T0213804

Lecture Notes in Computer Science   9443

Commenced Publication in 1973
Founding and Former Series Editors:
Gerhard Goos, Juris Hartmanis, and Jan van Leeuwen

More information about this series at http://www.springer.com/series/7412

Ana Fred · Maria De Marsico
Antoine Tabbone (Eds.)

Pattern Recognition Applications and Methods

Third International Conference, ICPRAM 2014
Angers, France, March 6–8, 2014
Revised Selected Papers

 Springer

Editors
Ana Fred
Instituto de Telecomunicações
Lisboa
Portugal

Maria De Marsico
Computer Science
Sapienza University of Rome
Rome
Italy

Antoine Tabbone
LORIA-UMR 7503
Vandoevre-lès-Nancy
France

ISSN 0302-9743   ISSN 1611-3349   (electronic)
Lecture Notes in Computer Science
ISBN 978-3-319-25529-3   ISBN 978-3-319-25530-9   (eBook)
DOI 10.1007/978-3-319-25530-9

Library of Congress Control Number: 2015953239

LNCS Sublibrary: SL6 – Image Processing, Computer Vision, Pattern Recognition, and Graphics

Springer International Publishing AG Switzerland is part of Springer Science+Business Media
(www.springer.com)

Preface

The present book includes extended and revised versions of a set of selected papers from the Third International Conference on Pattern Recognition Applications and Methods (ICPRAM 2014), held in Angers, France, during March 6–8, 2014. The conference was sponsored by the Institute for Systems and Technologies of Information, Control and Communication (INSTICC) in cooperation with the ACM Special Interest Group on Applied Computing (SIGAPP) and the Association for the Advancement of Artificial Intelligence (AAAI). ICPRAM 2014 was also technically co-sponsored by the Portuguese Association for Pattern Recognition (APRP).

Since its creation, the purpose of the International Conference on Pattern Recognition Applications and Methods has been to become a major point of contact between researchers, engineers, and practitioners in the different areas of pattern recognition, to foster a productive debate from both theoretical and application perspectives. The privileged focus was on contributions describing up-to-date applications of pattern recognition techniques to real-world problems, interdisciplinary research, and experimental and/or theoretical studies yielding new insights that advance pattern recognition methods. The final ambition was to spur new research lines and provide the occasion to start novel collaborations, most of all in interdisciplinary research scenarios.

ICPRAM 2014 received 179 submissions, from 44 countries in all continents, of which 14 % were presented at the conference as full papers, and their authors were invited to submit extended versions of their papers for this book. In order to evaluate each submission, a double-blind paper review was performed by the Program Committee. Finally, only the 18 best papers were included in this book.

We would like to highlight that ICPRAM 2014 included four plenary keynote lectures, given by internationally distinguished researchers, namely: Harry Wechsler (George Mason University, USA), Josef Kittler (University of Surrey, UK), John Shawe-Taylor (University College London, UK), and Robert Zlot (CSIRO, Australia). We must acknowledge the invaluable contribution of all keynote speakers who, as renowned researchers in their areas, presented cutting-edge work and thus contributed toward enriching the scientific content of the conference.

We especially thank the authors, whose research and development efforts are recorded here. The knowledge and diligence of the reviewers were essential to ensure the quality of the papers presented at the conference and published in this book. Finally, a special thanks to all members of the INSTICC team, whose involvement was fundamental for organizing a smooth and successful conference.

April 2015

Ana Fred
Maria De Marsico
Antoine Tabbone

Organization

Conference Chair

Ana Fred   Instituto de Telecomunicações, IST, University of Lisbon, Portugal

Program Chairs

Maria De Marsico   Sapienza Università di Roma, Italy
Antoine Tabbone   LORIA-Université de Lorraine, France

Organizing Committee

Helder Coelhas   INSTICC, Portugal
Bruno Encarnação   INSTICC, Portugal
Ana Guerreiro   INSTICC, Portugal
André Lista   INSTICC, Portugal
Andreia Moita   INSTICC, Portugal
Raquel Pedrosa   INSTICC, Portugal
Vitor Pedrosa   INSTICC, Portugal
Cláudia Pinto   INSTICC, Portugal
Susana Ribeiro   INSTICC, Portugal
Sara Santiago   INSTICC, Portugal
Mara Silva   INSTICC, Portugal
José Varela   INSTICC, Portugal
Pedro Varela   INSTICC, Portugal

Program Committee

Andrea F. Abate   University of Salerno, Italy
Ashraf AbdelRaouf   Cloudypedia, Dubai, UAE
Shigeo Abe   Kobe University, Japan
Gady Agam   Illinois Institute of Technology, USA
Mayer Aladjem   Ben-Gurion University of the Negev, Israel
Guillem Alenya   Institut de Robòtica i Informàtica Industrial, CSIC-UPC, Spain
Javad Alirezaie   Ryerson University, Canada
Francisco Martínez Álvarez   Pablo de Olavide University of Seville, Spain

Annalisa Appice	Università degli Studi di Bari Aldo Moro, Italy
Antonio Artés-Rodríguez	Universidad Carlos III de Madrid, Spain
Thierry Artières	Université Pierre et Marie Curie, France
Kevin Bailly	ISIR Institut des Systèmes Intelligents et de Robotique - UPMC/CNRS, France
Gabriella Sanniti di Baja	Institute of Cybernetics E. Caianiello, CNR, Italy
Emili Balaguer-Ballester	Bournemouth University, UK
Vineeth Nallure Balasubramanian	Indian Institute of Technology, India
Luis Baumela	Universidad Politécnica de Madrid, Spain
Jon Atli Benediktsson	University of Iceland, Iceland
Charles Bergeron	Union University of New York, USA
J. Ross Beveridge	Colorado State University, USA
Monica Bianchini	Università degli Studi di Siena, Italy
Marenglen Biba	University of New York Tirana, Albania
Anselm Blumer	Tufts University, USA
Joan Martí Bonmatí	Girona University, Spain
Mohamed-Rafik Bouguelia	LORIA, Lorraine University, France
Patrick Bouthemy	Inria, France
Francesca Bovolo	Remote Sensing Laboratory, University of Trento, Italy
Paula Brito	Universidade do Porto, Portugal
Hans du Buf	University of the Algarve, Portugal
Samuel Rota Bulò	University of Venice, Italy
Javier Calpe	Universitat de València, Spain
Rui Camacho	Universidade do Porto, Portugal
Francesco Camastra	University of Naples Parthenope, Italy
Virginio Cantoni	Università di Pavia, Italy
Ramón A. Mollineda Cárdenas	Universitat Jaume I, Spain
Marco La Cascia	Università Degli Studi di Palermo, Italy
Jocelyn Chanussot	Grenoble Institute of Technology, France
Amitava Chatterjee	Jadavpur University, India
Snigdhansu Chatterjee	University of Minnesota, USA
Imen Ben Cheikh	LaTICE lab, University of Tunis, Tunisia
Chi Hau Chen	University of Massachusetts Dartmouth, USA
Dmitry Chetverikov	MTA SZTAKI, Hungary
Seungjin Choi	Pohang University of Science and Technology, Republic of Korea
Ioannis Christou	Athens Information Technology, Greece
Pau-choo Chung	National Cheng Kung University, Taiwan
Miguel Coimbra	Universidade do Porto, Portugal
Tom Croonenborghs	KH Kempen University College, Belgium
Sergio Cruces	University of Seville, Spain
Justin Dauwels	Nanyang Technological University, Singapore

Jeroen Deknijf	Technical University Eindhoven, The Netherlands
Thorsten Dickhaus	Humboldt-Universität zu Berlin, Germany
Gianfranco Doretto	West Virginia University, USA
Bernadette Dorizzi	Telecom Sud Paris, France
Philippe Dosch	LORIA, France
Petros Drineas	Rensselaer Polytechnic Institute, USA
Gideon Dror	The Academic College of Tel-Aviv-Yaffo, Israel
Francisco Escolano	Universidad de Alicante, Spain
Yaokai Feng	Kyushu University, Japan
Mario Figueiredo	Technical University of Lisbon - IST, Portugal
Maurizio Filippone	University of Glasgow, UK
Gernot A. Fink	TU Dortmund, Germany
Vojtech Franc	Czech Technical University, Czech Republic
Damien François	Université catholique de Louvain, Belgium
Ana Fred	Instituto de Telecomunicações/IST, Portugal
Sabrina Gaito	Università degli Studi di Milano, Italy
Vicente Garcia	Institute of New Imaging Technologies/Universitat Jaume I, Spain
Azadeh Ghandehari	Islamic Azad University, Saveh Branch, Iran, Islamic Republic of
Giorgio Giacinto	University of Cagliari, Italy
Maryellen L. Giger	University of Chicago, USA
Bernard Gosselin	University of Mons, Belgium
Eric Granger	École de technologie supérieure, Canada
Giuliano Grossi	Università degli Studi di Milano, Italy
Marcin Grzegorzek	University of Siegen, Germany
Sébastien Guérif	University Paris 13 - SPC, France
Amaury Habrard	Laboratoire Hubert Curien, University of St. Etienne, France
Michal Haindl	Institute of Information Theory and Automation, Czech Republic
Barbara Hammer	Bielefeld University, Germany
Dongfeng Han	University of Iowa, USA
Makoto Hasegawa	Tokyo Denki University, Japan
Mark Hasegawa-Johnson	University of Illinois at Urbana-Champaign, USA
Pablo Hennings-Yeomans	Ontario Institute for Cancer Research, Canada
Tom Heskes	Radboud University Nijmegen, The Netherlands
Laurent Heutte	Université de Rouen, France
Anders Heyden	Lund Institute of Technology/Lund University, Sweden
Kouichi Hirata	Kyushu Institute of Technology, Japan
Bao-Gang Hu	Institute of Automation Chinese Academy of Sciences, China
Qinghua Huang	South China University of Technology, Guangzhou, China
Su-Yun Huang	Academia Sinica, Taiwan

Jose M. Iñesta	Universidad de Alicante, Spain
Akihiro Inokuchi	Kwansei Gakuin University, Japan
Yuji Iwahori	Chubu University, Japan
Alan J. Izenman	Temple University, USA
Saketha Nath Jagarlapudi	IIT-Bombay, India
Robert Jenssen	University of Tromso, Norway
Jiayan Jiang	Facebook, USA
Ata Kaban	University of Birmingham, UK
Heikki Kälviäinen	Lappeenranta University of Technology (LUT), Finland
Yasushi Kanazawa	Toyohashi University of Technology, Japan
Mohamed A. Khabou	University of West Florida, Pensacola, USA
Yunho Kim	Yale University, USA
Adams Kong	Nanyang Technological University, Singapore
Aryeh Kontorovich	Ben-Gurion University, Israel
Mario Köppen	Kyushu Institute of Technology, Japan
Walter Kosters	Universiteit Leiden, The Netherlands
Constantine Kotropoulos	Aristotle University of Thessaloniki, Greece
Sotiris Kotsiantis	Educational Software Development Laboratory, University of Patras, Greece
Konstantinos Koutroumbas	National Observatory of Athens, Greece
Adam Krzyzak	Concordia University, Canada
Gautam Kunapuli	University of Wisconsin-Madison, USA
Nojun Kwak	Seoul National University, Republic of Korea
Jaerock Kwon	Kettering University, USA
Raffaella Lanzarotti	Università degli Studi di Milano, Italy
Soo-Young Lee	KAIST, Republic of Korea
Xuejun Liao	Duke University, USA
Aristidis Likas	University of Ioannina, Greece
Hantao Liu	University of Hull, UK
Nicolas Loménie	Université Paris Descartes, France
Gaelle Loosli	Clermont Université, France
Jesús Malo	Universitat de Valencia, Spain
Alamin Mansouri	Université de Bourgogne, France
Francesco Marcelloni	University of Pisa, Italy
Stephane Marchand-Maillet	University of Geneva, Switzerland
Elena Marchiori	Radboud University, The Netherlands
Gian Luca Marcialis	Università degli Studi di Cagliari, Italy
Urszula Markowska-Kaczmar	Wroclaw University of Technology, Poland
Maria De Marsico	Sapienza Università di Roma, Italy
J. Francisco Martínez-Trinidad	Instituto Nacional de Astrofísica, Óptica y Electrónica, Puebla, Mexico

Sally Mcclean	University of Ulster, UK
Peter McCullagh	University of Chicago, USA
Majid Mirmehdi	University of Bristol, UK
Piotr Mirowski	Bell Labs (Alcatel-Lucent), USA
Delia Alexandrina Mitrea	Technical University of Cluj-Napoca, Romania
Giovanni Montana	King's College London, UK
Laurent Najman	Université Paris-Est, France
Yuichi Nakamura	Kyoto University, Japan
Michele Nappi	Università di Salerno, Italy
Claire Nédellec	MIG, INRA Centre de Jouy-en-Josas, France
Mikael Nilsson	Lund University, Sweden
Hasan Ogul	Baskent University, Turkey
Il-Seok Oh	Chonbuk National University, Republic of Korea
Ahmet Okutan	Isik University, Turkey
Yoshito Otake	Johns Hopkins University, USA
Vicente Palazón-González	Universitat Jaume I, Spain
Apostolos Papadopoulos	Aristotle University, Greece
Marcello Pelillo	University of Venice, Italy
Frederick Kin Hing Phoa	Academia Sinica (Taiwan ROC), Taiwan
Luca Piras	University of Cagliari, Italy
Fiora Pirri	University of Rome La Sapienza, Italy
Sylvia Pont	Delft University of Technology, The Netherlands
Philippe Preux	University of Lille 3, France
Lionel Prevost	University of French West Indies and Guiana, France
Hugo Proença	University of Beira Interior, Portugal
Subramanian Ramamoorthy	The University of Edinburgh, UK
Jan Ramon	K.U. Leuven, Belgium
Philippe Ravier	University of Orléans, France
Elisa Ricci	University of Perugia, Italy
Daniel Riccio	Univerity of Naples, Federico II, Italy
François Rioult	GREYC CNRS UMR6072 - Université de Caen Basse-Normandie, France
David Masip Rodo	Universitat Oberta de Catalunya, Spain
Marcos Rodrigues	Sheffield Hallam University, UK
Juan J. Rodríguez	University of Burgos, Spain
Lior Rokach	Ben-Gurion University of the Negev, Israel
Rosa María Valdovinos Rosas	Universidad Autonoma del Estado de Mexico, Mexico
Lorenza Saitta	Università del Piemonte Orientale, Italy

Antonio-José Sánchez-Salmerón	Universitat Politecnica de Valencia, Spain
Hector Satizabal	Haute Ecole d'Ingenierie et du Gestion du Canton de Vaud (HEIG-VD), Switzerland
Atsushi Sato	NEC, Japan
Michele Scarpiniti	Sapienza University of Rome, Italy
Paul Scheunders	University of Antwerp, Belgium
Friedhelm Schwenker	University of Ulm, Germany
Katsunari Shibata	Oita University, Japan
Bogdan Smolka	Silesian University of Technology, Poland
Tania Stathaki	Imperial College London, UK
Vassilios Stathopolous	University College London, UK
Mu-Chun Su	National Central University, Taiwan
Masashi Sugiyama	Tokyo Institute of Technology, Japan
Shiliang Sun	East China Normal University, China
Yajie Sun	Micron Tech/Aptina Imaging CA, USA
Zhenan Sun	Institute of Automation, Chinese Academy of Sciences (CASIA), China
Herbert Süße	Friedrich Schiller University of Jena, Germany
Johan Suykens	K.U. Leuven, Belgium
Kenji Suzuki	University of Chicago, USA
Antoine Tabbone	LORIA-Université de Lorraine, France
Alberto Taboada-Crispí	Universidad Central Marta Abreu de Las Villas, Cuba
Andrea Tagarelli	University of Calabria, Italy
Atsuhiro Takasu	National Institute of Informatics, Japan
Ichiro Takeuchi	Nagoya Institute of Technology, Japan
Toru Tamaki	Hiroshima University, Japan
Xiaoyang Tan	Nanjing University of Aeronautics and Astronautics, China
Lijun Tang	Microsoft Corp., USA
Anastasios Tefas	Aristotle University of Thessaloniki, Greece
Oriol Ramos Terrades	Centre de Visió per Computador - Universitat Autònoma de Barcelona, Spain
Fabien Torre	Lille University, Inria LNE and LIFL, France
Ricardo da S. Torres	University of Campinas (UNICAMP), Brazil
Andrea Torsello	Università Ca'Foscari Venezia, Italy
Genny Tortora	Università Degli Studi di Salerno, Italy
Godfried Toussaint	New York University Abu Dhabi, UAE
Kostas Triantafyllopoulos	University of Sheffield, UK
George Tsihrintzis	University of Piraeus, Greece
Maria Vanrell	Computer Vision Center, Universitat Autònoma de Barcelona, Spain
Antanas Verikas	Intelligent Systems Laboratory, Halmstad University, Sweden
Christian Viard-Gaudin	IRCCyN (UMR CNRS 6597) Universite de Nantes, France
Cinzia Viroli	University of Bologna, Italy

Jordi Vitrià	Universitat de Barcelona, Spain
Panayiotis Vlamos	Ionian University, Greece
Asmir Vodencarevic	Brunel GmbH, Germany
Yvon Voisin	University of Burgundy, France
Sviatoslav Voloshynovskiy	University of Geneva, Switzerland
Thomas Walsh	MIT, USA
Lei Wang	University of Wollogong, Australia
Jonathan Weber	Université de Lorraine, France
Harry Wechsler	George Mason University, USA
Joost van de Weijer	Autonomous University of Barcelona, Spain
David Windridge	University of Surrey, UK
Xianghua Xie	Swansea University, UK
Xin-Shun Xu	Shandong University, China
Jing-Hao Xue	University College London, UK
Nicolas Younan	Mississippi State University, USA
Pavel Zemcik	Brno University of Technology, Czech Republic
Huiyu Zhou	Queen's University Belfast, UK
Jovisa Zunic	University of Exeter, UK
Reyer Zwiggelaar	Aberystwyth University, UK

Additional Reviewers

Rene Grzeszick	TU Dortmund, Germany
Roberto Interdonato	DIMES - Università della Calabria, Italy
Kye-Hyeon Kim	POSTECH, Republic of Korea
Juho Lee	Pohang University of Science and Technology, Republic of Korea
Emilie Morvant	IST Austria, Austria
Arfika Nurhudatiana	Nanyang Technological University, Singapore
Jean-Philippe Peyrache	Jean Monnet University of Saint-Etienne, France
Marco Piccirilli	West Virginia University, USA
Yi Tao	ConocoPhillips, USA
Roi Weiss	Ben-Gurion University, Israel, Israel
Xijiong Xie	East China Normal University, China
Hengyi Zhang	Nanyang Technological University, Singapore
Gil Zohav	BGU, Israel

Invited Speakers

Harry Wechsler	George Mason University, USA
Josef Kittler	University of Surrey, UK
John Shawe-Taylor	University College London, UK
Robert Zlot	CSIRO, Australia

Contents

Theory and Methods

Multiple Image Segmentation

Jonathan Smets and Manfred Jaeger[✉]

Department for Computer Science, Aalborg University, Aalborg, Denmark
`jonathansmets@gmail.com, jaeger@cs.aau.dk`

Abstract. We propose a method for the simultaneous construction of multiple segmentations of images by combining a recently proposed "convolution of mixtures of Gaussians" model with a multi-layer hidden Markov random field structure. The resulting method constructs for a single image several, alternative segmentations that capture different structural elements of the image. We further introduce the notion of an image stack, by which we mean a collection of images with identical pixel dimensions. Here it turns out that the method is able to both identify groups of similar images in the stack, and to provide segmentations that represent the main structures in each group. We describe a variety of experimental results that illustrate the capabilities of the method.

Keywords: Segmentation · Multiple clustering · Probabilistic models

1 Introduction

Traditional clustering methods construct a single (possibly hierarchical) partitioning of the data. However, clustering when used as an explorative data analysis tool may not possess a single optimal solution that is characterized as the optimum of a unique underlying score function. Rather, there can be multiple distinct clusterings that each represent a meaningful view of the data. This observation has led to a recent research trend of developing methods for *multiple clustering* (or *multi-view clustering*). The general goal of these methods is to automatically construct several clusterings that represent alternative and complementary views of the data (see [12] for a recent overview, and the proceedings of the MultiClust workshop series for current developments).

The perhaps most typical application area for multiple clustering is document data (e.g. collections of news articles or web pages). For example, the standard benchmark WebKB dataset consists of university webpages that can be alternatively clustered according to page-type (e.g. personal homepage or course page), or the different universities the pages are taken from. Turning to image data, previously used benchmark sets are the CMU and the Yale Face Images data, which consists of portrait images of different persons in several poses, and accordingly can be clustered according to persons or poses [4,8]. In this setting, each image is a data-point, and (multiple) clustering means grouping images. When, instead, one views as a data-point a single image pixel, then multiple clustering becomes *multiple image segmentation.*

© Springer International Publishing Switzerland 2015
A. Fred et al. (Eds.): ICPRAM 2014, LNCS 9443, pp. 3–18, 2015.
DOI: 10.1007/978-3-319-25530-9_1

Relatively little work has been done on finding multiple, alternative image segmentations. Reference [10] developed a quite specific *factorial Markov random field* model in which an image is modeled as an overlay of several layers, and each layer corresponds to a binary segmentation. Reference [14] apply a general multiple clustering approach to a variety of datasets, including images. Their multiple clustering approach falls into the category of iterative multiple clustering, where given an initial (primary) clustering, a single alternative clustering is constructed. Our approach, on the other hand, falls into the category of simultaneous multiple clustering methods, where an arbitrary number of different clusterings is constructed at the same time, and without any priority ordering among the clusterings. Finally, [9] generate alternative segmentations based on color and texture features, respectively. However, the objective here is not to provide different, alternative segmentations, but to combine the two segmentations into a single one.

It is worth emphasizing that multiple clustering in the sense here considered is different from the construction of *cluster ensembles* [17]. In the latter, numerous clusterings are built in order to overcome the convergence to only locally optimal solutions of clustering algorithms, and to construct out of a collection of clusterings a single consensus clustering. The multiple segmentations in the sense of [6, 16] are segmentation analogues of cluster ensembles, not of multiple clusterings in our sense.

In this paper we develop a method for constructing multiple segmentations of images and *image stacks*, which we define as a collection of images with equal pixel dimensions. The most import type of image stacks are the collection of frames in a video sequence. However, we can also consider other such collections of pixel-aligned images. As we will see in the experimental section, multiple clustering of such image stacks can give results that combine elements of clustering at the image and at the pixel level. For the design of our method we build on the *convolution of mixtures of Gaussians* model of [8] which we customize for the segmentation setting by combining it with a Markov Random Field structure to account for the spatial dimension of the data.

Our approach is intended as a general method that can be applied to image data of quite different types, and that thereby is a quite general tool for explorative image data analysis. For more specialized application tasks, our general method may serve as a basis, but will presumably require additional modifications and adaptations.

2 The Convolutional Clustering Model

Probabilistic clustering approaches are based on *latent variable models* where a data point x is assumed to be sampled from a joint distribution $P(X, L \mid \theta)$ of an observed data variable X and a latent variable $L \in \{1, \ldots, k\}$, governed by parameters θ (throughout this paper we use bold symbols to denote tuples of variables, parameters, etc.; when talking about random variables, then uppercase letters stand for the variables, and lowercase letters for concrete values of the variables). Clustering then is performed by learning the parameters θ, and assigning x to the cluster with index i for which $P(X = x, L = i \mid \theta)$ is maximal.

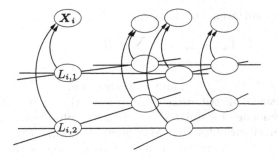

Fig. 1. Multi-layer Hidden Markov Random Field.

This probabilistic paradigm is readily generalized to multiple clustering models. One only needs to design a model $P(\boldsymbol{X}, \boldsymbol{L} \mid \boldsymbol{\theta})$ containing multiple latent variables $\boldsymbol{L} = L_1, \ldots, L_m$. Then the joint assignment $L_1 = i_1, \ldots, L_m = i_m$ (abbreviated $\boldsymbol{L} = \boldsymbol{i}$) maximizing $P(\boldsymbol{X} = \boldsymbol{x}, L_1 = i_1, \ldots, L_m = i_m \mid \boldsymbol{\theta})$ defines the cluster indices for $\boldsymbol{x}$ in m distinct clusterings. Models for multiple clustering that are based on multiple latent variables include the factorial Hidden Markov Model [5], the factorial Markov Random Fields of [10], convolution of mixtures of Gaussians [8], the latent tree models of [13], and the factorial logistic model of [7].

2.1 The Probabilistic Model

Our model is structurally identical to the factorial Markov Random Field model of [10]. Figure 1 shows the structure of such a *multi-layer hidden Markov random field*: with each pixel $i \in I$ (I the set of all pixels) are associated m latent variables $\boldsymbol{L}_{i,\bullet} = L_{i,1}, \ldots, L_{i,m}$ and a vector of observed variables $\boldsymbol{X}_i$. For $k = 1, \ldots, m$ the variables $\boldsymbol{L}_{\bullet,k} = L_{1,k}, \ldots, L_{|I|,k}$ take values in the set $\{1, \ldots, n_k\}$, so that the kth segmentation will consist of n_k segments.

For this paper we assume that in the case of single image analysis, $\boldsymbol{X}_i$ is simply the 3-dimensional vector (R_i, G_i, B_i) of *rgb*-values at pixel i. In the case of image stacks with N images, $\boldsymbol{X}_i$ will be a $3 \cdot N$-dimensional vector containing the *rgb*-values of all images in the stack. We denote with $\mid \boldsymbol{X} \mid_i$ the dimension of $\boldsymbol{X}_i$. Though we do not explore this in the current paper, we note that $\boldsymbol{X}_i$ could also contain differently defined observed features of pixel i.

For every $k = 1, \ldots, m$, the latent variables $\boldsymbol{L}_{\bullet,k}$ form a Markov random field with a square grid structure. The distribution of $\boldsymbol{X}_i$ depends conditionally on the latent variables $\boldsymbol{L}_{i,\bullet}$.

The marginal distribution $P(\boldsymbol{L} \mid \boldsymbol{\theta})$ is defined as a product of m Potts models defined by a common temperature parameter T:

$$P(\boldsymbol{L} = \boldsymbol{l} \mid \boldsymbol{\theta}) = P(\boldsymbol{L} = \boldsymbol{l} \mid T) = \frac{1}{Z} \prod_{k=1}^{m} e^{V(\boldsymbol{L}_{\bullet,k} = \boldsymbol{l}_{\bullet,k})/T}$$

where Z is the normalization constant, and

$$V(\boldsymbol{L}_{\bullet,k} = \boldsymbol{l}_{\bullet,k}) = \sum_{i,j:i\sim j} \mathbb{I}(l_{i,k} \neq l_{j,k})$$

with $\mathbb{I}(l_{i,k} \neq l_{j,k}) = 1$ if $l_{i,k} \neq l_{j,k}$, and $= 0$ otherwise.

For the conditional distribution $P(\boldsymbol{X} \mid \boldsymbol{L}, \boldsymbol{\theta})$ the model of Fig. 1 implies conditional independence for different pixels of the observed pixel features $\boldsymbol{X}_i$ given the latent pixel variables $\boldsymbol{L}_{i,\bullet}$. Moreover, we assume that the conditional model $P(\boldsymbol{X}_i \mid \boldsymbol{L}_{i,\bullet}, \boldsymbol{\theta})$ is identical for all i. It is defined as the convolution of m mixtures of Gaussians as follows. For $k = 1, \ldots, m$ and $j = 1, \ldots, n_k$ let $\mu_{k,j} \in \mathbb{R}^{|\boldsymbol{X}_i|}$. Writing $\boldsymbol{\mu}_k = \mu_{k,1}, \ldots, \mu_{k,n_k}$, we obtain for every k a distribution for a variable $\boldsymbol{Z}_{i,k}$ defined as a mixture of Gaussians

$$P(\boldsymbol{Z}_{i,k} \mid L_{i,k}, \boldsymbol{\mu}_k) = \sum_{j=1}^{n_k} N(\mu_{k,j}, \mathbf{1})\mathbb{I}(L_{i,k} = j),$$

where $\mathbf{1}$ stands for the unit covariance matrix. For two distributions $P(\boldsymbol{Y}), P(\boldsymbol{Z})$ of two k-dimensional real random variables $\boldsymbol{Y}, \boldsymbol{Z}$, we denote with $P(\boldsymbol{Y}) * P(\boldsymbol{Z})$ their convolution, i.e., the distribution of the sum $\boldsymbol{X} = \boldsymbol{Y} + \boldsymbol{Z}$. The final model for $\boldsymbol{X}_i$ now is defined as the m-fold convolution:

$$P(\boldsymbol{X}_i \mid \boldsymbol{L}_{i,\bullet}, \boldsymbol{\mu}_1, \ldots, \boldsymbol{\mu}_m) = P(\boldsymbol{Z}_{i,1} \mid L_{i,1}, \boldsymbol{\mu}_1) * \cdots * P(\boldsymbol{Z}_{i,m} \mid L_{i,m}, \boldsymbol{\mu}_m).$$

Combining the model for $\boldsymbol{L}$ and $\boldsymbol{X} \mid \boldsymbol{L}$, We now obtain

$$\log(P(\boldsymbol{L} = \boldsymbol{l}, \boldsymbol{X} = \boldsymbol{x} \mid \boldsymbol{\mu}, T)) \approx$$

$$-1/T \sum_{k=1}^{m} \sum_{i,j:i\sim j} \mathbb{I}(l_{i,k} \neq l_{j,k}) - \sum_{i \in I} \parallel \boldsymbol{x}_i - \sum_{k=1}^{m} \mu_{k,l_{i,k}} \parallel^2 \quad (1)$$

2.2 The Regularization Term

Maximizing the log-likelihood (1) alone is a sound approach to probabilistic multiple segmentation. However, [8] suggest to add to the likelihood the *regularization term*

$$-\lambda \sum_{\substack{k,k'=1,\ldots,m \\ k\neq k'}} \sum_{\substack{j=1,\ldots,n_k \\ j'=1,\ldots,n_{k'}}} (\mu_{k,j} \cdot \mu_{k',j'})^2 \quad (2)$$

Here $\lambda \geq 0$ is a weight parameter that regulates the strength of the influence of the regularization term. This penalty term is minimized when the means $\boldsymbol{\mu}_k, \boldsymbol{\mu}_{k'}$ corresponding to different segmentations lie in orthogonal subspaces. The rationale given for this regularization term is twofold. First, the likelihood function (1) does not have a unique maximum. Indeed, taking the case $m = 2$, the two solutions $(\mu_{1,1}, \ldots, \mu_{1,n_1}, \mu_{2,1}, \ldots, \mu_{2,n_2}, T)$ and $(\mu_{1,1} + c, \ldots, \mu_{1,n_1} + c, \mu_{2,1} - c, \ldots, \mu_{2,n_2} - c, T)$ $(c \in \mathbb{R}^3)$ define the same distribution, and therefore have the same likelihood score. Second, the likelihood alone does not give an explicit reward for the distinctness, or complementarity, of the resulting multiple clusterings. Following other approaches to multiple clustering, it is hoped that encouraging the means corresponding to different clusterings to lie in orthogonal subspaces will lead to a greater diversity of those clusterings.

We argue that the form and justification for this particular regularization term is slightly flawed, and that it should be replaced by a modified version. First, we note that the non-uniqueness of the optimal solution for (1) is not a real problem as long as two different optimal solutions define the same multiple segmentation. This, however, is exactly the case for the two solutions distinguished by the off-set vector c as described above. Second, regularization with (2) is not invariant under simple shifts of the coordinate system: adding a constant vector z to all data-points x_i should have no effect on the optimal segmentation, which should be characterized by also adding z to all model parameters $\mu_{k,j}$. Since (2) is not invariant under addition of a constant to all $\mu_{k,j}$, this is not the behavior one obtains with this regularization term. We therefore propose to modify (2) so as to reward means $\mu_k, \mu_{k'}$ to lie in orthogonal affine sub-spaces, rather than orthogonal linear sub-spaces. We therefore propose the following regularization term:

$$
-\lambda \sum_{\substack{k,k'=1,\ldots,m \\ k \neq k'}} \sum_{\substack{j,h=1,\ldots,n_k:j<h \\ j',h'=1,\ldots,n_{k'}:j'<h'}} \left(\frac{\mu_{k,j} - \mu_{k,h}}{\| \mu_{k,j} - \mu_{k,h} \|} \cdot \frac{\mu_{k',j'} - \mu_{k',h'}}{\| \mu_{k',j'} - \mu_{k',h'} \|} \right)^2 . \quad (3)
$$

Thus, we reward solutions in which normalized difference vectors between the means of different layers are orthogonal, rather than the means themselves. The term (3) now is invariant under adding, respectively subtracting, a constant vector c to all means of two different layers, and hence we again have the non-uniqueness of optimal solutions as for the pure likelihood (1). However, as argued above, we do not see this as a problem.

One small practical problem arises when we define our objective function as the sum of (1) and (3): the likelihood term (1) increases in magnitude linearly with the number of pixels. The regularization term, on the other hand, only increases as a function of the number of layers and the number of segments per layer. The choice of an appropriate tradeoff parameter λ between likelihood and regularization term, thus, would depend on the number of pixels. In order to get a more uniform scale for λ across different experiments, we therefore normalize the regularization term with the factor $|I|/K$, where K is the number of terms in the sum (3).

We remark that the probabilistic model (1) alone also has some built-in capability to encourage a diversity in the parameters μ_k for different layers, and hence, in the different segmentations. This is because having two layers with very similar means μ_k does not allow a much better fit to the data than a single layer with those means. Exploiting the full parameter space of the model to obtain a good fit to the data, thus, will tend to lead to some diversity in the parameters

μ_k. For this reason, in our experiments, we also pay particular attention to the case $\lambda = 0$, i.e., segmentation according to the pure probabilistic model (1).

The regularization terms (2) and (3) are intended to stimulate diversity in the computed segmentations, but they are not necessarily very meaningful, direct measurements for the diversity obtained. A common way to directly measure dissimilarity of two clusterings L_1, L_2 is *normalized mutual information*

$$NMI(L_1, L_2) = \frac{MI(L_1, L_2)}{\sqrt{H(L_1)H(L_2)}},$$

where MI is the mutual information and $H()$ the entropy of L_1, L_2, as determined by the empirical joint distribution of L_1, L_2 defined by the cluster assignments of the pixels. Low values of NMI indicate statistical independence, and hence dissimilarity of clusterings. Furthermore, a justification given by [8] for the regularization term (2) is that it induces a bias towards statistically independent clusterings. This justification carries over to our modified version (3). Therefore, the NMI as an evaluation measure is quite consistent with our objective function.

2.3 Clustering Algorithm

We take the model parameter $\beta := 1/T$ and the regularization parameter λ as user-defined inputs that may be varied in an iterative data exploration process. Large values of β mean that high emphasis is put on segmentations with large connected segments and smooth boundaries. Larger values of λ mean that diversity of segmentations as measured by the regularization term (3) is more strictly enforced.

Thus, the only model-parameters we have to fit are the mean vectors μ_k. Our goal, then, is to maximize a score function $S(\mu_1, \ldots, \mu_m, l)$ which is given as the sum of (1) and (3).

We use a typical 2-phase iterative process for this optimization: in a *MAP*-step we compute for a current setting of the μ_k the most probable assignment $L = l$ for the latent variables according to the likelihood function (1) (since (3) does not depend on l, we can ignore it in this phase). In a *M(aximization)*-step we recompute for the current setting $L = l$ the μ_k optimizing $S(\mu_1, \ldots, \mu_m, l)$. This well-known clustering approach (sometimes referred to as *hard EM*) has also been proposed for image segmentation in [3].

MAP-step. For the MAP-step we make use of the α-*expansion* algorithm of [1,2,11]. This algorithm provides solutions to segmentation problems characterized by an energy function E for segmentations s, which are of the form

$$E(s) = \sum_{i,j:i \sim j} V_{i,j}(s(i), s(j)) + \sum_i D_i(s(i)), \qquad (4)$$

where $s(i)$ is the segment label of pixel i, $V_{i,j}$ is a penalty function for discontinuities in s, and D_i is any non-negative function measuring the discrepancy of the label assignment $s(i)$ with the observed data for i. It is shown in [2] that if $V_{i,j}(s(i), s(j))$ is a metric on the label space, then the α-expansion algorithm is guaranteed to find a solution s that is within a constant factor of the globally minimal energy $E()$.

Up to a change of sign (and a corresponding change from a minimization to a maximization objective) our likelihood function (1) has the form (4) for the m-dimensional label space $\times_{k=1}^{m}\{1, \ldots, n_k\}$ (i.e. $s(i) = (l_{i,1}, \ldots, l_{i,m})$), with $V_{i,j}(s(i), s(j)) = \sum_{k=1}^{m} \mathbb{I}(l_{i,k} \neq l_{j,k})$ and $D_i(s(i)) = \parallel x_i - \sum_{k=1}^{m} \mu_{k,l_{i,k}} \parallel^2$.

Furthermore, it is straightforward to see that our $V_{i,j}$ is a metric on the m-dimensional label space.

To use the α-*expansion* algorithm we flatten our m-dimensional label space to a one-dimensional label space with $\prod_{k=1}^{m} n_k$ different labels. Thus, our method has a complexity that is exponential in the number of layers. On the other hand, the α-expansion algorithm in practice is quite efficient as a function of the number of pixels. It is reputed to show a linear complexity in practice [2], which was confirmed by what we observed in our experiments.

M-step. The M-step is performed by gradient ascent, leading to a local maximum of the score function given the current segmentation $L = l$.

Implementation. The algorithm is implemented in Matlab, using the α-expansion implementation provided by the gco-v3.0 library[1].

3 Experiments

In all our experiments we construct multiple segmentations with the same number of segments in each layer. We therefore refer to a multiple segmentation with m layers and k segments in each layer as a (m, k)-segmentation.

3.1 Single Images

Our first experiment establishes the baseline result that the segmentation methods works as intended when the input closely fits the underlying modeling assumption. To this end we construct the image shown in Fig. 2(c) as the overlay of the two images (a) and (b), and used our method to construct (2,3)-segmentations from the single input image (c). First setting $\lambda = \beta = 0$, we performed 200 runs of the algorithm with different random initializations. The highest-scoring solution that was found consists of the segmentations (d) and (e). In these figures, the color of the jth segment in the kth layer is set to $\tilde{\mu}_{k,j}$, where $\tilde{\mu}_{k,j}$ is obtained from $\mu_{k,j}$ by applying min-max normalization to re-scale the components of all the mean vectors μ_k $(k = 1, \ldots, m)$ into the interval $[0..255]$ of proper rgb-values. Essentially the same optimal result was found in

---

[1] http://vision.csd.uwo.ca/code/.

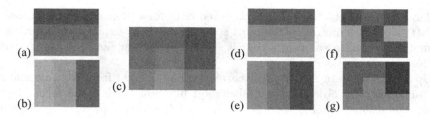

Fig. 2. Baseline: overlay image.

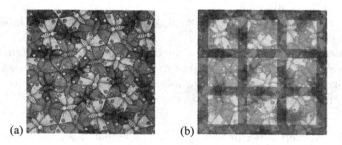

Fig. 3. Escher's butterflies: (a) original, (b) with added squares.

9 out of the 200 runs. In the remaining runs the algorithm converged to local maxima, an example of which is shown by (f) and (g). These results were clearly identified by the algorithm as sub-optimal by being associated with significantly lower score function values.

With increasing λ parameter the results in this experiment deteriorated. At $\lambda = 5000$ the "correct" solution was not found in 200 restarts. This is not very surprising, since for this image with $\lambda = \beta = 0$ the correct solution is clearly distinguished as the solution that can achieve a perfect score of 0 on the remaining Euclidean part of the likelihood term (1).

Next, we perform a series of experiments on the butterflies image by M.C. Escher, shown in Fig. 3(a), which has previously been used in [14]. The size of this image is 402×401 pixels.

We first compute (2,3)-segmentations with varying values of λ (and $\beta = 0$). Figure 4 shows the highest scoring results (in 20 restarts) obtained for $\lambda = 0, 1000, 10000$. In all cases, essentially the same two segmentations are computed: one that corresponds to the main colors of the three types of butterflies in the image, and one that captures the finer structure of the borders between the butterflies, as well as the shading inside the butterflies. The main effect of the regularization term here is not a difference in the segmentations, but only a difference in the means associated with the segments: for the high value $\lambda = 10000$, the means in the second segmentation all have a strong green component, whereas the means of the first component only have weak green components. This makes the means of the two components lie in near-orthogonal affine spaces. A similar color-separation does not appear at $\lambda = 0$.

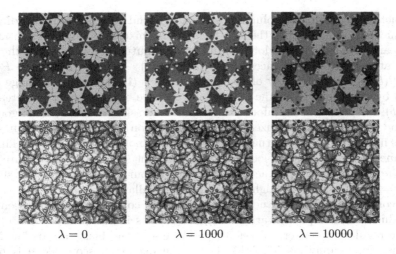

Fig. 4. Escher (2,3)-segmentations, varying λ (Color figure online).

(a)   (b)

Fig. 5. Mutual information vs. complementarity.

As discussed in Sect. 2.2, the regularization term is intended to stimulate complementarity of segmentations, whereas NMI would be used to actually measure complementarity. In this experiment the increasing λ-values place a higher weight on the regularization term, and the value of the regularization term decreases from $8.28 \cdot 10^6$ for the solution at $\lambda = 1000$ to $1.82 \cdot 10^6$ at $\lambda = 10000$ (at $\lambda = 0$ no regularization term is computed). However, the NMI values for the three solutions of Fig. 4 are $8.4 \cdot 10^{-3}, 5.4 \cdot 10^{-2}, 7.1 \cdot 10^{-2}$ for $\lambda = 0, 1000, 10000$, respectively. Thus, the NMI values are even slightly increasing for larger λ-values.

We note at this point that NMI values have to be used with caution when assessing dissimilarity of image segmentations (rather than other types of data clusterings): NMI is a function only of cluster membership of pixels. However, for segmentations one is perhaps more interested in the borders defined between segments, than in the global grouping of pixels into segments. To illustrate this issue we consider the modified butterfly image in Fig. 3(b), in which we have superimposed an additional square grid structure on the original image. Figure 5(a) shows a hypothetical (2,4)-segmentation (not computed by our method) of this image. Both segmentations identify the grid structure – the first one dividing the structure according to columns (and background), the second according to rows (and background). For the non-background pixels row and column membership are independent random variables. The mutual information of the two segmentations therefore reduces to $-P(b)\log P(b) - (1 - P(b))log(1 - P(b))$, where $P(b)$ is the probability of background pixels (i.e. the relative image area covered by background). In the limit where the size of the squares is increased, and $P(b) \to 0$, the mutual information of the two segmentations, thus, goes to zero (and so does the normalized mutual information). This shows that dissimilarity as measured by low mutual information need not correspond to the kind of complementarity we may be looking for in different segmentations. Figure 5(b) shows the (2,4)-segmentation actually obtained by our method. The result shown is for $\lambda = 0$, but results for higher λ-values are similar.

We conclude that neither need there be a good correspondence between low NMI values and complementarity of segmentations in the intuitive sense, nor does the regularization term necessarily induce a strong bias towards low NMI solutions. Fortunately, as Fig. 5(b) shows, the likelihood score alone is quite successful in producing segmentations that are complementary in an intuitively meaningful sense.

As a final experiment with the butterfly image, we do a (3,2)-segmentation with $\lambda = \beta = 0$. The result is shown in Fig. 6. The first segmentation again is based on the main underlying color distribution, isolating the blue butterflies

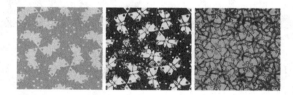

Fig. 6. Escher (3,2)-segmentation (Color figure online).

Fig. 7. Satellite image (Freiburg, Germany).

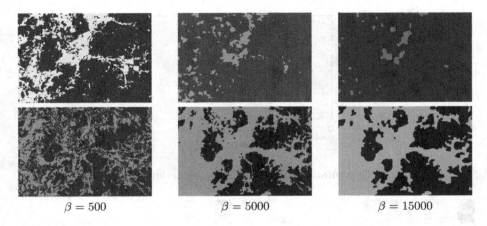

$\beta = 500$ $\beta = 5000$ $\beta = 15000$

Fig. 8. Satellite: results with varying β.

from the rest. The last segmentation again represents mostly the border structure and shading. Finally, the segmentation in the middle is mostly identifying the green butterflies, but also represents some structure. Reference [14] present a (2,2)-segmentation for the butterfly image obtained from their iterative clustering method. Their two segmentations are quite similar in nature to the first two in Fig. 6.

We next use the satellite image shown in Fig. 7 to investigate the influence of the β-parameter, as well as the scalability properties of our method. Figure 8 shows the result of (2,2)-segmentations with $\beta = 500, 5000, 15000$. We first observe that in all cases one segmentation mostly singles out the valley/city region against the rest (top row), whereas the second segmentation distinguishes the wooded area (bottom row). Increasing β-values have their primary intended effect to produce more coherent segments with smoother boundaries. At the same time, with increasing β the complementarity of the two segmentations here becomes rather more pronounced, and the valley/city segment shrinks to a segment more specifically identifying the city areas only. All results presented here are the top scoring results out of 10 random restarts for each setting of β.

The input image used for the experiment shown in Fig. 8 had a resolution of $500 \times 346 = 173.000$ pixels. Figure 9 shows the runtime per restart for varying resolutions of the same input image. We clearly observe a linear scaling of the runtime as a function of the image size, which, in particular, confirms the in practice linear behavior of the α-expansion algorithm.

3.2 Image Stacks

As a first experiment with an image stack, we used the collection of 25 flag-images shown in Fig. 10 (each at a resolution of 150×75 pixels).

Again setting $\lambda = \beta = 0$, the highest scoring (2,3)-segmentation is shown at the right of Fig. 10. Here we now depict the different segments using arbitrarily

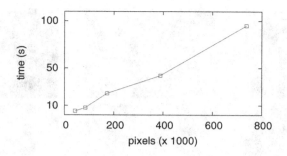

Fig. 9. Computation time for (2,2)-segmentations of Fig. 7.

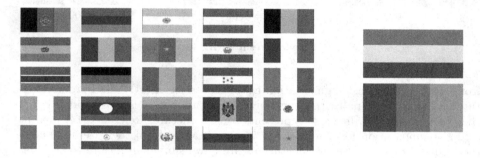

Fig. 10. Stack of flag images (Color figure online).

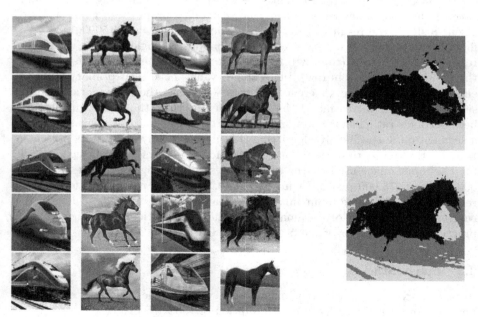

Fig. 11. Stack of Horse and Train images.

chosen greyscale values. The means $\mu_{k,j}$ characterizing segments now are $3 \cdot 25$ dimensional vectors that can be interpreted as an average color sequence for pixels in a segment. Taking for visualization the average over all colors in the sequence typically leads to all segments represented by very similar brownish colors (although, curiously, in this particular case the average colors for the segmentation with the vertical stripes yield a somewhat washed-out looking French flag). The same "correct" solution here was found in 9 out of 50 random restarts.

A second image stack we constructed consists of 10 images each of trains and horses, as shown in Fig. 11. We performed (2,3)-segmentation with $\lambda = 0$ and $\beta = 50$. The highest scoring result within 400 runs is shown at the right of Fig. 11. The method identifies the main structures in the two groups of images also in this somewhat more diverse collection of images. The results in the different runs were relatively stable, with other high-scoring solutions similar to the top-scoring one. Results with lower scores often separated the two groups of images less clearly, or contained segmentations in which one segment was reduced to very few pixels. The average runtime per restart in this experiment was about 1 min.

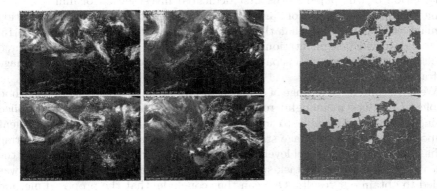

Fig. 12. Weather satellite image stack.

In a last image stack experiment we use a stack of 17 weather satellite images showing the cloud distribution over Europe on different days in the summer months June–August in years 2011–2014[2]. Figure 12 shows a representative 4 of the 17 input images, and the highest scoring result from 10 restarts of a (2,2)-segmentation. Interestingly, the top 5 solutions in the 10 restarts were visually indistinguishable from the one shown in Fig. 12, and achieved almost the same optimal score (note that even identical segmentations can have somewhat different scores, because the score is a function of the underlying model parameters $\mu_{k,j}$, not the segmentation alone). This robustness in the results under random restarts indicates that the found (2,2)-segmentation really shows relevant patterns in the input data, which one may cautiously try to interpret as patterns of cloud distributions.

---

[2] Image source: http://www.sat24.com.

In all our experiments results were quite robust under variations of the λ and β parameters. Good results are typically already obtained at the baseline setting $\lambda = \beta = 0$. Note that $\beta = 0$ means that the Markov random field structure of the model is ignored, and that the MAP step could be implemented in a much simplified manner. In applications where smooth and contiguous segments are required, settings of $\beta > 0$ will be needed. The impact of the λ parameter on the segmentations was rather small. It appears that larger values of λ affected the placement of the mean parameters representing the different segments, but not so much the resulting segmentations themselves.

4   Conclusions

We have introduced a method for constructing multiple segmentations of image stacks by combining the convolution of mixtures of Gaussians model [8] with a multi-layer Markov Random field. While novel in this form, the resulting model is a quite straightforward combination of existing components. The main original contribution of this paper is the first dedicated investigation of multiple clustering for image segmentation, and the introduction of (multiple) segmentation of image stacks. We note that the latter is different from cosegmentation [15] and standard video segmentation, where also "stacks" of images are segmented simultaneously, but where a separate segmentation is computed for each image (or frame).

We have conducted a range of experiments that demonstrate that the method is able to produce meaningful results in a broad variety of datasets. Applied to single images, it is able to identify the structures of multiple constituent components. Applied to image stacks, it can perform a simultaneous clustering at the image and at the pixel level. All these results were obtained using only the basic rgb pixel features. No task-specific preprocessing or feature engineering was needed to obtain our results. One can thus conclude, that the proposed method provides a useful baseline approach for explorative image analysis.

For more specific application purposes or data analysis objectives, it will be necessary to construct more specific pixel features. One possible such application domain is multiple segmentation of video sequences. The frames of a video can obviously be seen as an image stack. Using only the rgb pixel features our method is not very well adapted to video analysis, since it does not take into account the temporal order of the frames. New pixel features that capture some of the temporal dynamics of the pixel values can be constructed, for example, simply by considering the variance of the pixel's rgb values, or by constructing features that describe the trajectory of the pixel's rgb values in rgb-space. Performing multiple segmentation of video sequences based on such features is a topic for future work.

In this paper we have also tried to evaluate the usefulness of regularization terms along the lines proposed in [8] for stimulating diversity in the multiple segmentations. Our results lead to some doubts both with regard to the effectiveness of the regularization term to produce segmentations with low mutual information, and with regard of the usefulness of mutual information as a measure for diversity in image segmentations. On the other hand, our results indicate that the likelihood term (1) alone is quite capable of identifying the most relevant, distinct segmentations.

References

1. Boykov, Y., Kolmogorov, V.: An experimental comparison of min-cut/max- flow algorithms for energy minimization in vision. IEEE Trans. Pattern Anal. Mach. Intell. **26**(9), 1124–1137 (2004)
2. Boykov, Y., Veksler, O., Zabih, R.: Fast approximate energy minimization via graph cuts. IEEE Trans. Pattern Anal. Mach. Intell. **23**(11), 1222–1239 (2001)
3. Chen, S., Cao, L., Wang, Y., Liu, J., Tang, X.: Image segmentation by MAP-ML estimations. IEEE Trans. Image Process. **19**(9), 2254–2264 (2010)
4. Cui, Y., Fern, X., Dy, J.: Non-redundant multi-view clustering via orthogonalization. In: Proceedings of Seventh IEEE International Conference on Data-Mining (ICDM 2007), pp. 133–142 (2007)
5. Ghahramani, Z., Jordan, M.: Factorial hidden markov models. Mach. Learn. **29**(2–3), 245–273 (1997)
6. Hoiem, D., Efros, A., Hebert, M.: Geometric context from a single image. In: Tenth IEEE International Conference on Computer Vision (ICCV 2005), pp. 654–661 (2005)
7. Jaeger, M., Lyager, S.P., Vandborg, M.W., Wohlgemuth, T.: Factorial clustering with an application to plant distribution data. In: Proceedings of the 2nd Multi-Clust Workshop, pp. 31–42 (2011). Online proceedings http://dme.rwth-aachen. de/en/MultiClust2011
8. Jain, P., Meka, R., Dhillon, I.S.: Simultaneous unsupervised learning of disparate clusterings. Stat. Anal. Data Min. **1**(3), 195–210 (2008)
9. Kato, Z., Pong, T.-C., Qiang, S.G.: Unsupervised segmentation of color textured images using a multilayer MRF model. In: Proceedings of the IEEE International Conference on Image Processing (ICIP 2003), vol. 1, pp. 961–964. IEEE (2003)
10. Zabih, R., Kim, J.: Factorial Markov random fields. In: Heyden, A., Sparr, G., Nielsen, M., Johansen, P. (eds.) ECCV 2002, Part III. LNCS, vol. 2352, pp. 321–334. Springer, Heidelberg (2002)
11. Kolmogorov, V., Zabin, R.: What energy functions can be minimized via graph cuts? IEEE Trans. Pattern Anal. Mach. Intell. **26**(2), 147–159 (2004)
12. Müller, E., Günnemann, S., Färber, I., Seidl, T.: Discovering multiple clustering solutions: grouping objects in different views of the data. In: Proceedings of 28th International Conference on Data Engineering (ICDE 2012), pp. 1207–1210 (2012)
13. Poon, L.K.M., Zhang, N.L., Chen, T., Wang, Y.: Variable selection in model-based clustering: To do or to facilitate. In: Proceedings of the 27th International Conference on Machine Learning (ICML 2010), pp. 887–894 (2010)
14. Qi, Z., Davidson, I.: A principled and flexible framework for finding alternative clusterings. In: Proceedings of the 15th ACM SIGKDD International Conference on Knowledge Discovery and Data Mining (KDD 2009), pp. 717–725 (2009)

15. Rother, C., Minka, T., Blake, A., Kolmogorov, V.: Cosegmentation of image pairs by histogram matching-incorporating a global constraint into MRFs. In: 2006 IEEE Computer Society Conference on Computer Vision and Pattern Recognition, pp. 993–1000. IEEE (2006)
16. Russell, B., Freeman, W., Efros, A., Sivic, J., Zisserman, A.: Using multiple segmentations to discover objects and their extent in image collections. In: 2006 IEEE Computer Society Conference on Computer Vision and Pattern Recognition, pp. 1605–1614 (2006)
17. Strehl, A., Ghosh, J.: Cluster ensembles – a knowledge reuse framework for combining multiple partitions. J. Mach. Learn. Res. **3**, 583–617 (2003)

Aggregation of Biclustering Solutions for Ensemble Approach

Blaise Hanczar[✉] and Mohamed Nadif

LIPADE, University Paris Descartes, 45 rue des saint-peres, 75006 Paris, France
hanczar_blaise@parisdescartes.fr

Abstract. The biclustering methods have an increasing interest in the community of machine learning and data mining. These methods identify subsets of examples and features with interesting patterns. Recently ensemble approach has been applied to the biclustering problems with success. Their principle is to generate a set of different biclusters then aggregate them into only one. The crucial step of this approach is the consensus functions that compute the aggregation of the biclusters. We identify the main consensus functions commonly used in the clustering ensemble and show how to extend them in the biclustering context. We evaluate and analyze the performances of these consensus functions on several experiments based on both artificial and real data.

Keywords: Biclustering · Ensemble methods · Aggregation functions

1 Introduction

Biclustering, also called direct clustering [18], simultaneous clustering in [14,29] or block clustering in [15] is now a widely used method of data mining in various domains in particular in text mining and bioinformatics. For instance, in document clustering, in [7] the author proposed a spectral block clustering method which makes use of the clear duality between rows (documents) and columns (words). In the analysis of microarray data, where data are often presented as matrices of expression levels of genes under different conditions, the co- clustering of genes and conditions overcomes the problem encountered in conventional clustering methods concerning the choice of similarity. Cheng and Church [4] were the first to propose a biclustering algorithm for microarray data analysis. They considered that biclusters follow an additive model and used a greedy iterative search to minimize the mean square residue (MSR). Their algorithm identifies the biclusters one by one and was applied to yeast cell cycle data, and made it possible to identify several biologically relevant biclusters. Lazzeroni and Owen [20] have proposed the popular plaid model which has been improved by Turner et al. [29]. The authors assumed that biclusters are organized in layers and follow a given statistical model incorporating additive two way ANOVA models. The search approach is iterative: once $(K - 1)$ layers (biclusters) were identified, the K-th bicluster minimizing a merit function depending on all layers is selected. Applied to data

© Springer International Publishing Switzerland 2015
A. Fred et al. (Eds.): ICPRAM 2014, LNCS 9443, pp. 19–34, 2015.
DOI: 10.1007/978-3-319-25530-9_2

from the yeast, the proposed algorithm reveals that genes in biclusters share the same biological functions. In [11] the authors developed their localization procedure which improves the performance of a greedy iterative biclustering algorithm. Several other methods have been proposed in the literature, two complete surveys of biclustering methods can be found in [3,22].

Here we propose to use the ensemble methods to improve the performance of biclustering. It is important to note that we do not propose a new biclustering method in competition with the previously mentioned algorithms. We seek to adapt the ensemble approach to the biclustering problem in order to improve the performance of any biclustering algorithm. The principle of ensemble biclustering is to generate a set of different biclustering solutions, then aggregate them into only one solution. The crucial step is based on the consensus functions computing the aggregation of the different solutions. In this paper we have identified four types of consensus function commonly used in ensemble clustering and giving the best results. We show how to extend their use in the biclustering context. We evaluate their performances on a set of both numerical and real data experiments.

The paper is organized as follows. In Sect. 2, we review the ensemble methods in clustering and biclustering. In Sect. 3, we formalize the collection of biclustering solutions and show how to construct it from the Cheng and Church algorithm that we chose for our study. In Sect. 4, we extend four commonly used consensus functions to the biclustering context. Section 5 is devoted to evaluate these new consensus functions on several experimentations. Finally, we summarize the main points resulting from our approach.

2   Ensemble Methods

The principle of ensemble methods is to construct a set of models, then to aggregate them into a single model. It is well-known that these methods often perform better than a single model [9]. Ensemble methods first appeared in supervised learning problems. A combination of classifiers is more accurate than single classifiers [21]. A pioneer method boosting, the most popular algorithm which is adaboost, was developed mainly by Shapire [25]. The principle is to assign a weight to each training example, then several classifiers are learned iteratively and between each learning step the weight of the examples is adjusted depending on the classifier results. The final classifier is a weighted vote of classifiers constructed during the procedure. Another type of popular ensemble methods is bagging, proposed by Breiman [1]. The principle is to create a set a classifiers based on bootstrap samples of the original data. The random forests [2] are the most famous application of bagging. They are a combination of tree predictors, and have given very good results in many domains [8].

Several works have shown that ensemble methods can also be used in unsupervised learning. Topchy et al. [27] showed theoretically that ensemble methods may improve the clustering performance. The principle of boosting was exploited by Frossyniotis et al. [13] in order to provide a consistent partitioning of the data. The boost-clustering approach creates, at each iteration, a new training set using weighted random sampling from original data, and a simple clustering algorithm is applied to provide new clusters. Dudoit and Fridlyand [10] used bagging to improve the accuracy of clustering in reducing the variability of the PAM algorithm (Partitioning Around Medoids) results [19]. Their method has been applied to leukemia and melanoma datasets and made it possible to differentiate the different subtypes of tissues. Strehl et al. [26] proposed an approach to combine multiple partitioning obtained from different sources into a single one. They introduced heuristics based on a voting consensus. Each example is assigned to one cluster for each partition, an example has therefore as many assignments as number of partitions in the collection. In the aggregated partition, the example is assigned to the cluster to which it was the most often assigned. One problem with this consensus is that it requires knowledge of the cluster correspondence between the different partitions. They also proposed a cluster-based similarity partitioning algorithm. The collection is used to compute a similarity matrix of the examples. The similarity between two examples is based on the frequency of their co-association to the same cluster over the collection. The aggregated partition is computed by a clustering of the examples from the similarity matrix. Fern [12] formalized the aggregation procedure by a bipartite graph partitioning. The collection is represented by a bipartite graph. The examples and clusters of partitions are the two sets of vertices. An edge between an example and a cluster means that example has been assigned to this cluster. A partition of the graph is performed and each sub-graph represents an aggregated cluster. Topchy [28] proposed to modelize the consensus of the collection by a multinomial mixture model. In the collection, each example is defined by a set of labels that represents their assigned clusters in each partition. This can be seen as a new space in which the examples are defined, each dimension being a partition of the collection. The aggregated partition is computed from a clustering of examples in this new space. Since the labels are discrete variables, a multinomial mixture model is used. Each component of the model represents an aggregated cluster.

Some recent works have shown that the ensemble approach can also be useful in biclustering problems [17]. DeSmet presented a method of ensemble biclustering for querying gene expression compendia from experimental lists [5]. Actually the ensemble approach is performed only one dimension of the data (the gene dimension). Then biclusters are extracted from the gene consensus clusters. A bagging version of biclustering algorithms has been proposed and tested for microarray data [16]. Although this last method improves the performance of biclustering, in some cases it fails and returns empty biclusters, i.e. without examples or features. This is because the consensus function handles the sets of examples and features on the same dimension as in the clustering context. The consensus function must respect the structure of the biclusters. For this reason, the consensus functions mentioned above, can be applied to biclustering problems. In this paper we adapt these consensus functions to the biclustering context.

3   Biclustering Solution Collection

The first step of ensemble biclustering is to generate a collection of biclustering solution. Here we give the formalization of the collection and a method to generate it from the Cheng and Church algorithm that we have chosen for our study.

3.1   Formalization of the Collection

Let a data matrix be $\mathbb{X} = \{\mathbb{E}, \mathbb{F}\}$ where $\mathbb{E} = \{e_1, ..., e_N\}$ is the set of N examples represented by M-dimensional vectors and $\mathbb{F} = \{f_1, ..., f_M\}$ is the set of M features represented by N-dimensional vectors. A bicluster B is a submatrix of X defined by a subset of examples and a subset of features: $B = \{(E_B, F_B) | E_B \subseteq \mathbb{E}, F_B \subseteq \mathbb{F}\}$. A biclustering operator Φ is a function that returns a biclustering solution (i.e. a set of biclusters) from a data matrix: $\Phi(X) = \{B_1, ..., B_K\}$ where K is the number of biclusters. Let ϕ be the function giving for each point of the data matrix the label of the bicluster to which it belongs. The label is 0 for points belonging to no bicluster.

$$\phi(x_{ij}) = \begin{cases} k \ if \ e_i \in E_{B_k} \ and \ f_j \in F_{B_k} \\ 0 \ if \ e_i \notin E_{B_k} \ or \ f_j \notin F_{B_k} \ \forall k \in [1, K]. \end{cases}$$

A biclustering solution can be represented by a label matrix L giving for each point: $I_{ij} = \phi(x_{ij})$. In the following it will be convenient to represent this label matrix by an label vector indexed by u defined as $u = i * |\mathbb{F}| + (|\mathbb{F}| - j)$, where $|.|$ denotes the cardinality. $\mathbb{J}$ is the vector form of the matrix L: $\mathbb{J}_u = \mathbb{J}_{i*|\mathbb{F}|+(|\mathbb{F}|-j} = \phi(x_{ij})$.

Let's the true biclustering solution of the data set $\mathbb{X}$ represented by $\Phi(\mathbb{X})^*$, L^* and $\mathbb{J}^*$. An estimated biclustering solution is a biclustering solution returned by an algorithm from the data matrix, it is denoted by $\hat{\Phi}(\mathbb{X})$, $\hat{L}$ and $\hat{\mathbb{J}}$. The objective of the biclustering task is to find the closest estimated biclustering solution from the true biclustering solution. In ensemble methods, we do not use only one estimated biclustering solutions but we generate a collection of several solutions. We denote this collection of biclustering solutions as follows $\mathbb{C} = \{\hat{\Phi}(X)_{(1)}, ..., \hat{\Phi}(X)_{(R)}\}$. This collection can be represented by an $NM \times R$ matrix $\mathbb{J} = (\mathbb{J}_1^T, ..., \mathbb{J}_{NM}^T)^T$ by merging together all label vectors $\mathbb{J}_u = (J_{u1}, ..., J_{uR})^T$ where $J_{ur} = \phi(x_{ij})_{(r)}$ with $r \in [1, R]$. The objective of the consensus function is to form an aggregated biclustering solution, represented by $\overline{\Phi}(X)$, $\overline{L}$ and $\overline{\mathbb{J}}$, from the collection of estimated solutions. Each of these functions is illustrated with an example in Fig. 1.

3.2   Construction of the Collection

The key point of the generation of the collection is to find a good trade-off between the quality and diversity of the biclustering solutions of the collection. If all the generated solutions are the same, the aggregated solution is identical to the biclusters of the collection. Different sources of the diversity are possible. We can use a resampling method such as bootstrap or jacknife. In applying the biclustering operator to each resampled data, different solutions are produced. We can also include the source of diversity directly in the biclustering operator. In this case the algorithm is not deterministic and will produce different solutions from the same original data.

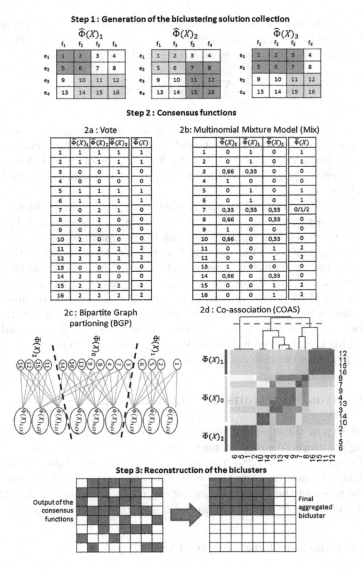

Fig. 1. Procedure of ensemble biclustering with the four consensus functions. (1) 3 different biclustering solutions with 2 biclusters for the same data matrix forming the collection. (2a) The collumns represents the labels of each data points obtained by the three biclustering solution. The last column represents the results of the VOTE consensus. (2b) The first three columns give the probability for each data point to be associated to the three labels of the mixture model. The last column represents the results of the MIX consensus. (2c) The bipartite graph representing all biclusters of the collection. The cuts of the graph give the results of the BGP consensus. (2d) The coassociation matrice of the data points. The 3 clusters obtained from this matrix represent the results of the COAS consensus. (3) An example of the reconstruction step of our methods.

In our experiments the biclustering operator is the Cheng and Church algorithm (CC) (algorithm 4 in the reference [4]). This algorithm returns a set of biclusters minimizing the mean square residue (MSR).

$$MSR(B_k) = \frac{1}{|B_k|} \sum_{i,j} z_{ik} w_{jk} (X_{ij} - \mu_{ik} - \mu_{jk} + \mu_k)^2,$$

where μ_k is the average of B_k, μ_{ik} and μ_{jk} are respectively the means of E_i and F_j belonging to bicluster B_k. z and w are the indicator functions of the examples and features. $z_{ik} = 1$ when the feature i belongs to the bicluster k, $z_{ik} = 0$ otherwise. $w_{jk} = 1$ when the example j belongs to the bicluster k, $w_{jk} = 0$ otherwise.

The CC algorithm is iterative and the biclusters are identified one by one. To detect each bicluster, the algorithm begins with all the features and examples, then it drops the feature or example minimizing the mean square residue (MSR) of the remaining matrix. This procedure is totally deterministic. We modified the CC algorithm by including a source of diversity in the computation of the bicluster. At each iteration, we selected the top $\alpha\%$ of the features and examples minimizing MSR of the remaining matrix. The element to be dropped was randomly chosen from this selection. Thus the parameter α controls the level of diversity of the bicluster collection; in our simulations $\alpha = 5\%$ seemed a good threshold. This modified version of the algorithm was used in all our experiments in order to generate the collection of biclustering solutions from a dataset.

4   Consensus Functions for Biclustering

The second step of the ensemble approach is the aggregation of the collection of biclustering solutions. We present here the extension of four consensus functions for biclustering ensemble. These methods assign a bicluster label to the $N \times M$ points of the data matrix. Note that even when the numbers of biclusters in the different solutions of the collection are not equal, these consensus functions can be used; it suffices to fix the final number of aggregated biclusters to K.

4.1   Co-association Consensus (COAS)

The idea of COAS is to group in a bicluster the points that are assigned together in the biclustering collection. This consensus is based on the bicluster assignation similarity between the points of the data matrix. The similarity between two points is defined by the proportion of times that they are associated to the same bicluster over the whole collection. All these similarities are represented by a distance matrix D defined by:

$$D_{uv} = 1 - \frac{1}{R} \sum_{r=1}^{R} \delta(J_{ur} = J_{vr}),$$

where $\delta(x)$ returns 0 when x is false and 1 when true. From this dissimilarity data matrix, $K + 1$ clusters are identified in using the Partitioning Around Medoids (PAM) algorithm [10]. The K clusters of points represent the K aggregated biclusters, the last cluster groups all the points that belongs to no bicluster.

4.2 Voting Consensus (VOTE)

This consensus function is based on the majority vote of the labels. Each point is assigned to the bicluster with which it has been assigned the most of the time in the biclustering collection. For each point of the data matrix, the consensus returns the most represented label in the collection of the biclustering solution. The main problem of this approach is that there is no correspondence between the labels of two different estimated biclustering solutions. All the biclusters of the collection have to be re-labeled according to their best agreement with some chosen reference solution. Any estimated solution can be used as reference, here we used the first one $\hat{\Phi}(X)_{(1)}$. The agreement problem can be solved in polynomial time by the Hungarian method [23] which relabels the estimated solution such the similarity between the solutions is maximized. The similarity between two biclustering solutions was computed by using the F-measure (details in Sect. 5.1). The label of the aggregated biclustering solution for a point is therefore defined by:

$$\overline{\mathbb{I}}_u = argmax_k \left(\sum_{r=1}^{R} \delta(\Gamma(J_{ur}) = k) \right).$$

where Γ is the relabelling operator performed by the Hungarian algorithm.

4.3 Bipartite Graph Partitionning Consensus (BGP)

In this consensus the collection of estimated solutions is represented by a bipartite graph where the vertices are divided into two sets: the point vertices and the label vertices. The point vertices represent the points of the data matrix $\{(e_i, f_j)\}$ while the set of label vertices represents all the estimated biclusters of the collection $\{\hat{B}_{k,(r)}\}$, for each estimated solution there is also a vertice that represents the points belonging to no bicluster. An edge links a point vertice to a label vertice if the point belongs to the corresponding estimated bicluster. The degree of each point is therefore R and the degree of each estimated bicluster represents the number of points that it contains. Finding a consensus consists in finding a partition of this bipartite graph. The optimal partition is the one that maximizes the numbers of edges inside each cluster of nodes and minimizes the number of edges between nodes of different clusters. This graph partitioning problem is a NP-hard problem, so we rely on a heuristic to an approximation of the optimal solution. We used a method based on a spin-glass model and simulated annealing [24] in order to identify the clusters of nodes. Each cluster of the partition represents an aggregated bicluster formed by all the points contained in this cluster.

4.4   Multivariate Mixture Model Consensus (MIX)

In [28], the authors have used the mixture approach to propose a consensus function. In the sequel we propose to extend it to our situation. In model-based clustering it is assumed that the data are generated by a mixture of under-lying probability distributions, where each component k of the mixture represents a cluster. Specifically, the $NM \times R$ data matrix $\mathbb{J}$ is assumed to be an $\mathbb{J}_1, \ldots, \mathbb{J}_u, \ldots, \mathbb{J}_{NM}$ i.i.d sample where $\mathbb{J}_u$ from a probability distribution with density

$$\varphi(\mathbb{J}_u|\Theta) = \sum_{k=0}^{K} \pi_k P_k(\mathbb{J}_u|\theta_k),$$

where $P_k(\mathbb{J}_u|\theta_k)$ is the density of label $\mathbb{J}_u$ from the kth component and the θ_ks are the corresponding class parameters. These densities belong to the same parametric family. The parameter π_k is the probability that an object belongs to the kth component, and K, which is assumed to be known, is the number of components in the mixture. The number of components corresponds to the number of biclusters minus one since one of the components represents the points belonging to no bicluster. The parameter of this model is the vector $\Theta = (\mathbf{p}i_0, \ldots, \mathbf{p}i_K, \theta_0, \ldots, \theta_K)$. The mixture density of the observed data $\mathbb{J}$ can be expressed as

$$\varphi(\mathbb{J}|\Theta) = \prod_{u=1}^{NM} \sum_{k=0}^{K} \pi_k P_k(\mathbb{J}_u|\theta_k).$$

The $\mathbb{J}_u$ labels are nominal categorical variables, we consider the latent class model and assume that all R categorical variables are independent, condition-nally on their memebership of a component;

$$P_k(\mathbb{J}_u|\theta_k) = \prod_{r=1}^{R} P_{k,(r)}(J_{ur}|\theta_{k,(r)}).$$

Note that $P_{k,(r)}(\mathbb{J}_u|\theta_{k,(r)})$ represents the probability to have $\mathbb{J}_u$ labels in the kth component for the estimated solution $\hat{\Phi}(X)_{(r)}$. If $\alpha_k^{r(j)}$ is the probability that the rth label takes the value j when an $\mathbb{J}_u$ belongs to the component k, then the probability of the mixture can be written $P_k(\mathbb{J}_u|\theta_k) = \prod_{r=1}^{R} \prod_{j=1}^{K} [\alpha_k^{r(j)}]^{\delta(J_{ur}=j)}$. The parameter of the mixture Θ is fitted in maximizing the likelihood function:

$$\Theta^* = argmax_\Theta \left(\log \left(\prod_{u=1}^{NM} P(\mathbb{J}_u|\theta) \right) \right).$$

The optimal solution of this maximization problem cannot generally be com-puted, we therefore rely on an estimation given by the EM algorithm [6].

In E-step, we compute the posterior probabilities of each label $s_{uk} \propto P_k(\mathrm{L}_u|\theta_k)$ and in the M-step we estimate the parameters of the mixture as follows

$$\pi_k = \frac{\sum_u s_{uk}}{NM} \text{ and } \alpha_k^{r(j)} = \frac{\sum_u s_{uk}\delta(J_{ur} = j)}{\sum_u s_{uk}}.$$

To limit the problems of local minimum during the EM algorithm, we performd the optimization process ten times with different initializations and kept the solution maximizing the log-likelihood. At the convergence, we consider that the largest π_k corresponds to labels representing the points belonging to no biclusters. The estimators of posterior probabilities give rise to a fuzzy or hard clustering using the maximum a posteriori principle (MAP). Then the consensus function consists in taking for each L_u the cluster such that k maximizing its conditional probability $k = argmax_{\ell=1,...,K}s_{u\ell}$, and we obtained the ensemble solution noted $\overline{\varPhi}(\mathbb{X})$.

4.5   Reconstruction of the Biclusters

The four consensus functions presented above, return a partition in $K+1$ clusters of the points of the data matrix. K of these clusters represent the K aggregated biclusters, the last one groups all the points that belong to no biclusters in the aggregated solution. The k aggregated biclusters are not actual biclusters yet. They are just sets of points that do not necessarily form submatrices of the data matrix. A reconstruction step has to be applied to each aggregated bicluster in order to transform it into a submatrix. This procedure consists in finding the submatrix containing the maximum of points that are in the aggregated bicluster and the minimum of points that are not in the aggregated bicluster. The k-th aggregated bicluster is reconstructed by minimizing the following function:

$$L(\overline{B_k}) = \sum_{i=1}^{N}\sum_{j=1}^{M} \delta(e_i \in \overline{E}_{B_k} \wedge f_i \in \overline{F}_{B_k})\delta(\overline{I}_{ij} \neq k)$$
$$+ \delta(e_i \notin \overline{E}_{B_k} \vee f_i \notin \overline{F}_{B_k})\delta(\overline{I}_{ij} = k).$$

This optimization problem is solved by a heuristic procedure. We started with all the examples and features involved in the aggregated bicluster. Then iteratively, we dropped the example or feature that maximizes the decrease of $L(\overline{B_k})$. This step was iterated until $L(\overline{B_k})$ did not decrease. Once the reconstruction procedure was finished, we obtained the final aggregated biclusters.

5   Results and Discussion

5.1   Performance of Consensus Functions

In our simulations, we considered six different data structures with $M = N = 100$ in which a true biclustering solution is included. The number of biclusters varies from 2 to 6 and their sizes from 10 examples by 10 features to 30 examples by 30

Fig. 2. The six data structures considered in the experiments.

features. We have defined six different structures of biclusters depicted in Fig. 2. For each data, from each true bicluster an estimated bicluster was generated, then a collection of estimated biclustering solutions was obtained. The quality of the collection is controlled by the parameters α_{pre} and α_{rec} that are the average precision and recall between estimated biclusters and their corresponding true biclusters. To generate an estimated bicluster we started with the true bicluster, then we randomly removed features/examples and have added features/examples that were not in the true bicluster in order to obtain the target precision α_{pre} and recall α_{rec}. Once the collection was generated, the four consensus functions were applied to obtain the aggregated biclustering solutions. Finally to evaluate the performance of each aggregated solution we computed the F-measure (noted Δ) between the obtained solution $\overline{\Phi}(\mathbb{X})$ and the true biclustering solution $\Phi(\mathbb{X})^*$;
$\Delta(\Phi(\mathbb{X})^*, \overline{\Phi}(\mathbb{X})) = \frac{1}{K} \sum_{k=1}^{K} M_{Dice}(B_k^*, \overline{B}_k)$ where $M_{Dice}(B_k^*, \overline{B}_k) = \frac{|B_k^* \cap \overline{B}_k|}{|B_k^*| + |\overline{B}_k|}$ is the Dice measure.

Figure 3 shows the performance of the different consensus in function on the size of the biclustering solution collection R with $\alpha_{pre} = \alpha_{rec} = 0.5$. Each of the six panels gives the results on the six data structures. The dot, triangle, cross and diamond curves represent respectively the F-measure in function of R for VOTE, COAS, BGP and MIX consensus. The full gray curve represents the mean of the performance of the biclustering collection. In the six panels, the performance of the collection is constantly around 0.5. That is be expected, since the performance of the collection does not depend on its size and by construction the theoretical performance of each estimated solution is 0.5. On the six dataset structures, from $R \geq 40$, all the consensus functions give much better performances than the estimated solutions of the collection. The performances of MIX in all the situations are strongly increasing with the size of the collection. Mix does not require a high value of R to record good results, for $R \geq 20$ it converges to their maximum and reaches 1 in all panels. The curves of BGP have the same shape, they begin with a strong increase then they converge to their maximums, but the increase phase is much longer than in MIX. It also worth noting that BGP begins with very low performances for small values of R, it is often lower than the performances of the collection. BGP reaches its best performances with $R \geq 60$, in four panels it obtains the second best results and the third on the two last panels. The performance of VOTE increases slowly and more or less linearly with the collection size. Even with very low values of R, the performance of the consensus is significantly better than the collection. VOTE gives the second best performances for S1 and S5 and the third best for the four other data structures.

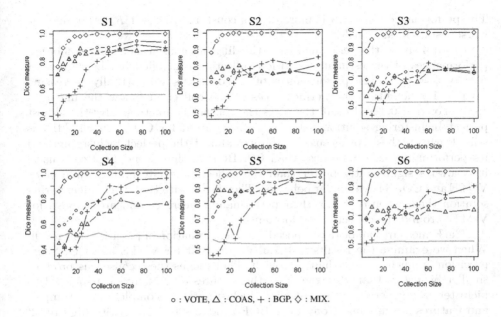

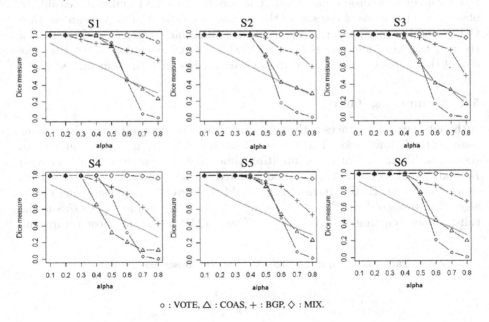

Fig. 3. Performance of the consensus in function of R (size of the biclustering solution collection) with $\alpha_{pre} = \alpha_{rec} = 0.5$.

Fig. 4. Performance of the consensus in function on the mean precision α_{pre} and recall α_{rec} of the biclustering solution collection with $\alpha = \alpha_{pre} = \alpha_{rec}$.

The performance of COAS is more or less constant whatever R; it obtains the worst results in all panels.

Figure 4 shows the performances of the different consensus in function of the performances of the estimated solution collection controlled by the parameter $\alpha = \alpha_{pre} = \alpha_{rec}$. The performances of all consensus are naturally decreasing with α. By definition the performances of the collection follow the line $y = 1 - x$. For $\alpha \leq 0.4$ and in all the cases the consensus functions give the almost perfect biclustering solution with $\Delta \approx 1$, expected for COAS in S4. MIX is still clearly the best consensus, it produces almost the perfect biclustering and its performances are never less than 0.9. BGP is the second best consensus, it is always significantly better than the collection whatever the value of α. VOTE and COAS have similar behavior. They begin with the perfect biclustering solution then, when $\alpha \geq 0.5$, their performances decrease and are at best, for VOTE, around the collection performance.

The F-measure can be decomposed into a combination of precision and recall. When we examine the results in detail we see that for VOTE and COAS the precision is much greater than the recall. That means these consensus produce smaller biclusters than the true ones, the features and examples associated to biclusters are generally good but these biclusters are incomplete i.e. examples and features are missing. Conversely BGP produces biclusters with high recall and low precision. The aggregated biclusters are generally complete but they also contain some extra wrong features and examples. MIX gives balanced biclusters with equal precision and recall. The experiment on S4 makes it possible to observe the influence of the size of the biclusters on the results. We can see that COAS obtains very bad performance on the small biclusters, since the recall on the two smallest biclusters is 0. MIX, VOTE, COAS are independent from the size of the biclusters, their performances are similar with the four biclusters.

5.2   Computing Time

Although the performances of the consensus functions are good, they also present some critical drawbacks. The use of these methods requires large amount of resources. Table 1 gives the computing time of each consensus function with $R = 50$. VOTE is the fastest method followed by MIX which is about ten times slower than VOTE, this inconvenient could be overcomed by using the eLEM algorithm proposed in [?] or the classification EM algorithm [?]. COAS is the third, about ten times slower than MIX and BGP needs the most computing

Table 1. Computing time (in s) of the consensus functions.

	S1	S2	S3	S4	S5	S6
VOTE	2.6	2.9	2.8	2.6	2.5	2.5
MIX	13.1	28.5	60.7	54	12.3	10.8
COAS	199.6	205.6	234.1	240.8	247.4	248.1
BGP	2502	3147.2	3345.1	3043.5	2806.6	2834.5

time, about ten times more than COAS. After observing S1, S2, S3 we can note that the number of biclusters has an impact of the computing time, specially for MIX. VOTE and MIX require loading an $NM \times R$ matrix than contains all the labels of the collection. BGP has to generate a graph containing $NM + R$ vertices while COAS requires computing resources for large distance matrices of size $NM \times NM$.

5.3   Results on Real Data

To evaluate our approach in terms of performance on real datasets, we used four datasets:

- Nutt: Gene expression data on the classification of gliomas in the brain.
- Pomeroy: Gene expression data on different types of tumors in the central nervous system.
- Sonar: Sonar signal from metal objects or rocks.
- Wdbc: Biological data on breast cancer.

The description of these datasets in terms of size is given in Table 2.

Table 2. Description of the four datasets.

Data sets	N	M
Nutt	50	500
Pomeroy	42	500
Sonar	208	60
Wdbc	569	30

Unlike numerical experiments and since we do not known the true biclustering solutions, the measures of performance can be based on external indices, like Dice score. Obviously, the quality of a biclustering solution can be measured by the AMSR i.e. the average of MSR computed from each bicluster belonging to the biclustering solution; the lower the AMSR, the better the solution. A problem with this approach is that the MSR is biased by the size of the biclusters. Indeed, the smallest biclusters favour AMSR. To remove this size bias we set the size of the biclusters in the parameters of the algorithms. All the methods will therefore return biclusters of the same size. The better solutions will be those minimizing AMSR. To compare the different consensus functions, we computed their gain which is the percentage of AMSR decreasing from the single biclustering solution i.e. the solution obtained by the classic CC algorithm without the ensemble approach. This is computed by:

$$Gain = 100 \frac{AMSR(\overline{\Phi}_{single}) - AMSR(\overline{\Phi}_{ensemble})}{AMSR(\overline{\Phi}_{single})},$$

Table 3. Gain of each consensus function on the four real datasets in function of the size of the biclusters.

Nutt dataset							
	50	100	200	300	400	600	800
VOTE	94	64	18	20	34	27	27
MIX	13	3	43	39	36	18	3
COAS	28	37	14	14	32	5	6
BGP	73	68	74	1	30	22	16

Pomeroy dataset							
	50	100	200	300	400	600	800
VOTE	79	85	79	69	32	63	60
MIX	84	83	69	52	37	75	74
COAS	69	78	21	36	30	43	39
BGP	68	80	21	22	30	46	51

Sonar dataset							
	50	100	200	300	400	600	800
VOTE	20	30	41	75	93	86	88
MIX	29	47	55	88	92	77	82
COAS	28	17	33	45	72	36	76
BGP	34	51	50	46	20	21	32

Wdbc dataset							
	50	100	200	300	400	600	800
VOTE	15	20	28	20	4	11	3
MIX	26	19	42	32	23	21	12
COAS	−4	−18	−15	−7	−17	−8	−25
BGP	6	13	37	31	2	10	−4

where $\overline{\Phi}_{single}$ and $\overline{\Phi}_{ensemble}$ are the biclustering solution returned respectively by the single and ensemble approaches.

Table 3 gives the gain of each consensus function for all the datasets in function on the size of the biclusters. We can observe that:

- In all the situations, all the consencus functions give an interesting gain, expected for COAS for Wdbc dataset. We know that in the merging process, once a cluster is formed it does not undo what was previously done; no modifications or permutations of objects are therefore possible. This disadvantage can be a handicap for COAS in some situations such as in Wdbc dataset.
- VOTE and MIX outperform BGP in most cases. In addition their behavior does not to depend on the size of biclusters. In Nutt and Sonar datasets, their performance has increased or decreased respectively.

- VOTE appears more efficient than MIX for the Nutt dataset which is the larger. However, the size of the biclusters seems unaffected MIX in other experiments.
- The difference of performance between VOTE/MIX and BGP/COAS is large. We observe that the size of the bicluster may impact the performance of the methods but there is no clear rule, it is only dependent on the data. Further investigation will be necessary.

In summary VOTE and MIX produce the best performances, the third is BGP and the last is COAS. Knowing that VOTE and MIX require less computing time than BGP, both appear therefore more efficient.

6   Conclusions

Unlike to the standard clustering contexts, biclustering considers both dimensions of the matrix in order to produce homogeneous submatrices. In this work, we have presented the approach of ensemble biclustering which consists in generating a collection of biclustering solutions then to aggregate them. First, we have showed how to use the CC algorithm to generate the collection. Secondly, concerning the aggregation of the collection of biclustering solutions, we have extended the use of four consensus functions commonly used in the clustering context. Thirdly we have evaluated the performance of each of them.

On simulated and real datasets, the ensemble approach appears fruitful. The results show that it improves significantly the performance of biclustering whatever the consensus function among VOTE, MIX and BGP. Specifically, VOTE and MIX give clearly the best results in all experiments and require less computing than BGP. We thus recommend to use one of these two methods for ensemble biclustering problems.

References

1. Breiman, L.: Bagging predictors. Mach. Learn. **24**, 123–140 (1996)
2. Breiman, L.: Random forests. Mach. Learn. **45**, 5–32 (2001)
3. Busygin, S., Prokopyev, O., Pardalos, P.: Biclustering in data mining. Comput. Oper. Res. **35**(9), 2964–2987 (2008)
4. Cheng, Y., Church, G.M.: Biclustering of expression data. In: Proceedings of the International Conference Intelligent Systems for Molecular Biology, vol. 8, pp. 93–103 (2000)
5. De Smet, R., Marchal, K.: An ensemble biclustering approach for querying gene expression compendia with experimental lists. Bioinformatics **27**(14), 1948–1956 (2011)
6. Dempster, A., Laird, N., Rubin, D.: Maximum likelihood from incomplete data via the em algorithm. J. R. Stat. Soc. Ser. B **39**(1), 1–38 (1977)
7. Dhillon, I.S.: Co-clustering documents and words using bipartite spectral graph partitioning. In: Proceedings of the Seventh ACM SIGKDD International Conference on Knowledge Discovery and Data Mining, KDD 2001, pp. 269–274 (2001)

8. Diaz-Uriarte, R., Alvarez de Andres, S.: Gene selection and classification of microarray data using random forest. BMC Bioinform. **7**(3) (2006)
9. Dietterich, T.G.: Ensemble methods in machine learning. In: Kittler, J., Roli, F. (eds.) MCS 2000. LNCS, vol. 1857, pp. 1–15. Springer, Heidelberg (2000)
10. Dudoit, S., Fridlyand, J.: Bagging to improve the accuracy of a clustering procedure. Bioinformatics **19**(9), 1090–1099 (2003)
11. Erten, C., Sözdinler, M.: Improving performances of suboptimal greedy iterative biclustering heuristics via localization. Bioinformatics **26**, 2594–2600 (2010)
12. Fern, X.Z., Brodley, C.E.: Solving cluster ensemble problems by bipartite graph partitioning. In: Proceedings of the Twenty-First International Conference on Machine Learning, ICML 2004, p. 36 (2004)
13. Frossyniotis, D., Likas, A., Stafylopatis, A.: A clustering method based on boosting. Pattern Recogn. Lett. **25**, 641–654 (2004)
14. Govaert, G.: Simultaneous clustering of rows and columns. Control Cybern. **24**(4), 437–458 (1995)
15. Govaert, G., Nadif, M.: Clustering with block mixture models. Pattern Recogn. **36**, 463–473 (2003)
16. Hanczar, B., Nadif, M.: Bagging for biclustering: application to microarray data. In: Balcázar, J.L., Bonchi, F., Gionis, A., Sebag, M. (eds.) ECML PKDD 2010, Part I. LNCS, vol. 6321, pp. 490–505. Springer, Heidelberg (2010)
17. Hanczar, B., Nadif, M.: Ensemble methods for biclustering tasks. Pattern Recogn. **45**(11), 3938–3949 (2012)
18. Hartigan, J.A.: Direct clustering of a data matrix. J. Am. Stat. Assoc. **67**(337), 123–129 (1972)
19. van der Laan, M., Pollard, K., Bryan, J.: A new partitioning around medoids algorithm. J. Stat. Comput. Simul. **73**(8), 575–584 (2003)
20. Lazzeroni, L., Owen, A.: Plaid models for gene expression data. Technical report, Stanford University (2000)
21. Maclin, R.: An empirical evaluation of bagging and boosting. In: Proceedings of the Fourteenth National Conference on Artificial Intelligence, pp. 546–551. AAAI Press (1997)
22. Madeira, S.C., Oliveira, A.L.: Biclustering algorithms for biological data analysis: a survey. IEEE/ACM Trans. Comput. Biol. Bioinform. **1**(1), 24–45 (2004)
23. Papadimitriou, C.H., Steiglitz, K.: Combinatorial Optimization: Algorithms and Complexity. Prentice-Hall Inc., Upper Saddle River (1982)
24. Reichardt, J., Bornholdt, S.: Statistical mechanics of community detection. Phys. Rev. E **74**, 016110 (2006)
25. Schapire, R.: The boosting approach to machine learning: an overview. In: Denison, D.D., Hansen, M.H., Holmes, C.C., Mallick, B., Yu, B. (eds.) Nonlinear Estimation and Classification, vol. 171, pp. 149–171. Springer, New York (2003)
26. Strehl, A., Ghosh, J.: Cluster ensembles - a knowledge reuse framework for combining multiple partitions. J. Mach. Learn. Res. **3**, 583–617 (2002)
27. Topchy, A.P., Law, M.H.C., Jain, A.K., Fred, A.L.: Analysis of consensus partition in cluster ensemble. In: Fourth IEEE International Conference on Data Mining, pp. 225–232 (2004)
28. Topchy, A., Jain, A.K., Punch, W.: A mixture model of clustering ensembles. In: Proceedings of the SIAM International Conference on Data Mining (2004)
29. Turner, H., Bailey, T., Krzanowski, W.: Improved biclustering of microarray data demonstrated through systematic performance tests. Comput. Stat. Data Anal. **48**(2), 235–254 (2005)

Fuzzy C-Means Stereo Segmentation

Michal Krumnikl[(⊠)], Eduard Sojka, and Jan Gaura

Faculty of Electrical Engineering and Computer Science,
Department of Computer Science, VŠB - Technical University of Ostrava,
17. listopadu 15, 708 33 Ostrava-Poruba, Czech Republic
{michal.krumnikl,eduard.sojka,jan.gaura}@vsb.cz
http://www.cs.vsb.cz/

Abstract. An extension to the popular fuzzy c-means clustering method
is proposed by introducing an additional disparity cue. The creation of
the fuzzy clusters is driven by a degree of the stereo match and thus it
enables to separate the objects not only by their different colours but also
on their different spatial depth. In contrast to the other approaches, the
clustering is not performed on the individual input images, but on the
stereo image pairs and takes into accounts the stereo matching properties known from the stereo matching algorithms. The proposed method is
capable of calculating the output segmentations, as well as the disparity
maps. The results of the algorithm show that the proposed method can
improve the segmentation in difficult settings. However, the drawback of
this approach is that it requires the stereo image pairs of the segmented
scenes that are not always easily obtainable.

Keywords: Fuzzy c-means · Segmentation · Stereo matching ·
Disparity

1 Introduction

In this paper we would like to propose an extension to the popular fuzzy
c-means clustering method by introducing an additional disparity cue. The reason for introducing the additional cues is to improve the segmentation. This can
be achieved by using the following approach, but only under the condition of
having the stereo image pair of the segmented scene. Beside the segmentation
with the additional depth constraints, our method is also capable of producing
the disparity map of the input image pair and hence can be also considered as
a form of the stereo matching algorithm.

The following text describes the adaptation of the fuzzy c-means algorithm
to perform the clustering in space extended by the dimension of the disparity.
The creation of the clusters will be driven by a degree of the stereo match (this
measure will be described later on). An attractive aspect of this strategy is that
we are able to take advantage of known number of depth levels or objects (if this
information is available).

© Springer International Publishing Switzerland 2015
A. Fred et al. (Eds.): ICPRAM 2014, LNCS 9443, pp. 35–49, 2015.
DOI: 10.1007/978-3-319-25530-9_3

The motivation for our work was to provide an algorithm that can separate objects based on their different colour and spatial depth. We regard this method as more suitable in specific cases (will be described later on) than the segmentation with the final disparity maps of the stereo matching algorithms. The distance, based on both dissimilarities (spatial and colour), provide more sensitive segmentation (especially on segment borders) than the segmentation performed on the filtered disparity maps which contain only the best matches, and do not take into account segment properties. The algorithm was originally developed for a very specific purpose – the segmentation of the moss clusters (as a part of a biological research involving these species). Therefore, we have tested and evaluated the algorithm mainly on the "Map" dataset, introduced in [26], as it strongly resembles the stone structures which are frequently covered by the moss layers. However, as we will show in the next paragraphs, the algorithm can be used in more general cases. The main domain of application of our algorithm is defined by the following constraints:

- The images should contain relatively small number of segments.
- The segments should be preferably planes or linear gradients.
- There should be no or minimal occlusions.

The clustering technique is usually described as a process of forming partitions from a data set on the basis of a performance function, also known as an objective function. The underlying idea of our algorithm is to consider the disparity space (e.g., in disparity maps) as a specific type of the data set, consisting of clusters representing the three dimensional objects of the scene. The fuzzy c-means algorithm has already been used to create the segmentations based on the depth information or disparity maps, e.g., [1,22], and was also adapted to incorporated the spatial neighbourhood information, e.g., [7,17,19], but in these approaches, the algorithms were run on the input data already containing the depth information. In contrast, the proposed algorithm does not need the depth information in advance, since it calculates it itself by means of the stereo matching.

The stereo matching problem itself is a multicriterion decision problem. The most common classification of the stereo matching algorithms is based on the size of the processed area. In this way, we recognize the local and global methods. In the local matching methods, the correspondence of a pixel is based on the similarity of its neighbourhood. The similarity itself can be computed using the measures such as the sum of the absolute differences (SAD), sum of the squared difference (SSD), normalized SSD, normalized cross-correlation etc. A comparison of the different similarity measures can be found in [9]. The global methods usually tend to minimize an energy function, e.g., by using the dynamic programming [8,30], graph cuts [6,16], Markov random fields [5] or belief propagation with segmentation [15].

The problem has been also solved by the fuzzy aggregation operators [31] or fuzzy relaxation technique [23]. The last method improves the matching in case of partially occluded objects. In [3], a fuzzy integral was introduced to improve the results obtained with the classical fuzzy averaging operators. The basic idea of using the clustering technique together with the stereo matching process was introduced in [4,28] and further developed in [18,27,29,32]. Compared to these, our approach differs in several aspects. The clustering is not performed on the individual input images, but on both stereo images simultaneously, and takes into account the matching properties. In each step, the clusters are adjusted to minimize the matching cost.

The paper is organized as follows. In Sect. 2 we briefly introduce the classical fuzzy c-means algorithm. Then, the extension of the fuzzy c-means is described in Sect. 3 in order to provide the depth segmentation based on the differences of the two stereo images. The experimental validation and the benchmark results are provided in Sect. 4. Finally, conclusions are presented in Sect. 5.

2 Fuzzy C-Means

Let us briefly introduce the original method. Fuzzy c-means is a widely used clustering technique, developed by [10] and improved by [2]. It is based on a standard least squared error model that generalizes an earlier and popular non-fuzzy c-means mode [20]. Fuzzy c-means can be generalized in many ways to include, e.g., Minkowski, Hamming, Canberrar or hybrid distances.

The fuzzy c-means algorithm attempts to partition a collection of n data points $\{x_k\}_{k=1}^n$ into a collection of c fuzzy clusters (represented by the cluster centres) on the basis of a distance d between the cluster centre and the data point. The algorithm is minimizing the objective function $J(U, V)$, where $V = (v_1, \ldots, v_c)$ is the set of cluster centres and $U = [u_{ki}]$ is the $n \times c$ membership matrix. The space of all possible values of U is denoted as U_f. The elements of the matrix U are organized as follows. The column i gives the membership of all n input data points (rows) in the cluster i for $i = 1 \ldots c$. The u_{ki} stands for the membership of the k-th point of the i-th cluster. The idea is that the closer the data point is to the cluster centre, the larger is its membership value towards that specific cluster. Consequently, the sum of all memberships of the data point across all clusters is equal to one. The fuzzy membership is formally given by the following constraint

$$U_f = \{U = (u_{ki}) : \sum_{j=1}^{c} u_{kj} = 1, 1 \leq k \leq n;$$

$$u_{ki} \in [0,1], 1 \leq k \leq n, 1 \leq i \leq c\}. \tag{1}$$

The minimized objective function $J(U, V)$ is defined as [2]

$$J(U, V) = \sum_{i=1}^{c} \sum_{k=1}^{n} (u_{ki})^m d(x_k, v_i), (1 \leq m \leq \infty), \tag{2}$$

where u_{ki} is a degree of membership of x_k in the cluster i, and v_i represents the centre of the cluster. The parameter m is called the weighting exponent of the model. For $m = 1$, the memberships converge to 0 or 1, producing a crisp partitioning. The best choice for m is probably in the interval $[1.5, 2.5]$, where $m = 2$ is the most common choice [24]. The distance $d(x_k, v_i)$ represents (usually) Euclidean distance between the k-th data point and the i-th cluster centre.

We should notice that the minimization of the objective function $J(U, V)$ is not an exact minimization but an iteration procedure of so called "alternate minimization". In essence, the algorithm is searching for a local optimal solution, which we will denote with stripe (e.g., $\bar{U}$). The overall iterative process may be summarised as follows.

Algorithm Steps

1. Initialize the matrix U by randomly generated u_{ki} membership coefficients for all cluster centres $\bar{V} = (\bar{v_1}, \ldots, \bar{v_c})$.
2. Find the optimal U by iteratively calculating $\bar{U} = \arg \min_{U \in U_f} J(U, \bar{V})$. The following solution can be derived using the Lagrange multiplier method [20]

$$\bar{u}_{ki} = \left[\sum_{j=1}^{c} \left(\frac{d(x_k, \bar{v}_i)}{d(x_k, \bar{v}_j)} \right)^{\frac{2}{m-1}} \right]^{-1}, (x_k \neq v_i). \qquad (3)$$

The solution for $(x_k = v_i)$ is obviously $\bar{u}_{ki} = 1$.
3. Find the optimal V by calculating $\bar{V} = \arg \min_V J(\bar{U}, V)$. The solution is computed by differentiating J with respect to V [20]:

$$\bar{v}_i = \frac{\sum_{k=1}^{n} (\bar{u}_{ki})^m x_k}{\sum_{k=1}^{n} (\bar{u}_{ki})^m}. \qquad (4)$$

4. Repeat from step 2 until $\bar{U}$ and $\bar{V}$ is convergent.

The convergence is achieved when $\max_{k,i} |\bar{u}_{ki} - u_{ki}| < \epsilon$, where $\bar{u}$ is the new solution, u is the value from the previous iteration and ϵ is a small positive number, the threshold. Alternatively, we can use $\max_{1 \leq i \leq c} \|\bar{v}_i - v_i\| < \epsilon$ as a convergence condition.

3   Introducing the Matching Constraint to Fuzzy C-Means

In a simplified way, we can say that the original fuzzy c-means algorithm (when used in image processing) is usually based only on the pixel positions and their intensities (colours). In our approach, we have extended this algorithm to include the matching constraints. First, by expanding the dimension of the data vector to include the disparity (depth), and then, by evaluating the dissimilarity of the stereo pair (which will be explained later).

As stated in Sect. 2, the algorithm attempts to partition the elements with respect to a given criterion, defined as a degree of belonging that is related inversely to the distance. However, for the depth segmentation, we need to add additional components measuring the intensity (colour) difference between the point and its supposed projection and the distance between the point disparity and the disparity of its supposed cluster. The sources of the spatial information are the small differences in the stereo images. In that way, we associate the clusters with the disparity space. Therefore, we have to define the vector of the cluster centre as

$$v = (v_X, v_Y, v_I, v_D), \tag{5}$$

where v_X, v_Y stand for the spatial position, v_I for the brightness and v_D for the disparity value. For the clarity, the capitalized subscripts, X, Y, I and D, are used to indicate the vector elements (e.g., v_X). The small subscripts will later be used to specify a particular vector from the set (e.g., x_k).

Fig. 1. Illustration of the rationale behind the algorithm. The left figure shows the coloured disparity levels of the dataset [25], while the right one depicts the original pixel colours. The algorithm is based on the observation that the objects share the similar disparity, as well as similar colour. This can be clearly seen on the red lamp in the foreground or the white statue on the left (Color figure online).

Our new membership function takes into account the dissimilarity of the left image pixel ($\phi_L(x_X, x_Y)$) and the right image pixel shifted by the average cluster disparity ($\phi_R(x_X + v_D, x_Y)$). Basically, we use the disparity in the similar fashion as the intensity, grouping the pixels sharing the same, or almost the same disparity value (see Fig. 1). For this, we need to adapt the membership function to penalize the pixels having the incorrect match (not similar to their projections on the other image) and provide the way of measuring the distance between the cluster centres and pixels with associated disparity value.

We propose the use of the extended vector space model with the additional dimensions reflecting the disparity and pixel dissimilarity in the stereo image pair. The distance in the proposed vector space is, for clarity, separated into the two components (d and d_s), described later on.

The proposed fuzzy stereo partitioning is carried out using the following membership function (the subscripts k, i, j are the indexes)

$$\bar{u}_{ki} = \left[\sum_{j=1}^{c} \left(\frac{d^2(x_k, \bar{v}_i) + d_s^2(x_k, \bar{v}_i)}{d^2(x_k, \bar{v}_j) + d_s^2(x_k, \bar{v}_j)} \right)^{\frac{1}{m-1}} \right]^{-1}, (x_k \neq v_i). \quad (6)$$

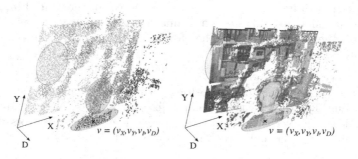

Fig. 2. Visualisation of the data points and their clusters taken from our experiments. The points on the left figure are coloured according to the disparity levels associated with them. The right figure shows their real colour. The both figures shows the depth levels as obtained from the calculations of the proposed modification of the fuzzy c-means algorithm (Color figure online).

The membership $\bar{u}_{ki}$ is related inversely to the distance between the processed point and the cluster centre (as calculated in the previous iteration). The new term d_s reflects the correctness of the stereo match between the pixel of the left (ϕ_L) and its projection on the right (ϕ_R) image (the subscripts X, Y, D denotes the vector elements):

$$d_s^2(x, v) = \lambda_m (\phi_L(x_X, x_Y) - \phi_R(x_X + v_D, x_Y))^2, \quad (7)$$

where x is the data point (vector) and v is the cluster centroid. The uppercase subscript of the vector denotes its component. The constant λ_m stands for the weight of the matching term. For ϕ_L and ϕ_R we assume the rectified images. It is possible to replace the difference $\phi_L(x_X, x_Y) - \phi_R(x_X + v_D, x_Y)$ by the difference of the aggregating windows (SAD, SSD, etc.), but as the aggregation of the pixels is inherently given by the fuzzy c-means, it does not provide any further advantage and even worsens the results by blurring the edges. The distance $d(x, v)$ is calculated (as in original method) using the Euclidean distance:

$$d^2(x, v) = \lambda_i (x_I - v_I)^2 + \lambda_d (x_D - v_D)^2 + \\ \lambda_s (x_X - v_X)^2 + \lambda_s (x_Y - v_Y)^2, \quad (8)$$

where x_X, x_Y are the pixel coordinates, x_I colour intensity and x_D is the disparity value. When compared to the original method, we have added the term measuring the disparity distance of the processed point and the cluster (see Fig. 2). For the pixel disparity x_D we can take an initial guess since, as we will show later, the algorithm is quite insensitive to this value. Basically, it only helps in the beginning to form the initial clusters. The values λ_i, λ_d and λ_s denote the intensity, disparity and spatial weights. The effects of these weights are discussed with results (Sect. 4).

The iteration steps remain the same as in Sect. 2. The outline of the algorithm can be summarized as follows: (i) choose the proper parameters, especially the number of clusters (discussed in Sect. 3.1), (ii) to each point assign random cluster membership coefficients, (iii) in each iteration compute the centroid for each cluster (Eq. 4), followed by the computation of the membership coefficients for all points (Eq. 6). Repeat this step until the algorithm has converged. Finally, create the output disparity map based on the cluster disparities (iv).

The algorithm was tested on several types of real images (depicting the processed botanical samples) and also on the standard dataset used for the evaluation of the stereo matching algorithms [25]. While our approach is not intended to be used as the general purpose stereo matching algorithm, we would like to give the reader an opportunity to examine the results in the standard stereo matching benchmark tests (see Sect. 4).

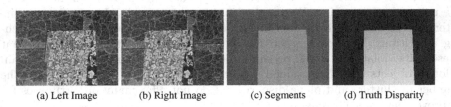

| (a) Left Image | (b) Right Image | (c) Segments | (d) Truth Disparity |

Fig. 3. The reference images (a, b), the ground truth disparity map (d), and its segments (c) used for the demonstration of the cluster count problem.

3.1 Cluster Count Problem

The disadvantage of the fuzzy c-means (as well as k-means) is the result dependency on the initial choice of weights. This is also true for our method. Despite the algorithm minimizes the intra-cluster variance, calculated minimum is still only a local minimum. But more serious problem of the fuzzy c-means algorithm is that it requires the number of clusters to be known in advance.

The correct choice of the cluster count is ambiguous, with interpretations depending on the shape and scale of the data point distribution in the input data set and the desired resolution. This may seem as a disadvantage for general settings, but may be an advantage for special cases, where the number of segments or number of disparity planes is already known. For example, in Fig. 3

the box is the only object in the foreground, and can be easily represented by only a small number of segments. As you can see (Fig. 4), with only a few clusters, we are able to acquire very precise disparity map and by increasing the number of the segments, we are able to capture even smaller changes in the disparity gradient (the box in the example is slightly tilted). We can say that by choosing the number of clusters, we can set, whether we are more interested in large segments covering the whole objects, or small fine-grained parts.

The results of our approach surpass (but only for the specific types of scenes, similar to the sample images) the performance of the majority of the standard state-of-the-art algorithms (see Table 1, "Map" column). However, due to the algorithm specialization, it is less suitable for the other types of scenes. But still, the additional cue improves the segmentation results.

4   Tests and Results

This section describes the experiments and shows the results confirming the anticipated segmentation features and proper depth discrimination.

First, we have performed the tests on the images fulfilling the assumptions, we made in the beginning – the scenes with only a few objects, each having almost the flat depth. The "Map" dataset (Fig. 4) complies with these requirements. The results for this specific dataset are very satisfactory (Table 1, column "Map"); however, the results for the other types of image pairs (from the dataset) are not very encouraging. We do not consider this as a disadvantage, since the intentions of this algorithm are different than the general purpose stereo matching algorithms. The explanation for the results on the other samples is that these pairs violate the initial presumptions of our algorithm; the scenes contain a lot of objects with fine-grained disparity. The limits of our algorithm – the number of clusters and plane disparities – do not offer many opportunities for improvements in such general cases.

To illustrate the algorithm performance on the images with optimal object configurations, we have chosen several samples from the Adobe Open Source Data Sets[1]. The data set contains stereo images and ground truth segmentation of the foreground object. The results of the selected images are visible in Fig. 6. We have to point out that these images illustrate the optimal cases.

Nevertheless, we have also performed the tests on the images that are not very suitable for our approach. The absolute results with the comparison of the other algorithms are shown in Table 1. The evaluation has been performed on the Middlebury dataset [25]. The full list of algorithms is available on the Middlebury stereo vision website. While our algorithm is not the typical stereo matching algorithm, due to the lack of more suitable, generally accepted dataset for segmenting the stereo images, we decided to perform the tests on these images. The parameters were maintained the same for all images – cluster count $n = 200$, $\lambda_d = 1.0$, $\lambda_s = 0.1$, $\lambda_i = 0.05$, and $\lambda_m = 0.1$.

---

[1] http://sourceforge.net/adobe/adobedatasets/.

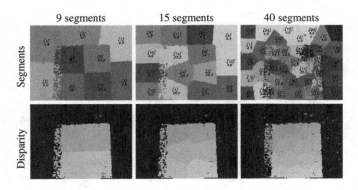

Fig. 4. The influence of the cluster count on the output disparity map. For better reading, the segments are coloured, numbered (number in brackets), and marked with their disparity values (the value below the number in brackets). The reference images are depicted in Fig. 3. The subfigures show the results of the modified fuzzy c-means algorithm set to 9, 15 and 40 segments. The top subfigures show the segments, while the bottom ones show the disparity maps obtained from the segments disparity values (Color figure online).

Table 1. The performance of the modified fuzzy c-means algorithms according the Middlebury stereo test bed [25]. The overall performance is measured by the percentage of bad pixels in the non-occluded areas (nocc). The performance measured on the whole image (all) is provided as well. Our algorithm is denoted as FZ. The total cluster count was set to 200. In order to give a better idea of the performance of our methods compared to the state-of-the-art algorithms, we have included the results of the selected algorithms from the Middlebury evaluation.

	Tsukuba		Venus		Teddy		Cones		Map		
Algorithm	nocc	all	nocc	all	nocc	all	nocc	all	nocc	all	avg
[12]	2.61	3.29	0.25	0.57	5.14	11.8	2.77	8.35	1.09	2.82	5.33
[15]	0.97	1.75	0.16	0.33	6.47	10.7	4.79	10.7	3.39	5.79	5.85
[11]	3.26	3.96	1.00	1.57	6.02	12.2	3.06	9.75	1.12	2.97	6.09
[14]	1.94	4.12	1.79	3.44	16.5	25.0	7.70	18.2	0.74	6.82	11.51
[8]	4.12	5.04	10.1	11.0	14.0	21.6	10.5	19.1	6.04	12.12	13.77
SSD	5.23	7.07	3.74	5.16	16.5	24.8	10.6	19.8	8.49	14.57	14.28
FZ	12.7	14.3	12.5	13.5	32.3	37.8	32.0	36.9	**0.72**	**7.17**	21.93

The proposed algorithm converges approximately after 15 iterations on all images of the given set. The outputs with 100 segments are displayed in Fig. 7 (evaluated outputs with 200 segments were not used for the illustration purposes, due to the hard distinguishability of the small clusters). The images in the upper row show the segments. The disparity maps obtained from the segment properties are displayed below. As you can see, the proposed algorithm is

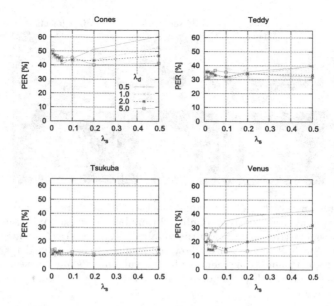

Fig. 5. The algorithm results achieved with different disparity and spatial weights (λ_d and λ_s). The algorithm was set to generate 100 segments. The different disparity weights (λ_d) are represented by the different line colours. The significant effects of the disparity weight (λ_d) can be seen only on the images containing the planar objects (e.g., the "Venus" pair) (Color figure online).

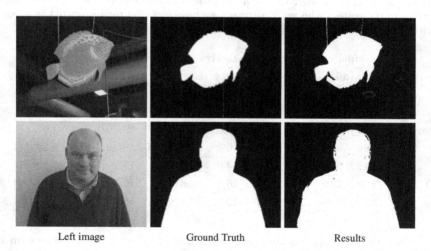

Left image   Ground Truth   Results

Fig. 6. Segmentation results of two samples from the Adobe Open Source Data Sets. These samples illustrate the ideal configurations for the proposed algorithm – raised flat foreground objects.

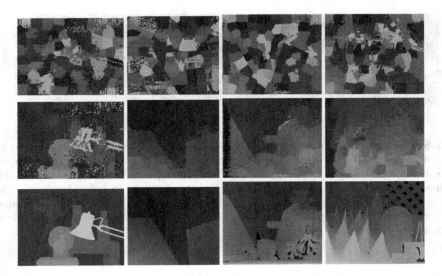

Fig. 7. The images show the disparity and cluster maps obtained for the default Middlebury dataset using our expanded fuzzy c-means algorithm (fz). The ground truth data are provided in the last row. As you can see, the output disparity maps are not as good as the results from the "Map" dataset (Fig. 4). The reason is that the Middlebury dataset contains images with a lot of details and a set of various objects, which contradicts the initial algorithm assumptions. In order to improve the performance it is necessary to significantly increase the number of the clusters, which consequently leads to a much longer processing time. Unfortunately, this still does not guarantee for all inputs the results comparable to the best algorithms.

capable of obtaining the disparity maps of more sophisticated scenes, but not at the level of detail as the generally used stereo matching approaches.

To increase the overall performance, it is possible to increase the number of clusters, which in result leads to a more grained segmentation, where each segment can have different disparity. The drawback of a huge number of clusters is the increasing computational time. At the certain level, the additional increasing of cluster count starts to be inefficient. We have used no more than 200 segments.

During the development, we have also performed several experiments to investigate the effects of the algorithm parameters on the segmentation performance. The parameter settings may vary from scenario to scenario, but generally, only two parameters appear to be particularly influential - the spatial and disparity weight (λ_s and λ_d). Figure 5 shows the influence of these weights on the output segmentation consisting of 100 segments. The experiment showed the significant effect of the disparity weight (λ_d) mainly on the images containing the planar objects (e.g., the "Venus" pair). This is a predicted behaviour as our algorithm favours planar disparities. On an example of "Venus" pair, you can see that the increasing disparity weight forces the algorithm to create segments with less disparity deviations from the cluster centroid, leading to the better results.

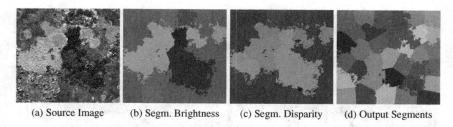

(a) Source Image   (b) Segm. Brightness   (c) Segm. Disparity   (d) Output Segments

Fig. 8. The segmentation results of the moss sample using the modified fuzzy c-means algorithm (Sect. 3). The figure shows (from left to right): the left image of the input pair depicting the moss layers on the stone base, the segments coloured according the average colour, the segments coloured according the disparity and the visualization of the clusters itself. As you can see, our modification of the fuzzy c-means still retains the properties of the original algorithm and in addition provides the disparity values (Color figure online).

However, for images not containing such objects (e.g., "Teddy" or "Tsukuba") the change in these parameters has only a small impact on the results. We have not evaluated all possible parameter configurations for all dataset images, but empirically, we can say that the best results were achieved with $\lambda_s = 0.1$. Increasing this value forced the algorithm to create too compact clusters and, vice versa, decreasing λ_s caused merging too distant pixels into one cluster.

In the application that the algorithm was originally developed for, it was important to separate the layer of the base (usually the stone) and the layer above, formed by the moss structures. An example is illustrated in Fig. 8. As you can see, the resulting segmentation strongly benefits from the inherit features of the algorithm. The design of the algorithm was strongly driven by the expected look of the captured samples.

5   Conclusions

In this paper, we have presented a modification of the fuzzy c-means algorithm. The fuzzy c-means algorithm is one of the most popular clustering techniques in image processing. In the past, it has been modified in many ways to take into account different constraints. In our case, we have added an additional disparity constraint and examined its impact on the segmentation performance and depth discrimination. In the context of the image segmentation, we see the advantage of the proposed joint analysis using brightness and depth constraints. We believe, such combination improves the segmentation by creating edges not only in places where brightness changes abruptly but also in places of the depth discontinuities. Objects of the similar colour in different depths may be connected by the classical algorithm but with an additional depth constraint they are separated correctly.

The motivation was to develop a segmentation technique that can be used in cases, where we have the possibility of obtaining the stereo images and, in such way, improve the segmentation by applying additional depth information. In the biological application (the segmentation of the moss layers), the method provided better results than the standard fuzzy c-mean algorithm. As the algorithm was intended for this specific application, we have mainly tested and evaluated the algorithm on the datasets that resemble stone structures (e.g., the standard "Map" dataset). For such cases, the algorithm provides very good results.

To sum up, the paper proposed the method that improves the segmentation in cases where the pixel intensities are not sufficient for correct segmentation and the stereo images are available. This area of research, however, still offers the space for improvements. The results can be further improved by tuning the distance weights. The goal is to create an algorithm that can automatically adapt the weight variables according to the input dataset. Similar approaches were already published for the closely related k-means clustering, e.g. [13,21], and should be applicable to the fuzzy c-means as well.

Acknowledgements. This work was partially supported by the SGS grant No. SP2014/ 170 of VŠB - Technical University of Ostrava, Faculty of Electrical Engineering and Computer Science.

References

1. Aik, L.E., Choon, T.W.: Enhancing passive stereo face recognition using pca and fuzzy c-means clustering. Int. J. Video Image Process. Netw. Secur. **11**(4), 1–5 (2011)
2. Bezdek, J.C.: Pattern Recognition with Fuzzy Objective Function Algorithms. Kluwer Academic Publishers, Norwell (1981)
3. Bigand, A., Bouwmans, T., Dubus, J.: A new stereomatching algorithm based on linear features and the fuzzy integral. Pattern Recogn. Lett. **22**(2), 133–146 (2001)
4. Bleyer, M., Gelautz, M.: A layered stereo algorithm using image segmentation and global visibility constraints. In: Proceedings of the IEEE International Conference on Image Processing, pp. 2997–3000 (2004)
5. Boykov, Y., Veksler, O., Zabih, R.: Markov random fields with efficient approximations. In: Proceedings of the IEEE Conference on Computer Vision and Pattern Recognition, pp. 648–655 (1998)
6. Boykov, Y., Veksler, O., Zabih, R.: Fast approximate energy minimization via graph cuts. IEEE Trans. Pattern Anal. Mach. Intell. **23**(11), 1222–1239 (2001)
7. Chuang, K.S., Tzeng, H.L., Chen, S., Wu, J., Chen, T.J.: Fuzzy c-means clustering with spatial information for image segmentation. Comput. Med. Imaging Graph. **30**(1), 9–15 (2006)
8. Cox, I.J., Hingorani, S.L., Rao, S.B., Maggs, B.M.: A maximum likelihood stereo algorithm. Comput. Vis. Image Underst. **63**, 542–567 (1996)
9. Cyganek, B.: An Introduction to 3D Computer Vision Techniques and Algorithms. Wiley, New York (2007)
10. Dunn, J.C.: A fuzzy relative of the ISODATA process and its use in detecting compact well-separated clusters. J. Cybern. **3**(3), 32–57 (1973)
11. Hirschmüller, H.: Accurate and efficient stereo processing by semi-global matching and mutual information. In: Proceedings of the IEEE Conference on Computer Vision and Pattern Recognition, pp. 807–814 (2005)

12. Hirschmüller, H.: Stereo vision in structured environments by consistent semi-global matching. In: Proceedings of the IEEE Conference on Computer Vision and Pattern Recognition, vol. 2, pp. 2386–2393. IEEE Computer Society (2006)

13. Huang, J., Ng, M., Rong, H., Li, Z.: Automated variable weighting in k-means type clustering. IEEE Trans. Pattern Anal. Mach. Intell. 27(5), 657–668 (2005)

14. Kim, J., Kolmogorov, V., Zabih, R.: Visual correspondence using energy minimization and mutual information. In: Proceedings of the IEEE International Conference on Computer Vision, vol. 2, pp. 1033–1040 (2003)

15. Klaus, A., Sormann, M., Karner, K.F.: Segment-based stereo matching using belief propagation and a self-adapting dissimilarity measure. In: Proceedings of the IEEE Conference on Computer Vision and Pattern Recognition, vol. 3, pp. 15–18 (2006)

16. Kolmogorov, V., Zabih, R.: Computing visual correspondence with occlusions using graph cuts. In: Proceedings of the IEEE International Conference on Computer Vision, vol. 2, pp. 508–515 (2001)

17. Liew, A.W.C., Leung, S.H., Lau, W.H.: Fuzzy image clustering incorporating spatial continuity. IEEE Proc. Vis. Image Sig. Process. 147, 185–192 (2000)

18. Liu, T., Zhang, P., Luo, L.: Dense stereo correspondence with contrast context histogram, segmentation-based two-pass aggregation and occlusion handling. In: Proceedings of the Pacific-Rim Symposium on Image and Video Technology, pp. 449–461 (2009)

19. Meena, A., Raja, R.: Spatial fuzzy c means pet image segmentation of neurodegenerative disorder. CoRR abs/1303.0647 (2013)

20. Miyamoto, S., Ichihashi, H., Honda, K.: Algorithms for Fuzzy Clustering: Methods in C-Means Clustering with Applications. Studies in Fuzziness and Soft Computing. Springer, Heidelberg (2008)

21. Modha, D., Spangler, S.: Feature weighting in k-means clustering. Mach. Learn. 52, 217–237 (2003)

22. Ntalianis, K. S., Doulamis, A., Doulamis, N., Kollias, S.: Unsupervised segmentation of stereoscopic video objects: investigation of two depth-based approaches. In: Proceedings of the 14th International Conference of Digital Signal Processing, vol. 2, pp. 693–696 (2002)

23. Ogawa, H.: A fuzzy relaxation technique for partial shape matching. Pattern Recogn. Lett. 15(4), 349–355 (1994)

24. Pal, N., Bezdek, J.: On cluster validity for the fuzzy c-means model. IEEE Trans. Fuzzy Syst. 3(3), 370–379 (1995)

25. Scharstein, D., Szeliski, R.: A taxonomy and evaluation of dense two-frame stereo correspondence algorithms. Int. J. Comput. Vis. 47(1–3), 7–42 (2002)

26. Szeliski, R., Zabih, R.: An experimental comparison of stereo algorithms. In: Triggs, B., Zisserman, A., Szeliski, R. (eds.) ICCV-WS 1999. LNCS, vol. 1883, pp. 1–19. Springer, Heidelberg (2000)

27. Taguchi, Y., Wilburn, B., Zitnick, C. L.: Stereo reconstruction with mixed pixels using adaptive over-segmentation. In: Proceedings of the IEEE Conference on Computer Vision and Pattern Recognition, pp. 1–8 (2008)

28. Tao, H., Sawhney, H.S., Kumar, R.: A global matching framework for stereo computation. In: Proceedings of the IEEE International Conference on Computer Vision, pp. 532–539 (2001)

29. Tombari, F., Mattoccia, S., di Stefano, L.: Segmentation-based adaptive support for accurate stereo correspondence. In: Proceedings of the Pacific-Rim Symposium on Image and Video Technology, pp. 427–438 (2007)

30. Veksler, O.: Stereo correspondence by dynamic programming on a tree. In: Proceedings of the IEEE Conference on Computer Vision and Pattern Recognition, vol. 2, pp. 384–390 (2005)
31. Zimmermann, H.: Fuzzy Set Theory and its applications. Kluwer Academic, Dordrecht (1991)
32. Zitnick, C.L., Kang, S.B.: Stereo for image-based rendering using image oversegmentation. Int. J. Comput. Vis. **75**(1), 49–65 (2007)

SCHOG Feature for Pedestrian Detection

Ryuichi Ozaki[✉] and Kazunori Onoguchi

Graduate School of Science and Technology, Hirosaki University, 3 Bunkyo-cho,
Hirosaki, Aomori 036-8561, Japan
h13ms504@stu.hirosaki-u.ac.jp, onoguchi@eit.hirosaki-u.ac.jp

Abstract. Co-occurrence Histograms of Oriented Gradients (CoHOG)
has succeeded in describing the detailed shape of the object by using
a co-occurrence of features. However, unlike HOG, it does not consider
the difference of gradient magnitude. In addition, the dimension of the
CoHOG feature is also very large. In this paper, we propose Similarity
Co-occurrence Histogram of Oriented Gradients (SCHOG) considering
the similarity and co-occurrence of features. Unlike CoHOG, SCHOG
quantize edge gradient direction to four directions. Therefore, the feature dimension for the co-occurrence between edge gradient direction
decreases greatly. In addition, the binary code representing the similarity between features is introduced. In spite of reducing the resolution
of the edge gradient direction, SCHOG realizes higher performance and
lower dimension than CoHOG by adding this similarity. In experiments
using the INRIA Person Dataset, SCHOG is evaluated in comparison
with the conventional CoHOG.

Keywords: Pedestrian detection · Co-occurrence histograms of oriented
gradients · Similarity · Support vector machine

1 Introduction

Recently, a pedestrian detection system have been put to practical use as a vehicle safety device [1]. Since features expressing characteristics of a person well is
important in these system, various features for pedestrian detection have been
proposed. T. Ojala et al. proposed the Local Binary Pattern (LBP) [2] representing the relation between the intensity of an interest pixel and the intensity
of eight adjacent pixels. This feature has been studied in various ways because
it's robust to illumination change and it's implemented easily. Y. Cao et al.
proposed the Advanced LBP [3] which is robust to noise and low intensity.
N. Dalal proposed HOG [4] feature which is robust to the change of the pedestrian's posture and the change of the illumination by generating the histogram
of the edge gradient orientation in each block and normalizing each block for
every cell. They also proposed the feature focusing on the gradient orientation
of the time series [5]. T. Watanabe et al. proposed CoHOG [6] feature that represented the co-occurrence of gradient orientation and showed high performance
for pedestrian detection. As other features using the co-occurrence, T. Kobayashi
et al. proposed a Gradient Local Auto-Correlation (GLAC) [7] which calculated

© Springer International Publishing Switzerland 2015
A. Fred et al. (Eds.): ICPRAM 2014, LNCS 9443, pp. 50–61, 2015.
DOI: 10.1007/978-3-319-25530-9_4

the autocorrelation of the position and edge gradient orientation. K. Yamaguchi et al. proposed a two-dimensional gradient orientation histogram using polar coordinates which can express small difference [8]. S. Walk et al. proposed the Color Self-Similarity (CSS) [9] feature using the similarity of HSV histogram in the local area. As mentioned above, the co-occurrence of feature is effective for improving the performance of pedestrian detection. However, there is a problem that the dimension of the feature increases significantly.

In this paper, we propose SCHOG which consists of the co-occurrence of edge gradient direction and the similarity. Although SCHOG quantizes edge gradient direction to the half of CoHOG, it can represent the shape of the object more finely than CoHOG by adding the similarity to the co-occurrence of edge gradient direction. Because the similarity is represented by the binary code, the dimension of SCHOG is a half of the conventional CoHOG in spite of adding the similarity. We evaluate three kind of similarity, such as the pixel intensity, the edge gradient magnitude and the edge gradient direction in the experiment. These values are not used directly in CoHOG. Therefore, the proposed feature can compensate information lost in CoHOG. In the experiment, the edge gradient magnitude showed the best performance.

Experimental result using the INRIA Person Dataset and the Support Vector Machine shows that the performance of SCHOG is better than the conventional CoHOG.

The rest of this paper is organized as follow. In Sect. 2, the proposed method is explained in detail and its extensibility is discussed. In Sect. 3, the performance of the SCHOG is evaluated by comparing with the conventional CoHOG. In Sect. 4, this paper is summarized.

2   Proposed Feature

Pedestrians show a various shape, e.g., standing, running, or shaking the hand. In addition, they wear clothes of various texture and color. Moreover, in the outdoor, illumination changes frequently and a lot of image noise occurs. CoHOG has solved these problems to some extent although it needs very large feature dimension. The proposed feature (SCHOG) improves CoHOG so that it can have better performance and lower feature dimension.

2.1   CoHOG

In this section, the outline of CoHOG is explained. CoHOG uses a two-dimensional histogram whose bin is a pair of edge gradient direction between the interest pixel and the offset pixel. The feature dimension becomes large since histograms are created for every combination of the interest pixel and the offset pixel. However, it can represent object shape finely and it is robust to the change of shape and illumination.

Fig. 1. Examples of gradient direction and gradient magnitude (Color figure online).

At first, the edge gradient magnitude (M) and the edge gradient direction (θ) are obtained from Eqs. (1) and (2).

$$\theta(x, y) = tan^{-1}\frac{f_y(x, y)}{f_x(x, y)} \tag{1}$$

$$M(x, y) = \sqrt{f_x(x, y)^2 + f_y(x, y)^2}, \tag{2}$$

where $f_x(x, y)$ and $f_y(x, y)$ denote edge gradient magnitude of horizontal direction and that of vertical direction in the pixel (x, y), which are calculated by Sobel operator. Gradient direction (θ) is quantized to eight directions by 45 degrees. Figure 1 shows the gradient direction image and the gradient magnitude image. The direction is represented by color and the magnitude is represented by the brightness. Same direction often appears around the contour of a pedestrian. CoHOG represents this characteristic by the co-occurrence of edge gradient direction between the interest pixel and the offset pixel.

31 offset pixels are set around the interest pixel as shown in Fig. 2. The interest pixel is included in offset pixels. Two dimensional histogram is created for every offset pixel. If the offset pixel corresponds the interest pixel, the histogram has eight bins because the gradient direction of each pixel is same. Except for this case, the histogram has $8 \times 8 = 64$ bins because the number of bins is a combination of gradient direction.

The input image is divided into several rectangular blocks as shown in Fig. 3. In each block, the 2D histogram is created for every offset pixels. Let (p, q) be

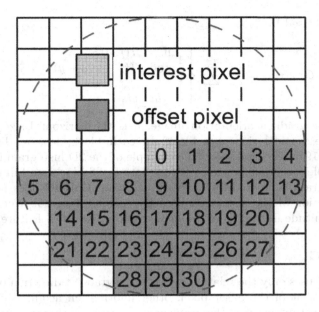

Fig. 2. The number and position of the offset pixel.

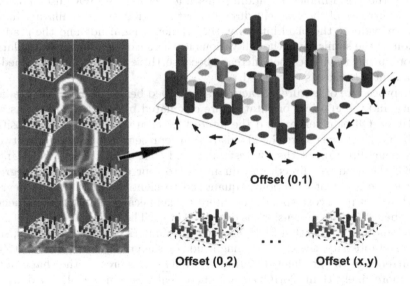

Fig. 3. Example of 2D histogram in CoHOG.

the image coordinate system whose origin is at the upper left of each block, (x, y) be a offset coordinate system whose origin is at the interest pixel and $C_{x,y}$ be the 2D histogram of an offset pixel (x, y). The bin $C_{x,y}(i, j)$ of 2D histogram

$C_{x,y}$ is incremented by

$$C_{x,y}(i,j) = \sum_{p=0}^{n-1}\sum_{q=0}^{m-1} \begin{cases} 1 & \begin{aligned} \text{if} \quad & I(p,q) == i \\ \text{and} \quad & I(p+x,q+y) == j \end{aligned} \\ \\ 0 & otherwise, \end{cases} \qquad (3)$$

where I is the gradient-orientation image, n is the horizontal size of a block and m is the vertical size of a block. In each block, feature dimension is $8 + 64 \times 30 = 1928$. Figure 3 shows the example of the 2D histogram in CoHOG. In this example, an input image is divided into 2×8 blocks and in each block, the 2D histogram is created for every offset pixels. CoHOG does not perform the normalization of the histogram because CoHOG does not accumulate the gradient magnitude in the bin of the histogram, unlike HOG feature.

2.2 SCHOG

Since CoHOG uses only the relation between the gradient direction of the interest pixel and that of the offset pixel, other information acquired on the way, such as the pixel intensity or the gradient magnitude, is thrown away. SCHOG improves the performance by adding this information. SCHOG uses not only the co-occurrence of the gradient direction but also that of the similarity. In this paper, we evaluate the pixel intensity, the gradient magnitude and the gradient direction as the similarity although various features can be use as the similarity. The computing time does not increase because these features are obtained as the gradient direction is calculated.

The procedure of feature extraction is described below. At first, the gradient intensity and the gradient orientation are calculated by Eqs. (1) and (2) as well as CoHOG. Offset pixels around the interest pixel are set as the same position as CoHOG. Next, we create the 2D histogram representing the relation between the gradient direction of the interest pixel and that of the offset pixel. Unlike CoHOG, the gradient direction is quantized to four directions by 90 degrees. However, the gradient direction is quantized to eight directions by 45 degrees when the offset pixel corresponds the interest pixel because this hardly influences the number of feature dimension, as described later. The main difference between SCHOG and CoHOG is that SCHOG adds the similarity between features, such as the pixel intensity, the gradient magnitude or the gradient orientation, to the co-occurrence of the gradient direction. SCHOG can represent the shape of the object more finely than CoHOG since these features which CoHOG does not use directly are incorporated. The similarity between the interest pixel and the offset pixel is given by

$$F_{sim1}(V_o, V_i) = \begin{cases} 0 \; if \quad T_1 < tan^{-1}\frac{V_i}{V_o} < T_2 \\ \\ 1 \; otherwise \end{cases} \qquad (4)$$

$$F_{sim2}(V_o, V_i) = \begin{cases} 0 & \begin{aligned} \text{if} & \quad T_3 < |V_o - V_i| \\ \text{or} & \quad T_4 > |V_o - V_i| \end{aligned} \\ 1 & otherwise, \end{cases} \tag{5}$$

where F_{sim1} is the similarity function for the pixel intensity or gradient magnitude, F_{sim2} is the similarity function for the gradient angle, V_i is the pixel intensity, the gradient magnitude or the gradient direction in the intensity pixel and V_o is the pixel intensity, the gradient magnitude or the gradient direction in the offset pixel. Thresholds T_1, T_2, T_3 and T_4 in Eqs. (4) and (5) were determined experimentally. The similarity returns 0 when features are similar and it returns 1 when features are different. The feature dimension is suppressed because the similarity is represented by the binary code.

Table 1. The name of SCHOG for each similarity.

Name	Similarity
SCHOG-pix	pixel intensity
SCHOG-gra	gradient magnitude
SCHOG-ang	gradient direction

We divide the input image into 6×12 blocks. Let (p, q) be the image coordinate system whose origin is at the upper left of each block, (x, y) be a offset coordinate system whose origin is at the interest pixel and $C_{x,y,s}$ be the histogram of an offset pixel (x, y) and similarity s. The bin $C_{x,y,s}(i, j, k)$ of histogram $C_{x,y,s}$ is incremented by

$$C_{x,y,s}(i, j, k) = \sum_{p=0}^{n-1} \sum_{q=0}^{m-1} \begin{cases} 1 & \begin{aligned} \text{if} & \quad I(p, q) == i \\ \text{and} & \quad I(p + x, q + y) == j \\ \text{and} & \quad F_{sim}(a, b) == k \end{aligned} \\ 0 & otherwise, \end{cases} \tag{6}$$

where I is gradient-orientation image, n and m represent the size of a block, a represents feature value at the offset pixel, b represents feature value at the interest pixel. k(0 or 1) represents the similarity. When the offset pixel corresponds the interest pixel, the dimension is 8. Since this case is not related to the co-occurrence, number of total feature dimension hardly increase even if the dimension is 8. The other offset pixel has 16 dimensions for a combination of 4 gradient directions and 2 dimensions for the similarity. Therefore, in each block, the total dimension of SCHOG is $8 + 16 \times 2 \times 30 = 968$. This is about a half of CoHOG. Figure 4 shows the example of the histogram representing the co-occurrence of the gradient direction and the gradient magnitude. There are two bins that represent the similarity for each combination of directions.

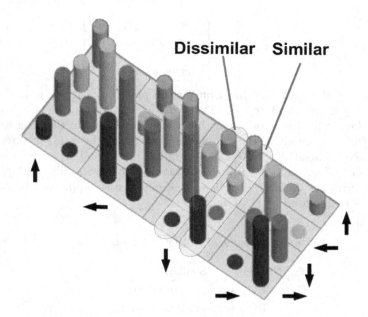

Fig. 4. Example of a histogram used in SCHOG.

In this paper, we use the pixel intensity, the gradient magnitude or the gradient direction as the feature for the similarity. However, the framework of SCHOG can easily introduce various features, avoiding the steep increase in a number of dimension because it uses the binary code to represent the similarity. The name described in Table 1 is attached for every kind of similarity. SCHOG-pix uses the pixel intensity as the similarity. SCHOG-gra uses the gradient magnitude as the similarity. Although this information is directly used in HOG, it is deleted in CoHOG. SCHOG-ang uses the gradient direction as the similarity. Since this similarity is calculated from the angle before quantization, it's expected that the finer relation between gradient directions can be expressed.

3   Experimental Results

We carried out experiments using a SVM classifier (SVMLight, Linear-kernel). The ROC curve, which shows the True Positive ratio for the vertical axis and shows the False Positive ratio for the horizontal axis, is used for evaluating the performance. It shows that performance is better, so that the curve goes to the upper left.

3.1   Dataset

We adopted INRIA Person Dataset that various previous paper have used for evaluation. Figure 5 shows some examples in this dataset. We used 2,416 positive

(a) positive image

(b) negative image

Fig. 5. Example of INRIA Person Dataset. (a) Person image (b) cropped negative.

images and 12,180 negative images for training. Ten regions randomly extracted from an image were used as negative images. The size of a positive image is 64×128 pixels and the size of a negative image is from 214×320 to 648×486. 1,132 positive images and 453 negative images are used for test. The size of a positive image is as same as an image for training and the size of a negative image is from 242×213 to 690×518. The dataset used in experiments is summarized in Table 2.

Table 2. Details of INRIA Person Dataset.

(a) Training data

image size	positive	64×128
	negative	214×320 - 648×486
number	positive	2416
	negative	$1218 \times 10 = 12180$

(b) Test data

image size	positive	64×128
	negative	242×213 - 690×518
number	positive	1132
	negative	453

3.2 Effect of Similarity

Figure 6(b) shows the performance of CoHOG, SCHOG-pix, SCHOG-gra and SCHOG-ang. SCHOG-pix, SCHOG-gra and SCHOG-ang use the pixel intensity,

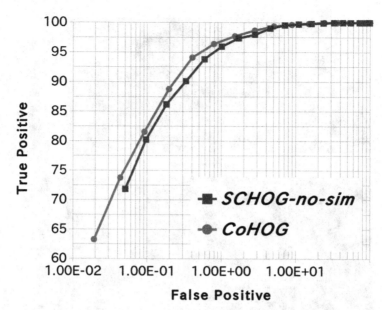

(a) Performance of CoHOG and SCHOG-no-sim.

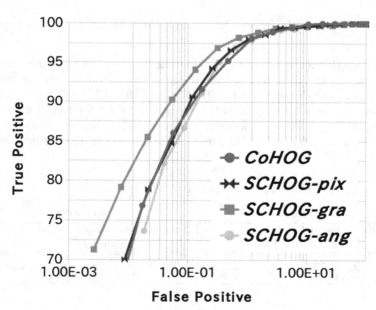

(b) Comparison of the proposed method and CoHOG.

Fig. 6. Performance of the proposed method.

the gradient magnitude and the gradient direction as the similarity respectively. The dimension of these features is a half of CoHOG. In Fig. 6(b), the True Positive ratio of SCHOG-pix, SCHOG-gra, SCHOG-ang and CoHOG is 90.12 %, 93.13 %, 87.37 % and 88.07 % respectively when the False Positive ratio is 0.1 %. SCHOG-pix shows the almost same performance as CoHOG although it uses the simple feature like the pixel intensity as the similarity. SCHOG-gra, which uses the gradient magnitude as the similarity, shows quite better performance than CoHOG. This result shows that the gradient magnitude which CoHOG omitted

Table 3. Summary of proposed features.

Name	Dimension per block	Similarity	Performance	TP (FP=0.1[%])
SCHOG-pix	$8 + 16 \times 2 \times 30 = 968$	pixel intensity	fair	90.1066
SCHOG-gra	$8 + 16 \times 2 \times 30 = 968$	magnitude	good	93.1310
SCHOG-ang	$8 + 16 \times 2 \times 30 = 968$	gradient direction	bad	87.3676
CoHOG	$8 + 8 \times 8 \times 30 = 1928$		bad	88.0733

(a) Examples to which both CoHOG and SCHOG-gra failed in detection.

(b) Examples to which only CoHOG failed in detection.

(c) Examples only SCHOG failed in detection.

Fig. 7. Examples of failure detection. Left is the original image, and right is gradient direction image.

is effective to improve performance of pedestrian detection. The performance of SCHOG-ang is slightly inferior to CoHOG. This result shows that the similar feature does not contribute to improvement in performance. In this experiment, it was shown that SCHOG whose similarity is the gradient magnitude can obtain better performance than CoHOG although the resolution of the gradient direction is a half of CoHOG. The summary of features used in this experiment is shown in Table 3.

Figure 7 shows failure examples of SCHOG-gra and CoHOG. Figure 7(a) shows examples to which both SCHOG-gra and CoHOG failed in detection. Pedestrians with low contrast to a background were not detected. Figure 7(b) shows examples to which only CoHOG failed in detection. CoHOG failed because the gradient direction around a pedestrian's contour is scattering, but SCHOG succeeded using the difference in the gradient magnitude between a pedestrian and a background. Figure 7(c) shows examples to which only SCHOG failed in detection. In these examples, pedestrians were not detected because the gradient magnitude around the pedestrian's contour is similar.

4 Conclusions

In this paper, we proposed the novel feature named SCHOG which improved CoHOG feature so that the detection performance might improve and the feature dimension might decrease. SCHOG consists of the co-occurrence of edge gradient direction and the similarity. Although SCHOG quantizes edge gradient direction to the half of CoHOG, it can represent the shape of the object more finely than CoHOG by adding the similarity to the co-occurrence of edge gradient direction. Because the similarity is represented by the binary code, the dimension of SCHOG is a half of the conventional CoHOG in spite of adding the similarity. Experimental results using INRIA Person Dataset showed that reducing quantization of the gradient direction hardly causes the fall of performance and SCHOG whose similarity is gradient magnitude have quite better performance than CoHOG.

As the similarity, the pixel intensity, the gradient magnitude and the gradient direction were evaluated in this paper. However, since the similarity is simply represented by the binary code, other various features such as color information or a combination of features are allowed as the similarity. Therefore, SCHOG can be applied to various application of object recognition. Presently, our method uses the same arrangement of offset pixels as CoOG. However, this arrangement is important for improving performance and the number of dimension can be reduced greatly if the number of offset pixels is reduced. In the future, we will examine the optimal arrangement and optimal number of offset pixels. Then, we clarify the performance by experiments using different data sets.

References

1. Hattori, H., Seki, A., Nishiyama, M., Watanabe, T.: Stereo-based pedestrian detection using multiple patterns. In: BMVC British Machine Vision Association (2009)

2. Ojala, T., Pietikäinen, M., Harwood, D.: A comparative study of texture measures with classification based on featured distributions. Pattern Recogn. **29**(1), 51–59 (1996)
3. Cao Yunyun, H.N.: Detecting pedestrians using an advanced local binary pattern histogram. In: 2011 Proceedings of 18th ITS World Congress, Orlando (2011)
4. Dalal, N., Triggs, B.: Histograms of oriented gradients for human detection. In: Schmid, C., Soatto, S., Tomasi, C. (eds.) International Conference on Computer Vision and Pattern Recognition. vol. 2, pp. 886–893. INRIA Rhone-Alpes, ZIRST-655, av. de l'Europe, Montbonnot-38334, June 2005. http://lear.inrialpes.fr/pubs/2005/DT05
5. Dalal, N., Triggs, B., Schmid, C.: Human detection using oriented histograms of flow and appearance. In: Leonardis, A., Bischof, H., Pinz, A. (eds.) ECCV 2006. LNCS, vol. 3952, pp. 428–441. Springer, Heidelberg (2006). http://dx.doi.org/10.1007/11744047_33
6. Tomoki, W., Satoshi, I., Kentaro, Y.: Co-occurrence histograms of oriented gradients for pedestrian detection. In: Wada, T., Huang, F., Lin, S. (eds.) PSIVT 2009. LNCS, vol. 5414. Springer, Heidelberg (2009). doi:10.1007/978-3-540-92957-4_4
7. Kobayashi, T., Otsu, N.: Image feature extraction using gradient local auto-correlations. In: Forsyth, D., Torr, P., Zisserman, A. (eds.) ECCV 2008, Part I. LNCS, vol. 5302, pp. 346–358. Springer, Heidelberg (2008)
8. Yamaguchi, K., Naito, T.: Two dimensional histograms of oriented gradients for pedestrian detection. IEICE Trans. Fundam. Electron. Commun. Comput. Sci. D, Inf. Syst. **94**(1), 365–373 (2011). http://ci.nii.ac.jp/naid/110008006578/
9. Walk, S., Majer, N., Schindler, K., Schiele, B.: New features and insights for pedestrian detection. In: Conference on Computer Vision and Pattern Recognition (CVPR). IEEE, San Francisco, June 2010

Learning Prior Bias in Classifier

Takumi Kobayashi$^{(\boxtimes)}$ and Kenji Nishida

National Institute of Advanced Industrial Science and Technology, 1-1-1 Umezono,
Tsukuba, Japan
{takumi.kobayashi,kenji.nishida}@aist.go.jp

Abstract. In pattern classification, a classifier is generally composed both of feature (vector) mapping and bias. While the mapping function for features is formulated in either a linear or a non-linear (kernel-based) form, the bias is simply represented by a constant scalar value, rendering prior information on class probabilities. In this paper, by focusing on the *prior bias* embedded in the classifier, we propose a novel method to discriminatively learn not only the feature mapping function but also the prior bias based on the extra prior information assigned to samples other than the class category, *e.g.*, the 2-D position where the local image feature is extracted. Without imposing specific probabilistic models, the proposed method is formulated in the framework of maximum margin to adaptively optimize the biases, improving the classification performance. We present a computationally efficient optimization approach for making the method applicable even to large-scale data. The experimental results on patch labeling in the on-board camera images demonstrate the favorable performance of the proposed method in terms of both classification accuracy and computation time.

Keywords: Pattern classification · Bias · Discriminative learning · SVM

1 Introduction

Prior information has been effectively exploited in the fields of computer vision and machine learning, such as for shape matching [1], image segmentation [2], graph inference [3], transfer learning [4] and multi-task learning [5]. Learning prior has so far been addressed mainly in the probabilistic framework on the assumption that the prior is defined by a certain type of generative probabilistic model [6,7]; especially, non-parametric Bayesian approach further considers the hyper priors of the probabilistic models [8].

In this paper, we focus on the (linear) classifier, $y = \boldsymbol{w}^\top \boldsymbol{x} + b$, and especially on the bias term, so called 'b' term [9]. The bias is also regarded as rendering the prior information on the class probabilities [10,11] and we aim to learn the *unstructured* prior bias b without assuming any specific models, while some transfer learning methods are differently built upon the prior of the weight $\boldsymbol{w}$ for effectively transferring the knowledge into the novel class categories [4,12] and the prior of $\boldsymbol{w}$ also induces a regularization on $\boldsymbol{w}$. While the bias b is generally

© Springer International Publishing Switzerland 2015
A. Fred et al. (Eds.): ICPRAM 2014, LNCS 9443, pp. 62–76, 2015.
DOI: 10.1007/978-3-319-25530-9_5

set as a constant across samples depending only on the class category, in this study we define it adaptively based on the extra prior information other than the class category. Note that, in the case of non-linear classification, the above classifier can be similarity formulated by $y = \boldsymbol{w}_\phi^\top \boldsymbol{\phi_x} + b$ where the feature vector $\boldsymbol{x}$ is simply replaced with the kernel-based feature vector $\boldsymbol{\phi_x}$ in the reproducing kernel Hilbert space defined by the kernel function $\mathrm{k}(\boldsymbol{x}_i, \boldsymbol{x}_j) = \boldsymbol{\phi}_{\boldsymbol{x}_i}^\top \boldsymbol{\phi}_{\boldsymbol{x}_j}$. Thereby, our focus also includes such kernel-based non-linear classifiers.

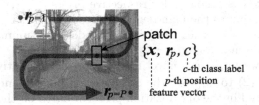

Fig. 1. The task of patch labeling is to predict the class labels c of the patches, each of which is represented by the appearance feature vector $\boldsymbol{x}$ and the prior position p. Note that there are P positions in total, $p \in \{1, \cdots, P\}$.

Suppose samples are associated with the extra prior information $p \in \{1, \cdots, P\}$ as well as the class category $c \in \{1, \cdots, C\}$, where P and C indicate the total number of the prior types and the class categories, respectively. For instance, in the task of labeling patches on the on-board camera images, each patch (sample) is assigned with the appearance feature vector $\boldsymbol{x}$, the class category c and the position (extra prior information) p, as shown in Fig. 1. The class category of the patch is effectively predicted by using not only the feature $\boldsymbol{x}$ but also the prior position p where the feature is extracted; the patches on an upper region probably belong to *sky* and the lower region would be *road*, even though the patches extracted from those two regions are both less textured, resulting in similar features.

The probabilistic structure that we assume in this study is shown in Fig. 2b with comparison to the **simple** model in Fig. 2a. By using generalized linear model [13], the standard classifier (Fig. 2a) is formulated to estimate the posterior on the class category c as[1]

$$\log \mathsf{p}(c|\boldsymbol{x}) \sim \log \mathsf{p}(\boldsymbol{x}|c) + \log \mathsf{p}(c) = \boldsymbol{w}_c^\top \boldsymbol{x} + b_c, \tag{1}$$

where $b_c = \log \mathsf{p}(c)$ indicates the class-dependent bias. On the other hand, the **proposed** model (Fig. 2b) using the extra prior p induces the following classifier;

$$\log \mathsf{p}(c|\boldsymbol{x}, p) \sim \log \mathsf{p}(\boldsymbol{x}|c) + \log \mathsf{p}(p|c) + \log \mathsf{p}(c) = \boldsymbol{w}_c^\top \boldsymbol{x} + b_c^{[p]}, \tag{2}$$

where the bias $b_c^{[p]} = \log \mathsf{p}(p|c) + \log \mathsf{p}(c)$ is dependent on both the class category c and the prior information p. Thus, if the bias could be properly determined,

---

[1] '$\sim$' in (1) means the equality in disregard of the irrelevant constant term $\log \mathsf{p}(\boldsymbol{x})$ or $\log \mathsf{p}(\boldsymbol{x}, p)$ in (2) and (3).

the classification performance would be improved compared to the standard classification model (1). One might also consider the `full-connected` model shown in Fig. 2c whose classifier is formulated by

$$\log p(c|\boldsymbol{x}, p) \sim \log p(\boldsymbol{x}|c, p) + \log p(p|c) + \log p(c) = \boldsymbol{w}_c^{[p]\top} \boldsymbol{x} + b_c^{[p]}, \qquad (3)$$

where the classifier weight $\boldsymbol{w}_c^{[p]}$ relies on the prior p as the bias $b_c^{[p]}$ does. This model is more complicated and consumes large memory storage since the classifier model $\{\boldsymbol{w}_c^{[p]}, b_c^{[p]}\}$ is prepared for respective priors $p = 1, \cdots, P$ and classes $c = 1, \cdots, C$. And, due to the high degree of freedom (D.O.F) of this model, it would be vulnerable to over-learning. These models are summarized in Table 1 and will be again discussed later.

In this paper, we propose a novel method for discriminatively learning the prior biases $b_c^{[p]}$ in (2) to improve the classification performance. The proposed method is formulated in the optimization problem based on the maximum margin criterion [14]. We also propose a computationally efficient approach for the optimization which contains large amount of samples drawn from all the priors $p \in \{1, \cdots, P\}$. Thereby, the proposed method is fast and applicable to large-scale data, while providing the high-performance classifier that exploits the extra prior information.

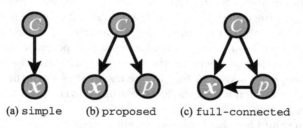

(a) `simple`   (b) `proposed`   (c) `full-connected`

Fig. 2. Graphical models to depict the probabilistic dependencies. The notations c, $\boldsymbol{x}$ and p denote the class category, the (appearance) feature vector and the extra prior information, respectively. The arrows show the probabilistic dependencies. (a) The feature $\boldsymbol{x}$ is simply drawn from the class category c in the `simple` model. (b) The `proposed` model incorporates the extra prior information p which is connected to $\boldsymbol{x}$ via c. (c) Those three variables are fully connected in the `full-connected` model.

2   Classifier Bias Learning

We detail the proposed method by first defining the formulation for learning the biases and then presenting the computationally efficient approach to optimize them. As we proceed to describe a general form regarding the prior biases p, for better understanding, it might be helpful to refer to the task of labeling patches shown in Fig. 1; the sample is represented by the appearance feature vector $\boldsymbol{x}$ and the prior position $p \in \{1, \cdots, P\}$.

Table 1. Classification methods for c-th class category. The dimensionality of the feature vector is denoted by D, $\boldsymbol{x} \in \mathbb{R}^D$, and the number of extra prior types is P.

Method	Model	D.O.F
simple	$y_c = \boldsymbol{w}_c^\top \boldsymbol{x} + b_c$	$D + 1$
proposed	$y_c = \boldsymbol{w}_c^\top \boldsymbol{x} + b_c^{[p]}$	$D + P$
full-connected	$y_c = \boldsymbol{w}_c^{[p]\top} \boldsymbol{x} + b_c^{[p]}$	$PD + P$

2.1 Formulation

We consider a binary classification problem for simplicity and take a one-vs-rest approach for multi-class tasks. Suppose we have P types of extra prior information, and let $\boldsymbol{x}_i^{[p]} \in \mathbb{R}^D$ denote the D-dimensional feature vector of the i-th sample ($i = 1, \cdots, n^{[p]}$) drawn from the p-th type of prior. As described in Sect. 1, we deal with the classification defined by

$$y = \boldsymbol{w}^\top \boldsymbol{x}^{[p]} + b^{[p]}, \tag{4}$$

where y denotes the classifier output which is subsequently thresholded by zero for performing binary classification, and $\boldsymbol{w}$ and $b^{[p]}$ are the classifier weight vector and the bias, respectively. Note again that the bias $b^{[p]}$ depends on the p-th type of prior, $p \in \{1, \cdots, P\}$. The classifier (4) can be optimized via the following formulation in the framework of maximum margin [14];

$$\min_{\boldsymbol{w}, \{b^{[p]}\}_p} \frac{1}{2}\|\boldsymbol{w}\|^2 + C \sum_p^P \sum_i^{n^{[p]}} \xi_i^{[p]} \tag{5}$$

$$s.t. \ \forall p \in \{1, \cdots, P\}, \ \forall i \in \{1, \cdots, n^{[p]}\}, \ y_i^{[p]}(\boldsymbol{w}^\top \boldsymbol{x}_i^{[p]} + b^{[p]}) \geq 1 - \xi_i^{[p]}, \ \xi_i^{[p]} \geq 0,$$

where C is the cost parameter. This is obviously convex and its Lagrangian is written by

$$L = \frac{1}{2}\|\boldsymbol{w}\|^2 + C \sum_p^P \sum_i^{n^{[p]}} \xi_i^{[p]} - \sum_p^P \sum_i^{n^{[p]}} \beta_i^{[p]} \xi_i^{[p]} - \sum_p^P \sum_i^{n^{[p]}} \alpha_i^{[p]} \{y_i^{[p]}(\boldsymbol{w}^\top \boldsymbol{x}_i^{[p]} + b^{[p]}) - 1 + \xi_i^{[p]}\}, \tag{6}$$

where we introduce the Lagrange multipliers $\alpha_i^{[p]} \geq 0$, $\beta_i^{[p]} \geq 0$. The derivatives of the Lagrangian are

$$\frac{\partial L}{\partial \boldsymbol{w}} = \boldsymbol{w} - \sum_p^P \sum_i^{n^{[p]}} \alpha_i^{[p]} y_i^{[p]} \boldsymbol{x}_i^{[p]} = \boldsymbol{0} \Rightarrow \boldsymbol{w} = \sum_p^P \sum_i^{n^{[p]}} \alpha_i^{[p]} y_i^{[p]} \boldsymbol{x}_i^{[p]} \tag{7}$$

$$\frac{\partial L}{\partial \xi_i^{[p]}} = C - \alpha_i^{[p]} - \beta_i^{[p]} = 0 \Rightarrow 0 \leq \alpha_i^{[p]} \leq C \tag{8}$$

$$\frac{\partial L}{\partial b^{[p]}} = \sum_i^{n^{[p]}} \alpha_i^{[p]} y_i^{[p]} = 0. \tag{9}$$

Thereby, the dual is finally obtained as

$$\min_{\{\alpha_i^{[p]}\}_{i,p}} \frac{1}{2} \sum_{p,q}^{P} \sum_{i}^{n^{[p]}} \sum_{j}^{n^{[q]}} \alpha_i^{[p]} \alpha_j^{[q]} y_i^{[p]} y_j^{[q]} {x_i^{[p]}}^{\top} x_j^{[q]} - \sum_{p}^{P} \sum_{i}^{n^{[p]}} \alpha_i^{[p]} \tag{10}$$

$$s.t. \ \forall p, \ \sum_{i}^{n^{[p]}} \alpha_i^{[p]} y_i^{[p]} = 0, \ \forall i, \forall p, \ 0 \le \alpha_i^{[p]} \le C.$$

This is a quadratic programming (QP) analogous to the dual of SVM [15] except that there exist P linear equality constraints with respect to $\boldsymbol{\alpha}^{[p]}$. The standard QP solver is applicable to optimize (10), though requiring substantial computation cost. For solving QP of the SVM dual, the method of sequential minimal optimization (SMO) [16] is successfully applied, but in this case, we can not employ it directly due to the multiple equality constraints. In what follows, we present a computationally efficient approach to optimize (10).

2.2 Optimization

A large number of variables $\{\alpha_i^{[p]}\}_{i,p}$ in the QP (10) are inherently partitioned into block-wise variables regarding the prior p; we obtain P blocks of $\boldsymbol{\alpha}^{[p]} = \{\alpha_i^{[p]}\}_{i=1,\cdots,n^{[p]}} \in \mathbb{R}^{n^{[p]}}$, $p = 1,\cdots,P$. According to those block-wise variables, (10) is decomposed into the following sub-problem as well:

$$\min_{\boldsymbol{\alpha}_i^{[p]}} \frac{1}{2} \sum_{i,j}^{n^{[p]}} \alpha_i^{[p]} \alpha_j^{[p]} y_i^{[p]} y_j^{[p]} {x_i^{[p]}}^{\top} x_j^{[p]} - \sum_{i}^{n^{[p]}} \alpha_i^{[p]} \left\{ 1 - y_i^{[p]} \sum_{q \neq p}^{P} \sum_{j}^{n^{[q]}} \alpha_j^{[q]} y_j^{[q]} {x_i^{[p]}}^{\top} x_j^{[q]} \right\} \tag{11}$$

$$s.t. \ \sum_{i}^{n^{[p]}} \alpha_i^{[p]} y_i^{[p]} = 0, \ \forall i, \ 0 \le \alpha_i^{[p]} \le C. \tag{12}$$

This is again a quadratic programming which resembles the SVM dual except for the linear term with respect to $\boldsymbol{\alpha}^{[p]}$ and thus is effectively optimized by using the SMO [16]. Therefore, the whole procedure for optimizing (10) consists of iteratively optimizing the sub-problem (11) with respect to every prior $p \in \{1,\cdots,P\}$ by means of SMO. According to [17], the order of the priors p to be optimized is randomly permuted. The detailed procedures are shown in Algorithm 1.

In order to discuss the convergence of the iterative optimization, we mention the KKT condition of (10) [18]. The optimizer $\alpha_i^{[p]}$ satisfies the following condition:

$$G_{i,p}(\boldsymbol{\alpha}) + b_i^{[p]} y_i^{[p]} = \lambda_i^{[p]} - \mu_i^{[p]}, \tag{13}$$

$$\lambda_i^{[p]} \alpha_i^{[p]} = 0, \ \mu_i^{[p]}(C - \alpha_i^{[p]}) = 0, \tag{14}$$

$$\lambda_i^{[p]} \ge 0, \ \mu_i^{[p]} \ge 0, \tag{15}$$

where $G_{i,p}(\boldsymbol{\alpha}) = y_i^{[p]} \boldsymbol{x}_i^{[p]^\top} \sum_q^P \sum_j^{n^{[q]}} \alpha_j^{[q]} y_j^{[q]} \boldsymbol{x}_j^{[q]} - 1$ is the derivative of the objective function in (10) with respect to $\alpha_i^{[p]}$. This condition is rewritten into

$$\alpha_i^{[p]} < C: \ G_{i,p}(\boldsymbol{\alpha}) + b_i^{[p]} y_i^{[p]} \geq 0, \tag{16}$$

$$\alpha_i^{[p]} > 0: \ G_{i,p}(\boldsymbol{\alpha}) + b_i^{[p]} y_i^{[p]} \leq 0, \tag{17}$$

and since $y_i^{[p]} \in \{+1, -1\}$, it results in

$$-y_i^{[p]} G_{i,p}(\boldsymbol{\alpha}) \begin{cases} \leq b_i^{[p]} \ i \in \mathbb{I}_+^{[p]} \\ \geq b_i^{[p]} \ i \in \mathbb{I}_-^{[p]} \end{cases}, \tag{18}$$

where

$$\mathbb{I}_+^{[p]} = \{i | (\alpha_i^{[p]} < C \wedge y_i^{[p]} = 1) \vee (\alpha_i^{[p]} > 0 \wedge y_i^{[p]} = -1)\}, \tag{19}$$

$$\mathbb{I}_-^{[p]} = \{i | (\alpha_i^{[p]} < C \wedge y_i^{[p]} = -1) \vee (\alpha_i^{[p]} > 0 \wedge y_i^{[p]} = 1)\}. \tag{20}$$

Therefore, we can conclude that $\alpha_i^{[p]}$ is a stationary point if and only if

$$\delta^{[p]} \triangleq \left[\max_{i \in \mathbb{I}_+^{[p]}} -y_i^{[p]} G_{i,p}(\boldsymbol{\alpha}) \right] - \left[\min_{i \in \mathbb{I}_-^{[p]}} -y_i^{[p]} G_{i,p}(\boldsymbol{\alpha}) \right] \leq 0. \tag{21}$$

On the basis of this measure, we can stop the iteration when $\max_p \delta^{[p]} < \epsilon$ with a small tolerance $\epsilon > 0$. At the optimum, the bias $b^{[p]}$ is retrieved by

$$b^{[p]} = \frac{1}{|\mathbb{I}^{[p]}|} \sum_{i \in \mathbb{I}^{[p]}} -y_i^{[p]} G_{i,p}(\boldsymbol{\alpha}), \text{ where } \mathbb{I}^{[p]} = \{i | 0 < \alpha_i^{[p]} < C\}, \tag{22}$$

since the right hand side in (13) equals zero for $i \in \mathbb{I}^{[p]}$.

2.3   Trivial Biases

Finally, we mention the trivial sub-problem for further reducing the computational cost in the optimization. From a practical viewpoint, the samples of the two class categories are not equally distributed across the priors $p = 1, .., P$ but are localized in limited number of priors. For instance, in the case of on-board camera images, the *road* never appears in upper regions where the *sky* usually dominates. That is, we occasionally encounter the following sub-problem;

$$\min_{\alpha_i^{[p]}} \frac{1}{2} \sum_{i,j}^{n^{[p]}} \alpha_i^{[p]} \alpha_j^{[p]} y_i^{[p]} y_j^{[p]} \boldsymbol{x}_i^{[p]^\top} \boldsymbol{x}_j^{[p]} - \sum_i^{n^{[p]}} \alpha_i^{[p]} \left\{ 1 - y_i^{[p]} \sum_{q \neq p}^P \sum_j^{n^{[q]}} \alpha_j^{[q]} y_j^{[q]} \boldsymbol{x}_i^{[p]^\top} \boldsymbol{x}_j^{[q]} \right\} \tag{23}$$

$$s.t. \sum_i^{n^{[p]}} \alpha_i^{[p]} y_i^{[p]} = 0, \ \forall i, \ 0 \leq \alpha_i^{[p]} \leq C, \ \forall i, \ \underline{y_i^{[p]} = 1} \ (\text{or } \forall i, \ y_i^{[p]} = -1). \tag{24}$$

The above QP is trivially optimized by $\boldsymbol{\alpha}^{[p]} = \mathbf{0}$ without exhaustive computation due to the constraint (24). Thus, by eliminating such priors that result in trivial optimization, we can reduce the computational burden of the whole procedure to optimize (10); see line 1 in Algorithm 1.

The only issue in this trivial case is how to determine the bias $b^{[p]}$. In this case, the bias is not uniquely determined but accepts any value satisfying the following;

$$y^{[p]} = +1 : b^{[p]} \geq \max_i -\boldsymbol{w}^\top \boldsymbol{x}_i^{[p]} + 1, \tag{25}$$

$$y^{[p]} = -1 : b^{[p]} \leq \min_i -\boldsymbol{w}^\top \boldsymbol{x}_i^{[p]} - 1. \tag{26}$$

Thus, we can provide three alternative ways for computing the bias on the trivial prior.

1. Tight bias:

$$b^{[p]} = \begin{cases} \max_i -\boldsymbol{w}^\top \boldsymbol{x}_i^{[p]} + 1 & (for\ y^{[p]} = +1) \\ \min_i -\boldsymbol{w}^\top \boldsymbol{x}_i^{[p]} - 1 & (for\ y^{[p]} = -1) \end{cases}. \tag{27}$$

Algorithm 1. Bias Learning.

Input: $\{\boldsymbol{x}_i^{[p]}, y_i^{[p]}\}$: feature vector and its class label of the i-th training sample from the p-th type of prior, $p = 1, .., P$, $i = 1, .., n^{[p]}$.

  $\epsilon > 0$: small tolerance for terminating the iteration.

1: $\mathbb{P} = \{p | \exists i, y_i^{[p]} = 1 \wedge \exists i, y_i^{[p]} = -1\}$
2: Initialization: $\forall p \in \{1, .., P\}$, $\boldsymbol{\alpha}^{[p]} = \mathbf{0}$
3: **repeat**
4:    Random permutation of $\{1, \cdots, P\}$: $\{\pi(1), \cdots, \pi(P)\}$
5:    **for** $i = 1$ to P **do**
6:      $p \leftarrow \pi(i)$
7:      Set $\boldsymbol{\alpha}^{[p]}$ as the optimizer of (11)
8:    **end for**
9: **until** $\max_{p \in \mathbb{P}} \delta^{[p]} < \epsilon$

Output: $\boldsymbol{w}$ computed by (7) and $\{b^{[p]}\}_{p=1,...,P}$ computed by (22) for $p \in \mathbb{P}$ and (27~29) for $p \notin \mathbb{P}$, using the optimizers $\{\boldsymbol{\alpha}^{[p]}\}_p$.

This gives the tight bias based on the above conditions (25, 26), which is computed by using only the samples $\boldsymbol{x}_i^{[p]}$ belonging to the prior p.

2. Mild bias:

$$b^{[p]} = \begin{cases} \max_{i,p} -\boldsymbol{w}^\top \boldsymbol{x}_i^{[p]} + 1 & (for\ y^{[p]} = +1) \\ \min_{i,p} -\boldsymbol{w}^\top \boldsymbol{x}_i^{[p]} - 1 & (for\ y^{[p]} = -1) \end{cases}, \tag{28}$$

By considering whole samples $\{\boldsymbol{x}_i^{[p]}\}_{i,p}$ across the priors, the bias is determined with a margin from the tight one, which might improve the generalization performance to some extent.

3. Extreme bias:

$$b^{[p]} = \begin{cases} +\infty & (for \ y^{[p]} = +1) \\ -\infty & (for \ y^{[p]} = -1) \end{cases},$$

(29)

By this bias, the samples from such a prior are definitely classified as positive (or negative) no matter how the appearance features of the samples are. In this case, the class category is solely dependent on the extra prior information via the bias $b^{[p]} \in \{+\infty, -\infty\}$.

These three ways are empirically compared in the experiments (Sect. 3.3).

2.4   Discussion

In the proposed method, all samples across all types of priors are leveraged to train the classifier, improving the generalization performance. In contrast, the `full-connected` method (Table 1) treats the samples separately regarding the priors, and thus the p-th classifier is learnt by using only a small amount of samples belonging to the p-th type of prior, which might degrade the performance. On the other hand, the `simple` method learning the classifier from the whole set of samples is less discriminative without utilizing the extra prior information associated with the samples. The proposed method effectively incorporate the prior information into the classifiers via the biases which are discriminatively optimized.

The proposed method is slightly close to the cross-modal learning [19,20]. The samples belonging to different priors are separated as if they are in different modalities, though the feature representations are the same in this case. The proposed method deals with them in a unified manner via the adaptive prior biases. Actually, the proposed method is applicable to the samples that are distributed differently across the priors; the sample distribution is shifted (translated) as $x^{[q]} = x^{[p]} + e$ and the prior bias can adapt to it by $b^{[q]} = b^{[p]} - w^\top e$ since $y^{[p]} = w^\top x^{[p]} + b^{[p]}$, $y^{[q]} = w^\top x^{[q]} + b^{[q]} = w^\top x^{[p]} + (b^{[q]} + w^\top e) = y^{[p]}$. Therefore, the samples of the different priors are effectively transferred into the optimization to improve the classification performance.

3   Experimental Results

We evaluated the proposed method on a task of patch labeling in on-board camera images by using CamVid dataset [21]. This patch labeling contributes to understand the scene surrounding the car.

3.1   Setting

The CamVid dataset [21] contains several sequences composed of fully labeled image frames as shown in Fig. 3: each pixel is assigned with one of 32 class labels including 'void'. Those labeled images are captured at 10 Hz. In this experiment, we employ the major 11 labels frequently seen in the image frames, *road, building,*

On-board image   Label image

Fig. 3. CamVid dataset [21].

sky, tree, sidewalk, car, column pole, sign symbol, fence, pedestrian and *bicyclist*, to form the 11-class classification task.

We extracted the GLAC image feature [22] from a local image patch of 20×40 pixels which slides at every 10 pixels over the resized image of 480×360. In this case, the feature vector $x \in \mathbb{R}^{2112}$ is associated with the 2D position of the patch as the extra prior information; the total number of prior types (grid points) is $P = 1551$. Thus, the task is to categorize the patch feature vectors extracted at 1511 positions into the above-mentioned 11 classes.

We used three sequences in the CamVid dataset, and partitioned each sequence into three sub-sequences along the time, one of which was used for training and the others were for test. This cross validation was repeated three times and the averaged classification accuracy is reported.

For comparison, we applied the methods mentioned in Sect. 1; `simple` and `full-connected` methods as listed in Table 1. The `simple` method is a standard classification using the weight w with the constant bias b in disregard of the prior information p. The `full-connected` method applies classifiers comprising $w^{[p]}$ and $b^{[p]}$ at respective priors $p = 1, \cdots, P$. This method requires tremendous memory storage for those P classifiers; in this experiment, 2112-dimensional weight vectors $w^{[p]}$ and scalar bias $b^{[p]}$ in 11 class categories are stored at each of 1511 positions. On the other hand, in the `proposed` method, the feature vectors are classified by using the identical weight w across the priors together with the adaptively optimized bias $b^{[p]}$ depending on the prior p. We consider the linear classification form $y = w^\top x + b$ in all these methods for fast computation time.

3.2   Computation Cost

We first evaluated the proposed method in terms of computation cost. The method trains the classifier by using all the samples across the priors, scale of which is as large as in the `simple` method. These methods are implemented by MATLAB on Xeon 3.4 GHz PC. In the proposed method, we apply libsvm [23]

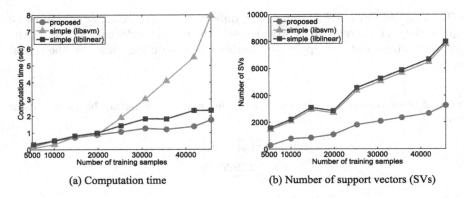

(a) Computation time   (b) Number of support vectors (SVs)

Fig. 4. Comparison of the `simple` and `proposed` methods in terms of (a) computation time as well as (b) number of support vectors (SVs).

to solve QP and efficiently compute the derivatives $G_{i,p}(\alpha)$ required for the linear term in the objective function (11) and $\delta^{[p]}$ in (21) by exploiting the linear classification form as in [24]. On the other hand, two types of solvers, libsvm and liblinear [24], are applied to the `simple` method.

Figure 4a shows the computation time on various sizes of training samples. The proposed method is significantly faster than the `simple` method using libsvm and competitive with that using liblinear. The time complexity of `simple` method which solves the standard SVM dual has been empirically shown to be $O(n^{2.1})$ [25]. The proposed optimization approach iteratively works on the block-wise subset into which the whole training set is decomposed (Sect. 2.2). The subset is regarded as the working set whose size is an important factor for fast computing QP [18]. In the `proposed` method, it is advantageous to inherently define the subset, *i.e.*, the working set, of adequate size according to the prior. Thus, roughly speaking, the time complexity of the `proposed` method results in $O(M\frac{n^{2.1}}{M^{2.1}}) = O(\frac{n^{2.1}}{M^{1.1}})$. Besides, the technique to skip the trivial subset (Sect. 2.3) empirically contributes to further reduce the computational cost. The computation time essentially depends on the (resultant) number of support vectors (SVs); Fig. 4b shows the number of support vectors produced by those two methods. The `proposed` method provides a smaller number of support vectors, which significantly contributes to reduce the computation time. As a result, the proposed optimization approach works quite well from the viewpoint of computation time. This result shows the favorable scalability of the `proposed` method, especially compared to the standard `simple` method.

3.3   Trivial Biases

As described in Sect. 2.3, in the proposed method, it is necessary to compute the biases on the trivial priors rather heuristically, since they are not theoretically determined. We presented three ways, *tight*, *mild* and *extreme* ones, for computing those biases (Sect. 2.3), and the classification performances are compared in Table 2. We can not find significant differences in performance across those approaches; the *extreme* way provides slightly better performance and in the following experiments we apply this method for computing biases on the trivial priors.

Table 2. Performance comparison (%) on the ways of computing biases on trivial priors.

Tight	Mild	Extreme
52.22	52.19	52.25

Table 3. Performance comparison.

(a) Classification accuracy (%)

	simple	full-connected	proposed-linear	proposed-kernel
road	93.10	93.80	**94.92**	95.43
building	75.90	72.96	**78.70**	80.51
sky	**90.52**	82.21	90.25	90.64
tree	70.49	77.59	**79.95**	83.15
sidewalk	77.06	78.43	**81.36**	83.11
car	53.84	58.64	**65.16**	73.82
column pole	9.53	**16.15**	12.85	27.32
sign symbol	**1.73**	1.62	1.70	8.76
fence	5.23	11.09	**13.48**	19.47
pedestrian	17.26	30.69	**31.52**	42.78
bicyclist	17.09	18.49	**24.88**	31.17
avg.	46.52	49.24	**52.25**	57.83

(b) Confusion matrix (%)

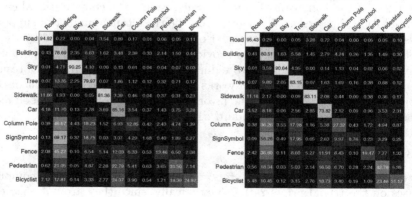

proposed-linear   proposed-kernel

3.4   Classification Performance

We finally compared the classification performance of the three methods, `simple`, `full-connected` and `proposed` (listed in Table 1); for reference, we also apply the kernel-based extension of the proposed method by using Gaussian kernel $k(\boldsymbol{x}_i, \boldsymbol{x}_j) = \exp(-\frac{\|\boldsymbol{x}_i - \boldsymbol{x}_j\|}{\gamma})$ where γ is mean of the pair-wise distances. Table 3 shows the overall performance, demonstrating that the `proposed` method outperforms the others. It should be noted that the `full-connected` method individually applies the classifier specific to the prior $p \in \{1, \cdots, P\}$, requiring a plenty of memory storage and consequently taking large classification time due to loading the enormous memory. The `proposed` method renders as fast classification as the `simple` method since it enlarges only the bias. By discriminatively optimizing the biases for respective priors, the performance is significantly improved

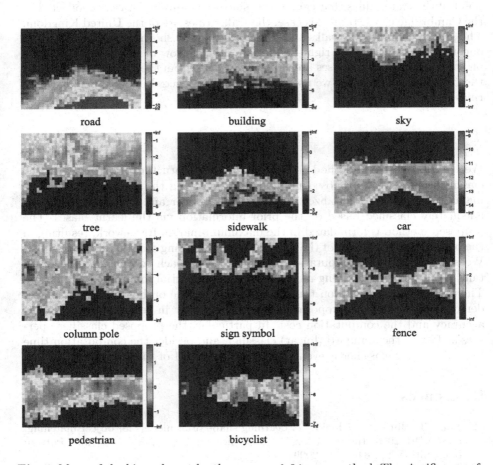

Fig. 5. Maps of the biases learnt by the `proposed-linear` method. The significance of the biases are shown by using pseudo colors from (dark) blue to (dark) red. This figure is best viewed in color (Color figure online).

in comparison to the `simple` method; the improvement is especially found at the categories of *car, pedestrian* and *bicyclist* that are composed of patch parts similar to other categories but are associated with the distinct prior positions.

The kernel-based method (`proposed-kernel`) further improves the performance on the foreground object categories, such as *column pole* and *pedestrian*. Those foreground objects exhibit large variations in appearance due to viewpoint changes and within-class variations themselves, and the kernel-based method produces more discriminative feature mapping function compared to the linear method.

Finally, we show in Fig. 5 the biases learnt by the `proposed` method; the biases $\{b^{[p]}\}_p$ are folded into the form of image frame according to the x-y positions. These maps of the biases reflect the *prior* probability over the locations where the target category appears. These seem quite reasonable from the viewpoint of the traffic rules that cars obeys; since the CamVid dataset is collected at the Cambridge city [21], in this case, the traffic rules are of the United Kingdom. The *pedestrian* probably walks on the *sidewalk* mainly shown in the left side. The oncoming *car* runs on the right-hand road, and the row of the *building* is found on the roadside. These biases are adaptively learnt from the CamVid dataset and they would be different if we use other datasets collected under different traffic rules.

4   Conclusions

We have proposed a method to discriminatively learn the prior biases in the classification. In the proposed method, for improving the classification performance, all samples are utilized to train the classifier and the input sample is adequately classified based on the prior information via the learnt biases. The proposed method is formulated in the maximum-margin framework, resulting in the optimization problem of the quadratic programming form similarly to SVM. We also presented a computationally efficient approach to optimize the resultant quadratic programming along the line of sequential minimal optimization. The experimental results on the patch labeling in the on-board camera images demonstrated that the proposed method is superior in terms of classification accuracy and the computation cost. In particular, the proposed classifier operates as fast as the standard (linear) classifier, and besides the computation time for training the classifier is even faster than the SVM of the same size.

References

1. Jiang, T., Jurie, F., Schmid, C.: Learning shape prior models for object matching. In: CVPR 2009, the 22nd IEEE Conference on Computer Vision and Pattern Recognition, pp. 848–855 (2009)
2. El-Baz, A., Gimel'farb, G.: Robust image segmentation using learned priors. In: ICCV 2009, the 12nd International Conference on Computer Vision, pp. 857–864 (2009)

3. Cremers, D., Grady, L.: Statistical priors for efficient combinatorial optimization via graph cuts. In: Pinz, A., Leonardis, A. (eds.) ECCV 2006. LNCS, vol. 3953, pp. 263–274. Springer, Heidelberg (2006)

4. Jie, L., Tommasi, T., Caputo, B.: Multiclass transfer learning from unconstrained priors. In: ICCV 2011, the 13th International Conference on Computer Vision, pp. 1863–1870 (2011)

5. Yuan, C., Hu, W., Tian, G., Yang, S., Wang, H.: Multi-task sparse learning with beta process prior for action recognition. In: CVPR 2013, the 26th IEEE Conference on Computer Vision and Pattern Recognition, pp. 423–430 (2013)

6. Wang, C., Liao, X., Carin, L., Dunson, D.: Classification with incomplete data using dirichlet process priors. J. Mach. Learn. Res. **11**, 3269–3311 (2010)

7. Kapoor, A., Hua, G., Akbarzadeh, A., Baker, S.: Which faces to tag: adding prior constraints into active learning. In: ICCV 2009, the 12th International Conference on Computer Vision, pp. 1058–1065 (2009)

8. Ghosh, J., Ramamoorthi, R.: Bayesian Nonparametrics. Springer, Berlin (2003)

9. Poggio, T., Mukherjee, S., Rifkin, R., Rakhlin, A., Verri, A.: b. Technical report CBCL Paper#198/AI Memo #2001-011, Massachusetts Institute of Technology, Cambridge (2001)

10. Bishop, C.M.: Neural Networks for Pattern Recognition. Oxford University Press, New York (1995)

11. Van Gestel, T., Suykens, J., Lanckriet, G., Lambrechts, A., De Moor, B., Vandewalle, J.: Bayesian framework for least squares support vector machine classifiers, gaussian processes and kernel fisher discriminant analysis. Neural Comput. **15**, 1115–1148 (2002)

12. Gao, T., Stark, M., Koller, D.: What makes a good detector? – structured priors for learning from few examples. In: Fitzgibbon, A., Lazebnik, S., Perona, P., Sato, Y., Schmid, C. (eds.) ECCV 2012, Part V. LNCS, vol. 7576, pp. 354–367. Springer, Heidelberg (2012)

13. Bishop, C.M.: Pattern Recognition and Machine Learning. Springer, Berlin (2006)

14. Smola, A.J., Bartlett, P., Schölkopf, B., Schuurmans, D.: Advances in Large-Margin Classifiers. MIT Press, Cambridge (2000)

15. Vapnik, V.: Statistical Learning Theory. Wiley, New York (1998)

16. Platt, J.: Fast training of support vector machines using sequential minimal optimization. In: Schölkopf, B., Burges, C., Smola, A. (eds.) Advances in Kernel Methods - Support Vector Learning, pp. 185–208. MIT Press, Cambridge (1999)

17. Hsieh, C.J., Chang, K.W., Lin, C.J., Keerthi, S.S., Sundararajan, S.: A dual coordinate descent method for large-scale linear svm. In: ICML 2008, the 25th International Conference on Machine Learning, pp. 408–415 (2008)

18. Fan, R.E., Chen, P.H., Lin, C.J.: Working set selection using second order information for training support vector machines. J. Mach. Learn. Res. **6**, 1889–1918 (2005)

19. Kan, M., Shan, S., Zhang, H., Lao, S.: Multi-view discriminant analysis. In: Fitzgibbon, A., Lazebnik, S., Perona, P., Sato, Y., Schmid, C. (eds.) ECCV 2012, Part I. LNCS, vol. 7572, pp. 808–821. Springer, Heidelberg (2012)

20. Sharma, A., Jacobs, D.: Bypassing synthesis: pls for face recognition with pose, low-resolution and sketch. In: CVPR 2011, the 24th IEEE Conference on Computer Vision and Pattern Recognition (CVPR), pp. 593–600 (2011)

21. Fauqueur, J., Brostow, G.J., Shotton, J., Cipolla, R.: Segmentation and recognition using structure from motion point clouds. In: Torr, P., Forsyth, D., Zisserman, A. (eds.) ECCV 2008, Part I. LNCS, vol. 5302, pp. 44–57. Springer, Heidelberg (2008)

22. Kobayashi, T., Otsu, N.: Image feature extraction using gradient local auto-correlations. In: Torr, P., Forsyth, D., Zisserman, A. (eds.) ECCV 2008, Part I. LNCS, vol. 5302, pp. 346–358. Springer, Heidelberg (2008)
23. Chang, C.C., Lin, C.J.: LIBSVM: a library for support vector machines (2001). Software available at http://www.csie.ntu.edu.tw/~cjlin/libsvm
24. Fan, R., Chang, K., Hsieh, C., Wang, X., Lin, C.: Liblinear: a library for large linear classification. J. Mach. Learn. Res. **9**, 1871–1874 (2008). Software available at http://www.csie.ntu.edu.tw/~cjlin/liblinear
25. Joachims, T.: Making large-scale svm learning practical. In: Schölkopf, B., Burges, C., Smola, A. (eds.) Advances in Kernel Methods - Support Vector Learning, pp. 169–184. MIT Press, Cambridge (1999)

Intra-class Variance Among Multiple Samples of the Same Person's Fingerprint in a Cooperative User Scenario

Vedrana Krivokuća$^{(\boxtimes)}$ and Waleed Abdulla

The University of Auckland, Auckland, New Zealand
{vedrana.krivokuca, w.abdulla}@auckland.ac.nz

Abstract. A significant challenge in the development of automated fingerprint recognition algorithms is dealing with intra-class variance among multiple samples of the same fingerprint. A major contributor to this intra-class variance is the inconsistency with which a finger is presented to the fingerprint scanner across multiple authentication attempts. This paper investigates the consistency of cooperative users in placing their finger on a typical fingerprint scanner, in terms of the amount of translation and rotation of the finger on the scanner surface and the percentage of reference minutiae that are present in the query fingerprint during each authentication attempt. A database of 800 fingerprint samples from 100 cooperative participants was collected for this purpose. Analysis of this database resulted in a median horizontal translation of 13 pixels (0.66 mm), a median vertical translation of 17 pixels (0.86 mm), a median rotation of 2°, and a median minutiae repeatability of 96.1 %.

Keywords: Biometrics · Fingerprints · Fingerprint recognition · Intra-class variance · Translation · Rotation · Alignment · Partial fingerprints · Minutiae · Missing minutiae

1 Introduction

The development of automated fingerprint recognition systems is a non-trivial task. The biggest challenge lies in dealing with the unavoidable variability between different samples of the same fingerprint, which is referred to as intra-class variance [1]. Multiple images of the same fingerprint, acquired at different times, will always exhibit some amount of intra-class variance, mainly due to the way in which a user interacts with the fingerprint scanner. The more consistent the user is in placing their finger onto the scanner, the more similar the resulting fingerprint images will be. The similarity between different samples of the same fingerprint can, therefore, be used as a measure of the consistency with which a user places their finger onto the scanner. Two fingerprint images may be considered to be the effectively the same if they are aligned and contain the same fingerprint minutiae. In the literature, it is generally assumed that two fingerprint images will never be the same, i.e., they are likely to be misaligned and are unlikely to contain all of the same minutiae. For this reason, over the years there has been extensive focus on the development of a variety of fingerprint alignment

© Springer International Publishing Switzerland 2015
A. Fred et al. (Eds.): ICPRAM 2014, LNCS 9443, pp. 77–92, 2015.
DOI: 10.1007/978-3-319-25530-9_6

techniques (e.g., [2–7]) and partial fingerprint matching methods (e.g., [8–10]) to address these two causes of intra-class variance.

While the development of fingerprint alignment and matching techniques is certainly important, the nature and complexity of these algorithms should be suited to the expected intra-class variance among the types of fingerprint images to which these algorithms will be applied. The currently available public fingerprint databases are suitable for testing fingerprint recognition algorithms that target uncooperative fingerprint capture scenarios (e.g., the FVC series [11]); however, these databases do not fairly represent the types of fingerprint images that are likely to be acquired from cooperative users in civilian fingerprint recognition applications. Consequently, the literature lacks information on the amount of intra-class variance that may reasonably be expected in such scenarios, which makes it difficult to reliably evaluate fingerprint recognition algorithms specifically designed to operate in those scenarios. Therefore, to aid in the development and evaluation of these techniques, this paper investigates the consistency with which cooperative users in civilian fingerprint recognition applications may be expected to place their fingers onto a fingerprint scanner during multiple authentication attempts. The consistency is measured in terms of the horizontal and vertical translations and rotations between multiple samples of the same fingerprint, and the percentage of reference minutiae repeating in a query sample of the same fingerprint. This investigation is important for the development and testing of fingerprint recognition algorithms that target cooperative-user civilian applications, because it provides insight into how users of such applications are likely to interact with the fingerprint scanner. The results of this investigation can also be used in the development of fingerprint alignment and matching algorithms, since they provide practical limits on the inconsistencies that may reasonably be expected between different samples of the same fingerprint. To the best of our knowledge, this type of investigation has not been undertaken before.

Section 2 of this paper explains the reason and methodology behind constructing our own database for this investigation. Section 3 analyses the new database to empirically quantify the consistency of the database participants in placing their fingers on the provided fingerprint scanner. Section 4 summarizes the findings of this investigation and offers suggestions on how user consistency may be further improved.

2   Fingerprint Database Collection

Commonly used public fingerprint databases (e.g., the FVC series [11]), have been constructed by asking the participants to exaggerate the inconsistency with which they place their finger onto the provided fingerprint scanner [12]. Figure 1 shows an example of three samples of the same fingerprint from the FVC2002 DB1_A database: the first image was acquired when the user's finger was placed on the scanner in a cooperative manner, and the second and third images are deliberately rotated and heavily translated samples of the same fingerprint.

The nature of these databases makes them suitable for testing fingerprint recognition algorithms designed for deployment in uncooperative user scenarios, e.g., forensics, where the latent prints are usually partial and of poor quality; border security,

Fig. 1. Three samples of the same fingerprint from FVC2002 DB1_A.

where a person may attempt to avoid being recognized as a criminal on a "wanted" list, etc. However, they are not representative of fingerprint samples that would be acquired from cooperative users in civilian fingerprint recognition applications. In such applications, it is in the users' best interests to be recognized, so it is fair to assume that they would be fairly consistent in the way in which they present their fingers to the fingerprint scanner.

The aim of this investigation was to quantify the expected consistency of cooperative users in civilian fingerprint recognition applications. This consistency was measured in terms of the translation and rotation of a person's finger on the fingerprint scanner surface across multiple authentication attempts, and the percentage of reference minutiae that are present in a query sample of the same person's fingerprint. At first, the FVC2006 public fingerprint database [13], which was collected by asking the participants to place their fingers on the scanner naturally, appeared suitable for our purposes. However, the construction of this database did not involve a quality check on the acquired fingerprint images. In our investigation, a quality check was important for two reasons. Firstly, since one of our measures of user consistency is the percentage of reference minutiae that are present in a query sample of the same fingerprint, and since we were interested in evaluating minutiae persistence based on user consistency *alone*, we had to eliminate the fingerprint quality factor from the database. This means that fingerprint images acquired from the same finger had to be of approximately the same quality. Secondly, our investigation targets civilian fingerprint recognition applications, which usually perform a quality check on the captured fingerprint images [14]. This helps to improve the chances of a correct authentication decision by ensuring that the acquired fingerprint images are all of a sufficiently high quality for subsequent processing. For this reason, using fingerprint images of very variable quality was irrelevant to our investigation. Hence, the FVC2006 database was an unsuitable testing platform for our purposes and it was necessary to collect our own fingerprint database. Sects. 2.1 to 2.3 describe our database collection procedure in detail.

2.1 Scanner Specifications

The images in our fingerprint database were acquired using the Futronic FS88 optical fingerprint scanner [15]. Futronic provides a simple user interface, which shows a live video of the user's fingerprint when it is placed on the scanner's surface. Scanning is quick and easy, producing an 8-bit grey level fingerprint image with a resolution of 320 × 480 pixels, 500dpi.

A crucial property of electronic fingerprint scanners, which sets them apart, is their underlying sensor technology. Since optical sensors are a popular choice in fingerprint scanner design [16] and since these types of scanners generally exhibit similar user interfaces, the FS88 scanner may be considered to be "typical". This means that the results of our investigation are not limited to this particular scanner.

2.2 Participant Selection

Our database consists of fingerprints provided by 100 volunteers. The requirement for participation to be voluntary was the first step in ensuring that the database represents cooperative users. The participants consisted of adults of both genders, from diverse ethnic backgrounds and of various ages in the range [18, 60] (although most of the participants were young adults).

2.3 Methodology

The participants were invited to play the part of cooperative users in a fingerprint-based computer login application. They were asked to sit down at a typical computer station with the scanner positioned on the desk approximately where the computer mouse would be. Participants were free to move the scanner around and position it in whichever way was most comfortable for them (as long as it stayed flat on the desk). Each user was asked to choose a finger, which they would use to authenticate them-selves in a fingerprint-based computer login application. The only guidance that the users received regarding the proper placement of their finger on the scanner was that the line of the first joint from the top of the finger should approximately lie on the line just below the glass platen on the fingerprint scanner, such that the maximum finger-print area is captured (see Fig. 2).

Fig. 2. Illustration of the proper placement of a finger on the Futronic FS88 scanner: the horizontal lines inside the red rectangles should approximately align (Color figure online).

Participants were then asked to find a comfortable position on the scanner, which they felt that they could naturally repeat for future scans. Prior to the first scan, the quality of their chosen fingerprint was examined on the live scan video. People with dry skin were asked to rub their fingers on the side of their noise or onto their forehead to apply some grease to the fingerprint, and those with excessively wet or greasy fingerprints were asked to dab their fingers on a piece of clothing to remove the excess moisture. A fingerprint image was deemed to be of sufficiently good quality when the difference between the ridges and valleys was clear. The participant's chosen fingerprint was then scanned 8 times.

Note that fingerprint databases are often constructed by acquiring multiple samples of the same person's fingerprint over several days. The purpose of this is to simulate natural variability between the samples; e.g., on some days a person's finger may be drier than on other days. However, since our investigation required elimination of the quality factor, simulating this natural variability was unnecessary. So, we elected to collect each of a participant's 8 fingerprint samples on the same day. To simulate multiple authentication attempts, after each scan the participant was asked to remove their finger from the scanner while their previous fingerprint image was manually saved by a human operator. The images were saved manually to deliberately introduce some delay in between the scans and to 'distract' the participant, thereby simulating different authentication attempts. Once the scanning started, the human operator did not guide the user in the placement of their finger on the scanner.

The participants were observed to be careful in the way in which they placed their fingers on the scanner. They also became very aware of what a good quality fingerprint image should look like after the initial quality check, and most controlled this quality on their own for subsequent scans, without prompting by the operator. This suggests that users are both capable and willing to be cooperative in a scenario in which they *want* to be recognized.

3   Analysis of Finger Placement Consistency

The consistency with which the participants placed their chosen finger on the scanner was analyzed in terms of three factors: translation, rotation, and captured fingerprint minutiae. These factors are further described and analyzed in Sects. 3.1 to 3.3.

3.1   Translation

Translation refers to the horizontal and vertical offsets between multiple samples of the same fingerprint. A horizontal translation occurs when the user moves their finger to the left or right on the scanner surface, and a vertical translation occurs when the user moves their finger up or down. The more consistent a user is in placing their finger on the scanner, the smaller these translations will be.

To measure the translation between each person's 8 fingerprint samples, a reference point inside each fingerprint was first chosen. A reference point is a feature that is present in all 8 of a person's fingerprint samples. The most commonly used reference point in practice is the *core point*, which has traditionally been defined as the center of the north-most loop-type pattern in a fingerprint image, or for fingerprints that do not contain loops the core usually corresponds to the point of maximum ridge line curvature [17]. We thus decided to use the core point as the common reference point between all 8 samples of each person's fingerprint.

The (x, y) location (corresponding to the (column, row) pixel indices in the fingerprint image) of the core point in every fingerprint was extracted using VeriFinger 6.7 [18]. To ensure that we were working with ground-truth data, each fingerprint was manually inspected to confirm that its core location was correctly determined. If the core in a particular fingerprint sample was detected in the wrong location, but it was correct in other samples of the same fingerprint, then those other samples were used as a guide in manually identifying the location of the core point in the former fingerprint. If the core was not detected in any samples of the same fingerprint, which was often the case for Arch type fingerprints, then the point of highest curvature was selected as the core point. The horizontal and vertical translations between *every pair* of a person's 8 fingerprint samples were then calculated using Eqs. (1) and (2), respectively:

$$HT_{ij} = |x_i - x_j| \tag{1}$$

$$VT_{ij} = |y_i - y_j| \tag{2}$$

In Eqs. (1) and (2), HT_{ij} and VT_{ij} denote the horizontal and vertical translations (in pixels), respectively, between fingerprint samples i and j from a single person. Further, (x_i, y_i) and (x_j, y_j) represent the x- and y-coordinates of the core point in the same two sample images (i and j, respectively). The absolute value brackets in Eqs. (1) and (2) suggest that we are only interested in the *quantities* of the translations, rather than their specific directions. By "specific directions", we mean directions within the larger class of horizontal and vertical translations, i.e., left or right for horizontal translations, and up or down for vertical translations. The reason that we are not interested in these more specific directions is simply because they are arbitrary depending on which of a pair of sample images is chosen to be i and which is chosen to be j in Eqs. (1) and (2).

Equations (1) and (2) were applied to our cooperative user fingerprint database to calculate the horizontal and vertical translation, respectively, between each pair of fingerprint samples originating from the same finger. Figure 3 compares the box and whisker plots corresponding to the horizontal and vertical translation distributions resulting from applying Eqs. (1) and (2) to all 100 people in our fingerprint database.

From Fig. 3, we can see that the median *horizontal* translation is 13 pixels, with an interquartile range of 17 pixels (upper quartile of 23 − lower quartile of 6) and a range of 48 pixels (upper whisker of 48 − lower whisker of 0). Similarly, Fig. 3 indicates that the median *vertical* translation is 17 pixels, with an interquartile range of 23 pixels (upper quartile of 30 − lower quartile of 7) and a range of 64 pixels (upper whisker of 64 − lower whisker of 0).

The fact that the median horizontal translation is slightly smaller than the median vertical translation makes sense, because the height of the scanning surface of the Futronic FS88 (which was used to acquire the fingerprint images for our database) is 1.5 times its width (this would have been a conscious design decision to approximately replicate the shape of a finger). This means that the user has more freedom to move their finger up and down than they do to move it left and right. Consequently, we would typically expect vertical translations to be larger than horizontal translations. Similarly, the fact that both the interquartile range and range of the horizontal translation distribution were found to be larger than the corresponding statistics pertaining to

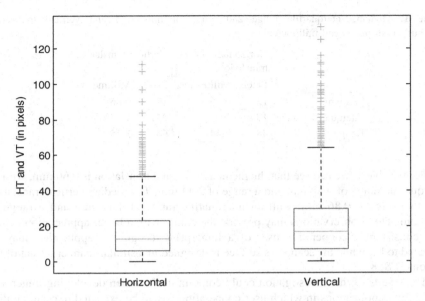

Fig. 3. Box and whisker plots comparing the horizontal and vertical translation distributions.

the vertical translation distribution suggests that there is more variability in the vertical translations between different samples of the same fingerprint than there is in the horizontal translations. So, not only are vertical translations more likely to be larger (larger median), but there is also likely to be more variation in the actual values of those vertical translations due to the greater degree of freedom in the vertical placement of the finger on the scanner. Since, for reasons outlined in Sect. 2.1, the Futronic FS88 may be considered a "typical" fingerprint scanner, we may conclude that the observations from Fig. 3 extend beyond the Futronic FS88 scanner.

The results in Fig. 3 provide a fair estimation of the amount of horizontal and vertical translation that we may expect, in pixels, between multiple images of the same fingerprint, when the fingerprints are captured from cooperative users. To gain a better appreciation of the significance of these translation amounts, Table 1 shows the pixel values of the median, interquartile range and range of the horizontal and vertical translation distributions from Fig. 3 in millimeters. Note that the fingerprint images in our database all measure 320 pixels in width and 480 pixels in height, and the scanning surface of the Futronic FS88 fingerprint scanner measures 16.26 mm in width and 24.38 mm in height. Taking these dimensions into account, Eq. (3) was used to convert a horizontal translation from pixels to millimeters, and Eq. (4) was used to convert a vertical translation from pixels to millimeters. Table 1 summarizes the results.

$$HT^{mm} = \left(HT^{pix} \div 320\right) \times 16.26 \qquad\qquad (3)$$

$$VT^{mm} = \left(VT^{pix} \div 480\right) \times 24.38 \qquad\qquad (4)$$

Table 1. Median, interquartile range and range of horizontal and vertical translation distributions in pixels and millimeters.

	Horizontal translation		Vertical translation	
	Pixels	Millimeters	Pixels	Millimeters
Median	13	0.66	17	0.86
Interquartile range	17	0.86	23	1.17
Range	48	2.44	64	3.25

From Table 1, we can see that the median horizontal translation is 0.66 mm, with an interquartile range of 0.86 mm and a range of 2.44 mm. The median vertical translation was found to be 0.86 mm, with an interquartile range of 1.17 mm and a range of 3.25 mm. These observations may provide the reader with a better appreciation of just how consistent a cooperative user of a fingerprint recognition application may be expected to be, when the scanner's surface is designed in a similar manner to that of the Futronic FS88.

The results of this investigation could come in useful when developing fingerprint recognition applications in which user cooperation would be expected (e.g., in civilian fingerprint recognition applications, such as computer login, for which it would be in the users' best interests to be recognized). The most obvious use for these results would be in testing the suitability of the Futronic FS88 scanner, or a scanner with a similar user interface, for a particular application. Since the results indicate a high level of user consistency in terms of finger translation on the scanner surface, we may conclude that this type of scanner would be suitable for an application in which user consistency is important. It is our hope that this investigation will inspire developers of fingerprint recognition systems to conduct similar experiments when evaluating the suitability of a particular scanner for their applications, as well as encouraging designers of fingerprint scanners to consider how these results may influence scanner surface design.

Another use for the results of this investigation would be in the development of fingerprint alignment algorithms. For example, our results in Fig. 3 and Table 1 provide developers with a realistic approximation of the amount of horizontal and vertical translation that is likely to occur among cooperative users (when the Futronic FS88 or a similar scanner is employed). The results indicate that a suitable alignment algorithm must be capable of resolving translations within quite a small range, which immediately suggests that a coarse alignment algorithm may be unsuitable for this purpose. If translational offsets between two different samples of the same fingerprint are to be corrected using a common reference point, the results of our investigation could be applied in speeding up the search for a common reference point between the two fingerprints. For example, our results indicate that the locations of the core point in two different samples of the same fingerprint should not differ by more than ± 48 pixels in the horizontal direction and ± 64 pixels in the vertical direction. So, once the core point is found in the first fingerprint sample, we may use the results of our investigation to approximate the likely location of the core point in the second fingerprint sample. Then, the search algorithm can be applied to the selected area first.

3.2 Rotation

Rotation refers to the difference in orientation between multiple samples of the same fingerprint. The more consistent a user is in placing their finger on the scanner, the more similar the orientations of their fingerprint samples will be, and thus the smaller the rotation will be.

To calculate the rotation between each person's 8 fingerprint samples, it was first necessary to establish the ground truth orientation of each fingerprint. Since the orientation of a fingerprint is commonly represented by the orientation of the core point, we decided to adopt this method for calculating the ground-truth orientation of each fingerprint in our database. As for the translation calculations in Sect. 3.1, the core point from each fingerprint was extracted using VeriFinger 6.7 [18]; however, this time we were only interested in the core *angle*, rather than its *location*. In order to ensure that we were working with ground-truth fingerprint orientations, the angle of each core point was manually inspected and corrected if necessary. The rotation between *every pair* of a person's 8 fingerprint samples was then calculated using Eq. (5), where θ_i and θ_j denote the orientations of fingerprints i and j, respectively, from the same person:

$$\phi_{ij} = \min(|\theta_i - \theta_j|, 360° - |\theta_i - \theta_j| \tag{5}$$

Figure 4 depicts the resulting rotation distribution in terms of a box and whisker plot.

From Fig. 4, we may conclude that cooperative users of a fingerprint recognition application may be expected to be very consistent in placing their finger onto the scanner, such that the median rotation between multiple samples of the same fingerprint should be 2°, with an interquartile range of 5° (upper quartile of 5° – lower quartile of

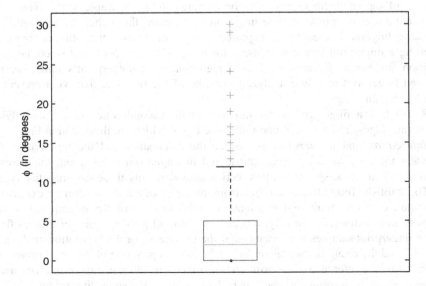

Fig. 4. Box and whisker plot of the rotation distribution.

0°) and a range of 12° (upper whisker of 12° – lower whisker of 0°). Note that this consistency would also be influenced by the scanner's design. The scanning surface of the Futronic FS88, which was used for our fingerprint database collection, has been designed to approximately replicate the shape of a finger; so, there is not much room for rotating the finger when it is placed on the scanning surface. We may thus assume that the results from this experiment extend to all other scanners with a similar design for the scanner surface as that of the Futronic FS88.

As for the investigation on translation in Sect. 3.1, the results from the investigation on rotation in the current section would be useful in the development of fingerprint recognition systems. For example, our results in Fig. 4 indicate that, in order to correct rotational differences between two different samples of the same fingerprint acquired from cooperative users, the adopted alignment algorithm must be capable of resolving small rotations. This suggests that a coarse alignment algorithm may be unsuitable. Furthermore, our results indicate that it may be unnecessary to check for rotations of more than ± 12°, which would be helpful in speeding up alignment methods that exhaustively check every possible rotation until the correct one is found (e.g., [6]). Finally, as suggested in our investigation on translation in Sect. 3.1, the results of our evaluation on rotation in this section would also be suitable for assessing the suitability of the Futronic FS88 scanner (and other scanners with a similar user interface) for a particular application, and it is hoped that both the results of this investigation and its methodology will come in useful for developers of fingerprint scanners.

3.3   Captured Fingerprint Minutiae

The area of the fingerprint that is captured during each scan is extremely important for fingerprint recognition. Depending on how a finger is placed on the scanner, different portions of the same fingerprint may be captured during multiple scans. The more consistent a user is in placing their finger on the scanner, the higher the probability of the same fingerprint area being captured every time. Since the ultimate point of acquiring a fingerprint image is to use it for recognition purposes, and since the most common fingerprint features used in recognition are a fingerprint's *minutiae*, the captured fingerprint area was analyzed in terms of the minutiae that were present in each fingerprint image.

Recall that a fingerprint's *minutiae* are small discontinuities in the underlying fingerprint ridges. The most common minutiae types, which are illustrated in Fig. 5, are the *bifurcation* and the *termination*. Since the bifurcation and the termination are generally the only minutiae types considered in fingerprint recognition, our investigation on minutiae persistence in this section considers only these two minutiae types.

To establish ground truth data, free from the errors of automated feature extractors, the minutiae (bifurcations and terminations only) in each of the fingerprints in our database were extracted manually. For each person, all possible pairs of images from their 8 fingerprint samples were then established (same as for the translation analysis in Sect. 3.1 and the rotation analysis in Sect. 3.2). In each pair, one of the fingerprints was chosen to be the reference fingerprint and the other was the test fingerprint. To ensure fairness, each fingerprint in each pair had a turn at being the reference. The

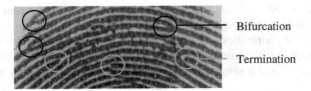

Bifurcation

Termination

Fig. 5. The most common fingerprint minutiae types.

corresponding minutiae between the reference and test fingerprints were then established. In order to avoid faulty or missing minutiae correspondences, which may be the result of using automated minutiae matchers, minutiae correspondences were determined manually. For each reference-test fingerprint pair, the percentage of reference minutiae that were paired up with a minutia in the corresponding test fingerprint was then calculated, and this was repeated for all 100 people in our fingerprint database. Figure 6 depicts the resulting distribution corresponding to the percentage of reference minutiae that are present in a test sample of the same fingerprint, in terms of a box and whisker plot.

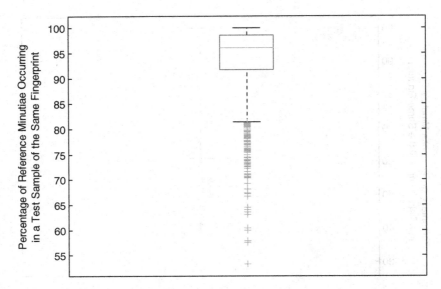

Fig. 6. Box and whisker plot of the distribution corresponding to the percentage of reference minutiae occurring in a test sample of the same fingerprint, when the minutiae are extracted and matched manually.

The box in the box and whisker plot from Fig. 6 suggests that, 50 % of the time, we may expect cooperative users of a fingerprint recognition application to be consistent enough in placing their finger on the fingerprint scanner to ensure that between 91.7 % and 98.6 % of the minutiae present in their reference fingerprint (acquired during enrolment) will be captured in their query fingerprint (acquired during authentication), with the median being 96.1 %. The whiskers indicate that this percentage would be expected to lie within the range [81.4 %, 100 %]. Note that this range encompasses 96.8 % of the entire distribution, which was used to generate the box and whisker plot in Fig. 6; therefore, we may conclude that, 96.8 % of the time, we may expect between 81.4 % and 100 % of the same minutiae to be captured during every authentication attempt.

The results presented in this section would be useful for the development of automated fingerprint recognition algorithms intended for deployment in cooperative user scenarios. For instance, knowing the percentage of reference minutiae that may be expected to occur in a query sample of the same fingerprint would help in an estimation of the likelihood that a false non-match is the user's fault (i.e., caused by user inconsistency in capturing the same minutiae). Such an analysis would be useful for honing in on the most problematic modules in a fingerprint recognition system. Although this point is dealt with in the dedicated investigation on minutiae persistence in our complimentary publication [19], here we provide the reader with an example. Consider the box and whisker plot in Fig. 7, which was generated by extracting and matching the minutiae from each fingerprint in our database *automatically* (using VeriFinger 6.7 [18]) instead of *manually*.

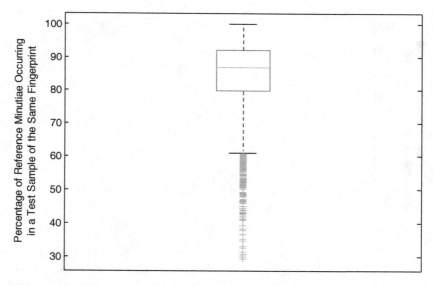

Fig. 7. Box and whisker plot of the distribution corresponding to the percentage of reference minutiae occurring in a test sample of the same fingerprint, when the minutiae are extracted and matched automatically.

Table 2 compares the box and whisker plot quantities from Fig. 6 to their corresponding statistics from Fig. 7.

From Table 2, we can see that, when the minutiae are extracted and matched *automatically*, a reference minutia is less likely to appear in a query sample of the same

Table 2. Comparison of box and whisker plot quantities for minutiae repeatability when the minutiae are extracted and matched manually versus automatically.

	Manual minutiae extraction and matching	Automatic minutiae extraction and matching
Lower whisker	81.4 %	60.9 %
Lower quartile	91.7 %	79.6 %
Median	96.1 %	86.8 %
Upper quartile	98.6 %	92.2 %
Upper whisker	100 %	100 %

fingerprint than when the minutiae are extracted and matched *manually*. This is because the minutiae repeatability results for the scenario in which *manual* minutiae extraction and matching are adopted is influenced by user consistency alone; however, when the minutiae are extracted and matched *automatically*, then minutiae repeatability is influenced by user consistency *and* potential minutiae extraction errors *and* potential matching errors. This hints at an interesting way in which the results of our investigation on minutiae repeatability may be applied in the development and testing of automated fingerprint recognition algorithms. For example, considering the *median* results in Table 2, we can see that the median percentage of reference minutiae occurring in a query sample of the same fingerprint when the minutiae are extracted and matched automatically is 9.3 % lower than the median percentage obtained when the minutiae are extracted and matched manually. We may thus conclude that 9.3 % of the reference minutiae are falsely not identified as being present in a query sample of the same fingerprint due to errors in the adopted automatic minutiae extractor and matcher alone. This tells us that, although minutiae repeatability is most significantly influenced by user consistency, in this case there are some errors in the automated minutiae extractor and matcher. Analysis of this sort would be useful for zoning in on the most problematic modules in a fingerprint recognition system, which would enable the development of more effective solutions.

Our results on minutiae repeatability would also be useful for gauging the suitability of the Futronic FS88 scanner (and other scanners with a similar user interface) for a particular application. Furthermore, these results may prove beneficial when applied towards the development of user-friendly fingerprint scanners, since the results indicate the ease with which the Futronic FS88 scanner (and others like it) allows its users to be consistent in capturing the same minutiae across multiple scans of the same fingerprint. A similar investigation conducted on different scanner designs would enable the respective developers to draw sensible conclusions regarding the usability of their product.

Another interesting application of our results on minutiae persistence would be towards an evaluation of the practicality of certain minutiae-based fingerprint template protection schemes. For example, we proposed a novel fingerprint template protection scheme based on compact minutiae patterns in [20]. This method has been specifically designed for deployment in cooperative user civilian fingerprint recognition applications, and it relies on the same subset of minutiae being present in every scan of a user's fingerprint. It would thus be valuable to apply the results of our investigation on minutiae persistence towards supporting the practicality of this method.

While the results in Fig. 6 are very encouraging in terms of supporting the expected consistency of cooperative users, the percentage of reference minutiae repeating in a test sample of the same fingerprint may be further improved. For example, note that the reference minutiae that were missing from a test fingerprint were always those minutiae that were close to the edges of the reference fingerprint. This suggests that edge minutiae should not be relied upon to be present during every scan. Thus, a simple, yet effective, way of increasing the proportion of reference minutiae that are present in another scan of the same fingerprint would be to combine multiple samples of the same fingerprint during enrolment to filter out only the most "reliable" minutiae. For example, we could ask the user to scan their finger 3 times during enrolment, extract the minutiae from each of those 3 fingerprints and ignore any minutiae that do not appear in *all* 3 fingerprints. The remaining minutiae would be considered the most likely to appear in another sample of the same fingerprint, and thus only those minutiae should constitute the reference minutiae set. We would expect that, the more reference fingerprints that are employed for reference minutiae filtering during enrolment, the greater our confidence that a reference minutia will be present in a query sample of the same fingerprint. A complete investigation into this matter is thoroughly detailed in our complementary publication [19], which is dedicated to empirically evaluating minutiae persistence among multiple samples of the same person's fingerprint in a cooperative user scenario.

4   Conclusions and Future Work

This paper investigates the intra-class variance between different samples of the same fingerprint, when the fingerprints are acquired in a cooperative user scenario. In particular, the focus of the paper is on empirically evaluating the consistency with which such users place their finger on a fingerprint scanner. The consistency is measured in terms of the translation and rotation between a reference fingerprint and a query sample of the same fingerprint, as well as the percentage of reference minutiae that are captured in the query fingerprint. Analysis of a specially constructed cooperative user fingerprint database resulted in a median horizontal translation of 13 pixels (0.66 mm), a median vertical translation of 17 pixels (0.86 mm), a median rotation of 2°, and a median minutiae repeatability of 96.1 %.

The obtained results suggest that we may expect cooperative users in a civilian fingerprint authentication application to be very consistent in the way in which they present their fingers to the fingerprint scanner – certainly much more consistent than would be implied by typical public fingerprint databases, which have been deliberately constructed to exaggerate user inconsistency. In light of these findings, it may be useful to provide public fingerprint databases that are more representative of the types of fingerprints that would actually be obtained in cooperative civilian fingerprint recognition applications. This would enable more realistic evaluations of fingerprint recognition algorithms that target such applications.

The results from this investigation may be useful in several ways. Perhaps the most obvious application is towards gauging the suitability of the Futronic FS88 scanner (and other scanners with a similar user interface) for a particular application. Furthermore, the results may be beneficial for estimating the user-friendliness of a similar fingerprint scanner during its design process, and it is hoped that the methodology of this entire investigation may inspire fingerprint scanner designers in general to conduct similar experiments to evaluate the usability of their product. Another use for the results of our investigation pertains to the development of fingerprint alignment algorithms intended for deployment in cooperative user scenarios. For example, the results obtained for the typical translation and rotation amounts and their respective ranges may be applied towards establishing the required resolution (i.e., the smallest translation and rotation amounts that must be able to be sensed) and computational complexity (e.g., the maximum translation and rotation that must be tested in an exhaustive search) of corresponding alignment algorithms. Finally, our results on the percentage of reference minutiae that are likely to be captured in a query sample of the same fingerprint would be helpful in the development and testing of automated fingerprint recognition systems and certain minutiae-based fingerprint template protection schemes by estimating the likelihood of a false non-match being the user's fault.

While the results obtained from this investigation are very encouraging, it is possible to further improve the consistency with which users place their fingers on a fingerprint scanner. A simple way of improving the percentage of reference minutiae that are present in another sample of the same fingerprint would be to filter out only the most reliable minutiae during enrolment. This could be done by capturing several samples of the user's fingerprint and then storing only those minutiae that appear in *all* of those scans. Another interesting extension to this investigation could be the implementation of a user guidance algorithm. For instance, to ensure that a user's finger is placed in approximately the same location on the scanner surface during each authentication attempt, an outline of the enrolled fingerprint could show up on the live scan. The users would then try to place their finger in the same location each time. This method could be useful in ensuring that the same minutiae are captured during each scan, by guiding the user into scanning the same fingerprint area every time. It would be interesting to test the capability of such a user guidance algorithm in decreasing the intra-class variance between multiple samples of the same person's fingerprint.

Acknowledgements. The authors wish to express their gratitude to the volunteers who donated their fingerprints to our database, without whom this investigation would not have been possible. We would also like to thank Vytautas Pranckenas from Neurotechnology's technical support team for his assistance in getting various components of the VeriFinger 6.7 SDK up and running.

References

1. Maltoni, D., et al.: Handbook of Fingerprint Recognition: Fingerprint Matching. Springer, New York (2009)
2. Qi, J., et al.: Fingerprint matching combining the global orientation field with minutia. Pattern Recogn. Lett. **26**, 2424–2430 (2005)

3. Tico, M., Kuosmanen, P.: Fingerprint matching using an orientation-based minutia descriptor. IEEE Trans. Pattern Anal. **25**, 1009–1014 (2003)
4. Xudong, J., Wei-Yun, Y.: Fingerprint minutiae matching based on the local and global structures. In: Proceedings of the 15th International Conference on Pattern Recognition ICPR 2000, pp. 1038–1041. IEEE (2000)
5. Jain, A., et al.: On-line fingerprint verification. IEEE Trans. Pattern Anal. **19**, 302–314 (1997)
6. Ratha, N.K., et al.: A real-time matching system for large fingerprint databases. IEEE Trans. Pattern Anal. **18**, 799–813 (1996)
7. Krivokuća, V., Abdulla, W.: Fast fingerprint alignment method based on minutiae orientation histograms. In: Proceedings of the 27th Conference on Image and Vision Computing New Zealand, pp. 486–491. ACM (2012)
8. Hrechak, A.K., McHugh, J.A.: Automated fingerprint recognition using structural matching. Pattern Recogn. **23**, 893–904 (1990)
9. Zhao, Q., et al.: High resolution partial fingerprint alignment using pore–valley descriptors. Pattern Recogn. **43**, 1050–1061 (2010)
10. Jea, T.-Y., Govindaraju, V.: A minutia-based partial fingerprint recognition system. Pattern Recogn. **38**, 1672–1684 (2005)
11. Databases. http://biolab.csr.unibo.it/databasesoftware.asp
12. Maio, D., et al.: FVC2002: second fingerprint verification competition. In: Proceedings of the 16th International Conference on Pattern Recognition, pp. 811–814. IEEE (2002)
13. FVC 2006 Fingerprint Verification Competition. http://bias.csr.unibo.it/fvc2006/databases. asp
14. Maltoni, D., et al.: Handbook of Fingerprint Recognition: Introduction. Springer, New York (2009)
15. FS88 FIPS201/PIV Compliant USB2.0 Fingerprint Scanner. http://www.futronic-tech.com/product_fs88.html
16. Jain, A.K., et al.: Introduction to Biometrics: Fingerprint Recognition. Springer US, New York (2011)
17. Maltoni, D., et al.: Handbook of Fingerprint Recognition. Springer, New York (2003)
18. VeriFinger SDK. http://www.neurotechnology.com/verifinger.html
19. Krivokuća, V., et al.: Minutiae persistence among multiple samples of the same person's fingerprint in a cooperative user scenario. In: Proceedings of the 3rd International Conference on Pattern Recognition Applications and Methods, pp. 76–86. SCITEPRESS, Portugal (2014)
20. Krivokuća, V., et al.: A non-invertible cancellable fingerprint construct based on compact minutiae patterns. Int. J. Biometrics. **6**, 125–142 (2014)

Kernel Matrix Completion for Learning Nearly Consensus Support Vector Machines

Sangkyun Lee$^{(\boxtimes)}$ and Christian Pölitz

Fakultät für Informatik, LS VIII, Technische Universität Dortmund,
44221 Dortmund, Germany
{sangkyun.lee,christian.poelitz}@tu-dortmund.de

Abstract. When feature measurements are stored in a distributed fashion, such as in sensor networks, learning a support vector machine (SVM) with a full kernel built with accessing all features can be pricey due to required communication. If we build an individual SVM for each subset of features stored locally, then the SVMs may behave quite differently, being unable to capture global trends. However, it is possible to make the individual SVMs behave nearly the same, using a simple yet effective idea we propose in this paper. Our approach makes use of two kernel matrices in each node of a network, a local kernel matrix built with only locally stored features, and an estimate of remote information (about "local" kernels stored in the other nodes). Using matrix completion, remote information is fully recovered with high probability from a small set of sampled entries. Due to symmetric construction, each node will be equipped with nearly identical kernel matrices, and therefore individually trained SVMs on these matrices are expected to have good consensus. Experiments showed that such SVMs trained with relatively small numbers of sampled remote kernel entries have competent prediction performance to full models.

Keywords: Support vector machines · Kernel methods · Matrix completion · Multiple kernel learning · Distributed features

1 Introduction

Training the support vector machines (SVMs) [2] in distributed environments has been an interesting topic in machine learning research. Considering distributed storage of data, the topic can be largely divided into two categories depending on whether examples or features are distributed. When examples are distributed (in this case, features are usually not assumed to be distributed), an SVM can be trained using a distributed optimization algorithm [3] but incurring potentially a good amount of communication, or individual SVMs can be trained locally with extra constraints to reduce their disparity to other SVM models in a network [7] with less amount of information exchange. Alternatively, local SVMs can be trained completely independently on data partitions and then combined to produce a more stable and accurate model than individual ones [6,10].

© Springer International Publishing Switzerland 2015
A. Fred et al. (Eds.): ICPRAM 2014, LNCS 9443, pp. 93–109, 2015.
DOI: 10.1007/978-3-319-25530-9_7

On the other hand, when features are distributed (examples are not distributed) and they are not agglomerated for analysis, learning machineries have to deal with a set of partitioned feature spaces, figuring out how they can accurately predict global trends overall feature spaces. Although this scenario has potential use for emerging applications such as sensor network [14], it has not been studied much in machine learning research for obvious difficulties. This paper focuses on this scenario and proposes a simple but effective method for training SVMs for distributed features, with a major difference to the existing methods [12,20] that no central coordination will be involved in learning. Part of this paper has been published in a conference [11], and this paper extends the previous one with updated results on projection error and matrix completion, and new discussions on classification error bounds of using approximate kernels.

Figure 1 shows a sketch of our proposed method. Each local feature storage node (represented as a circle) creates a local kernel matrix using only locally stored features. It also creates an empty matrix to store remote information, for which some entries are sampled from kernels stored in the other nodes (built with only features stored in them). The unseen entries of this partially observed remote kernel matrix are recovered via matrix completion, where the recovery is perfect with high probability if samples are noise-free. Then, each node makes use of a combination of the two matrices, a local and a recovered remote kernels, to train an SVM predictor as if the node has seen all features over a network.

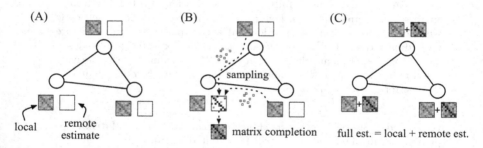

Fig. 1. A sketch of the proposed method. (A) each local feature storage node (a circle) create a kernel matrix for local features, and a null estimate of remote information, (B) each node obtains sampled entries of "local" kernels in the other nodes, and then recovers unseen entries by matrix completion, (C) each node makes use of its local kernel matrix together with recovered remote information to create its own SVM predictor.

In the following, we first introduce two decompositions of kernel matrices that enable us to approximate an original full-feature kernel matrix with separated kernels corresponding to local and remote features. Then we discuss the idea of matrix completion, a new observation of support vectors, and generalization error bounds of our method. Our description focuses on SVM classifiers, however it can be generalized to other kernel-based methods. We denote the Euclidean norm of vectors by $\| \cdot \|$ and the cardinality of a finite set A by $|A|$ throughout the paper.

2 Decomposition of Kernel Matrices

Consider local feature storages represented as nodes $n = 1, 2, \ldots, N$ in a network, where each node stores its features in a vector $\mathbf{x}_i[n]$ of length p_n. Here i is an index for examples, $i = 1, 2, \ldots, m$, and we assume that all nodes can observe the same examples (but through different set of features) and their label $y_1, \ldots, y_m$. For simplicity, we allow for communication between any pair of nodes. Then the collection of all features can be written as a single vector $\mathbf{x}_i = (\mathbf{x}_i[1]^T, \mathbf{x}_i[2]^T, \ldots, \mathbf{x}_i[N]^T)^T$ of length $p := \sum_{n=1}^{N} p_n$ (this vector is never created in our method).

2.1 Support Vector Machines

The dual formulation of the SVM, in a form where a bias term is augmented in a weight vector, is described as follows [13],

$$\min_{0 \le \alpha \le C1} \frac{1}{2} \alpha^T \mathbf{Q} \alpha - \mathbf{1}^T \alpha, \tag{1}$$

where α is a vector of length m, $\mathbf{1} := (1, 1, \ldots, 1)^T$, $\mathbf{y} := (y_1, y_2, \ldots, y_m)^T$, and $C > 0$ is a given constant. The $m \times m$ matrix $\mathbf{Q}$ is a scaled kernel matrix, that is, $\mathbf{Q} := \mathbf{YKY}$ for a positive semidefinite kernel matrix $\mathbf{K}$, where $\mathbf{Y} := \text{diag}(\mathbf{y})$ is the diagonal matrix with labels from $\mathbf{y}$. A typical SVM is built with $\mathbf{K}_{ij} = \langle \phi(\mathbf{x}_i), \phi(\mathbf{x}_j) \rangle$, where $\mathbf{x}_i$ and $\mathbf{x}_j$ are "full" feature vectors and $\phi : \mathfrak{R}^p \to \mathcal{H}$ is a map from the space of feature vectors to a Hilbert space [18].

2.2 Schur and MKL Kernels

Our goal is to build an SVM for each local feature storage node in a network, as if we have accessed all features but without explicitly doing so. This becomes possible by the observations we introduce here, about how to decompose the original kernel matrices into parts corresponding to local and remote features.

The construction is surprisingly simple. For instance, the expression of the Gaussian kernel [18] can be rewritten as follows,

$$
\begin{aligned}
\text{(Schur Kernel)} \quad [\mathbf{K}]_{ij} &= e^{-\gamma \|\mathbf{x}_i - \mathbf{x}_j\|} = \prod_{n=1}^{N} e^{-\gamma \|\mathbf{x}_i[n] - \mathbf{x}_j[n]\|^2} \\
&= \underbrace{e^{-\gamma \|\mathbf{x}_i[n] - \mathbf{x}_j[n]\|^2}}_{[\mathbf{S}_n]_{ij} \,:\, \text{local for node } n} \underbrace{\prod_{n' \ne n} e^{-\gamma \|\mathbf{x}_i[n'] - \mathbf{x}_j[n']\|^2}}_{[\mathbf{S}_{-n}]_{ij} \,:\, \text{remote for node } n}.
\end{aligned} \tag{2}
$$

Here $\gamma > 0$ is a given parameter. As we can see, the Gaussian kernel matrix can be decomposed into a matrix $\mathbf{S}_n$ created using only the features stored in a node n, and another matrix $\mathbf{S}_{-n}$ that captures information about features not in the node n. Since this kernel matrix is written as an elementwise product of matrices, known as the Schur (or Hadamard) product, we rename this kernel as the Schur kernel to distinguish itself from local Gaussian kernels such as $\mathbf{S}_n$.

We also consider another type of kernels constructed by a simple linear combination of multiple local Gaussian kernels. This kernel typically appears in the multiple kernel learning [9] scenarios, so we name it as the MKL kernel. This kernel is obviously decomposable:

$$
\text{(MKL Kernel)} \quad [\mathbf{K}]_{ij} = \frac{1}{N} \sum_{n=1}^{N} e^{-\gamma_n \|\mathbf{x}_i[n] - \mathbf{x}_j[n]\|^2}
$$

$$
= \underbrace{\frac{1}{N} \, e^{-\gamma_n \|\mathbf{x}_i[n] - \mathbf{x}_j[n]\|^2}}_{[\mathbf{M}_n]_{ij}:\ \text{local for node } n} + \underbrace{\frac{1}{N} \sum_{n' \neq n} e^{-\gamma_{n'} \|\mathbf{x}_i[n'] - \mathbf{x}_j[n']\|^2}}_{[\mathbf{M}_{-n}]_{ij}:\ \text{remote for node } n} \;.
$$

$$(3)$$

Similarly to the previous case, this kernel is decomposed into a local ($\mathbf{M}_n$) and a remote $\mathbf{M}_{-n}$ parts. Compared to the Schur kernel, the MKL kernel requires to specify N positive kernel parameters $\gamma_1, \gamma_2, \ldots, \gamma_N$, rather than a single parameter $\gamma > 0$. Despite of this potential inconvenience, MKL kernels have an advantage that we can reduce the size of kernel matrices and therefore speed up the process of matrix completion as discussed later in Sect. 3.3.

3   Recovery of Unseen Kernel Elements

In our model, each node n creates a matrix $\mathbf{S}_n$ or $\mathbf{M}_n$ depending on the type of kernel it use (Schur or MKL), from only locally stored features. Also, each node n creates an empty matrix for $\mathbf{S}_{-n}$ or $\mathbf{M}_{-n}$, in which each element is just a product (Schur) or a sum (MKL) of "local" kernel matrices stored in the other nodes, where only a few of these entries are observed by the node n by uniform random sampling.

3.1   Sampling

The key elements in our sampling strategy are that (i) the size of sample should be minimized to reduce communication cost, and at the same time (ii) the sample size should be large enough to guarantee the recovery of unseen entries with accuracy. As we see later, the theory of matrix completion make it possible to achieve both goals, surprisingly enough, with simple uniform random sampling.

For a node n, we denote the index pair set of observed entries of remote kernel matrices ("local" in the other nodes) by

$$
\Omega_n \subset \{(i, j) : 1 \leq i, j \leq m\},
$$

where the pairs (i, j) are chosen uniformly at random[1]. Then, all the other nodes n' transfer the entries of their local kernel matrix corresponding to Ω_n, i.e.,

$$\text{Node } n \longleftarrow \begin{cases} [\mathbf{S}_{n'}]_{\Omega_n} & (\text{Schur}) \\ [\mathbf{M}_{n'}]_{\Omega_n} & (\text{MKL}) \end{cases} \longleftarrow \text{Node } n'.$$

This requires that the sample index pair set Ω_n should be known to all nodes[2], which can be done once by exchanging such information when a pier recognizes other piers in a network. Or, we can fix all sets Ω_n for $n = 1, 2, \ldots, N$ the same, so that no exchange will be necessary if the set is pre-determined.

Given Ω_n, the communication cost of this type of transfer will be $\mathcal{O}((N - 1)|\Omega|)$, if a node n is connected to all the other nodes. As discussed later, matrix completion requires the size Ω_n relatively small, $\mathcal{O}(m \log^6 m)$, to guarantee a perfect recovery in high probability.

When a node n receives the information, it stores the entries to the corresponding storage as follows (left: Schur, right: MKL):

$$[\mathbf{S}_{-n}]_{\Omega_n} = \prod_{n' \neq n} [\mathbf{S}_{n'}]_{\Omega_n}, \qquad [\mathbf{M}_{-n}]_{\Omega_n} = \sum_{n' \neq n} [\mathbf{M}_{n'}]_{\Omega_n}.$$

The next step is to recover the unobserved entries in the matrix $\mathbf{S}_{-n}$ or $\mathbf{M}_{-n}$.

3.2 Low-Rank Matrix Completion with Kernel Constraints

To recover the full matrix of $\mathbf{S}_{-n}$ or $\mathbf{M}_{-n}$ from few observed entries indexed by Ω_n, we use the low-rank matrix completion [17]. Some modifications are required, however, to deal with the constraints that $\mathbf{S}_{-n}$ or $\mathbf{M}_{-n}$ should be a valid kernel matrix. In particular, $\mathbf{S}_{-n}$ and $\frac{1}{N-1}\mathbf{M}_{-n}$ must be symmetric and positive semi-definite matrices to satisfy the Mercer's theorem [18].

To describe a formal framework for matrix completion, suppose that $\mathbf{X} \in \Re^{m \times m}$ is a matrix we wish to recover, and that the observed entries of $\mathbf{X}$ (corresponding to a sample index pair set Ω_n) are stored in a data matrix $\mathbf{D}$ so that $\mathbf{D}_{\Omega_n} = [\mathbf{S}_{-n}]_{\Omega_n}$ for Schur kernels or $\mathbf{D} = [\mathbf{M}_{-n}]_{\Omega_n}$ for MKL kernels. Matrix completion recovers the full matrix of $\mathbf{X}$ as a solution of the following convex optimization problem [17],

$$\min_{\mathbf{X} \in \Re^{m \times m}} \sum_{(i,j) \in \Omega_n} (\mathbf{X}_{ij} - \mathbf{D}_{ij})^2 + \lambda \|\mathbf{X}\|_*, \quad \|\mathbf{X}\|_* := \sum_{k=1}^{m} \sigma_k(\mathbf{X}),$$

where $\|\mathbf{X}\|_*$ is called the *nuclear norm* of $\mathbf{X}$, which is the summation of singular values $\sigma_k(\mathbf{X})$ of $\mathbf{X}$ and penalizes the rank of $\mathbf{X}$ in effect. When the rank of $\mathbf{X}$ is r, then the nuclear norm simplifies to the expression [16,17],

$$\|\mathbf{X}\|_* = \min_{\mathbf{X} = \mathbf{L}\mathbf{R}^T} \frac{1}{2}\|\mathbf{L}\|_F^2 + \frac{1}{2}\|\mathbf{R}\|_F^2,$$

---

[1] We require Ω_n to be symmetric, that is, if $(i, j) \in \Omega_n$ then $(j, i) \in \Omega_n$, and also the values corresponding to the entries are the same.

[2] Not actual values, but the positions to be sampled from.

where $\mathbf{L}$ and $\mathbf{R}$ are $m \times r$ matrices, and $\|A\|_F := \left(\sum_{ij} A_{ij}^2\right)^{1/2}$ is the Frobenius norm of a matrix A. Using this property, the above optimization can be rewritten for rank-r matrix completion,

$$\min_{\mathbf{L},\mathbf{R} \in \Re^{m \times r}} \sum_{(i,j) \in \Omega_n} (\mathbf{L}_{i\cdot} \mathbf{R}_{j\cdot}^T - \mathbf{D}_{ij})^2 + \frac{\lambda}{2}\|\mathbf{L}\|_F^2 + \frac{\lambda}{2}\|\mathbf{R}\|_F^2. \tag{4}$$

A solution of this optimization can be obtained using the JELLYFISH algorithm [17] for example, which is a highly parallel incremental gradient descent algorithm using the fact that the gradient of the objective function depends on row vectors $\mathbf{L}_{i\cdot}$ and $\mathbf{R}_{j\cdot}$ and therefore updates of iterates can be distributed for the pairs (i,j) in Ω_n.

Projection to the Mercer Kernel Space. The outcome $\mathbf{X}^* = \mathbf{L}^*(\mathbf{R}^*)^T$ of low-rank matrix completion (4) may not be a valid Mercer kernel matrix. To make it a valid one, we project $\mathbf{X}^*$ to the space of Mercer kernels [18], which is a convex set $\mathcal{K}$ of symmetric positive semidefinite rank-r matrices in our case,

$$\mathcal{K} := \{\mathbf{X} : \mathbf{X} \succeq 0, \ \mathbf{X}^T = \mathbf{X}, \ \text{rank}(\mathbf{X}) = r\} = \{\mathbf{Z}\mathbf{Z}^T : \mathbf{Z} \in \Re^{m \times r}\}.$$

Here $\mathbf{X} \succeq 0$ means that $\mathbf{X}$ is a positive semi-definite matrix which satisfies $\mathbf{w}^T \mathbf{X} \mathbf{w} \geq 0$ for any $\mathbf{w} \in \Re^m$. The last equality implies that an element $\mathbf{X} \in \mathcal{K}$ must have the form $\mathbf{Z}\mathbf{Z}^T$ for an $m \times r$ matrix $\mathbf{Z}$, which can be easily verified using the eigen-decomposition [21] of symmetric $\mathbf{X}$.

To project $\mathbf{X}^* = \mathbf{L}^*(\mathbf{R}^*)^T$ onto $\mathcal{K}$, we use a simple projection for which a closed-form solution exists (derivable from the optimality conditions),

$$\mathbf{Z}^* = \arg\min_{\mathbf{Z} \in \Re^{m \times r}} \frac{1}{2}\|\mathbf{Z} - \mathbf{L}^*\|_F^2 + \frac{1}{2}\|\mathbf{Z} - \mathbf{R}^*\|_F^2, \quad \mathbf{Z}^* = \frac{\mathbf{L}^* + \mathbf{R}^*}{2}. \tag{5}$$

The next result shows that the gap between a recovered matrix from matrix completion $\mathbf{L}^*(\mathbf{R}^*)^T$ and a valid kernel matrix obtained after projection $\mathbf{Z}^*(\mathbf{Z}^*)^T$ is bounded and becomes small whenever $\mathbf{L}^* \approx \mathbf{R}^*$, which is indeed likely to happen since we require that the sample index pair set Ω is symmetric.

Lemma 1. *The distance between* $\mathbf{Z}^*(\mathbf{Z}^*)^T$ *and* $\mathbf{L}^*(\mathbf{R}^*)^T$ *is bounded as follows,*

$$\|\mathbf{Z}^*(\mathbf{Z}^*)^T - \mathbf{L}^*(\mathbf{R}^*)^T\|_F \leq \frac{\|\mathbf{L}^* - \mathbf{R}^*\|_F}{4} + \frac{\|\mathbf{L}^*(\mathbf{R}^*)^T - \mathbf{R}^*(\mathbf{L}^*)^T\|_F}{2}.$$

Proof. The result follows from the definition of $\mathbf{Z}^*$ and the triangle inequality.

After training SVMs, we apply the same technique for new test examples to build a test kernel matrix. This usually involves a smaller matrix completion problem corresponding to the support vectors and test examples.

3.3   Reduction with MKL Kernels Using Support Vectors

The matrix completion optimization (4) for recovering full kernel matrices involves m^2 variables, and therefore it requires more computational resource for larger m. It turns out that we can work on a smaller number of variables than m^2, using a property we discovered for *support vectors* (SVs) in case of the MKL kernel. To remind, a support vector is an example indexed by $i \in \{1, 2, \ldots, m\}$ for which the optimal solution α_i^* of the SVM problem (1) is strictly positive.

Let us consider a "full-information" SVM problem with the MKL kernel built with accessing all features, and its set of SVs S^*,

$$\alpha^* := \underset{0 \leq \alpha \leq C1}{\arg\min} \; \frac{1}{2}\alpha^T \mathbf{Y} \left[\frac{1}{N} \sum_{n=1}^{N} \mathbf{M}_n \right] \mathbf{Y}\alpha - \mathbf{1}^T\alpha, \;\; S^* := \{i : 1 \leq i \leq m, [\alpha^*]_i > 0\}.$$

And, we also consider the corresponding "local" SVM problems and their SV sets for each node n, $n = 1, 2, \ldots, N$,

$$\alpha_n^* := \underset{0 \leq \alpha \leq C1}{\arg\min} \; \frac{1}{2}\alpha^T \mathbf{Y}[\mathbf{M}_n]\mathbf{Y}\alpha - \mathbf{1}^T\alpha, \;\; S_n^* := \{i : 1 \leq i \leq m, [\alpha_n^*]_i > 0\}.$$

The number of SVs, $|S^*|$, is often much smaller than the total number of examples m, and it is well known that an SVM predictor is fully determined by the SVs [18]. Therefore, if we estimate S^* without too much cost, then we can focus on variables corresponding to S^* for recovery in matrix completion rather than considering all m^2 variables. Our next theorem shows that an estimation of S^* is possible without solving the full-information problem, simply by the union of local SV sets (the proof is in Appendix).

Theorem 1. *The support vector sets of the full-information and local SVM problems above with the MKL kernel satisfy*

$$S^* \subseteq \bigcup_{n=1}^{N} S_n^*.$$

Using the theorem, matrix completion (4) can be solved more efficiently with $|\cup_n S_n^*|^2$ variables. In our experiments, the number was much smaller than m^2: the ratio $|\cup_n S_n^*|^2 / m^2$ was in the range of $0.31 \sim 0.98$, where in the half of the cases the ratio was below 0.5.

4   Matrix Recovery and Classification Error Bounds

To decide the size of the sample index set Ω, it is important to understand when a (perfect) recovery of unseen matrix elements is possible from only a few observed entries. It is also closely related how well an SVM trained with a recovered kernel matrix will perform in classification, compared to the case of using full-information kernel matrices.

4.1   Conditions for Matrix Recovery

The technique of matrix completion guarantees the perfect recovery of a partially observed matrix with high probability, under a certain condition called the *strong incoherence property* [4].

To describe this property, we define a parameter $\mu > 0$ as follows. Suppose that we try to recover all unseen entries of a matrix $\mathbf{D} \in \Re^{m \times m}$ from a few observed entries $\mathbf{D}_{ij}$, $(i,j) \in \Omega$. The rank of $\mathbf{D}$ is assumed to be r, so that the reduced singular value decomposition of $\mathbf{D}$ can be written as,

$$\mathbf{D} = \mathbf{U}\mathbf{\Sigma}\mathbf{V}^T, \quad \mathbf{U}^T\mathbf{U} = \mathbf{I}, \quad \mathbf{V}^T\mathbf{V} = \mathbf{I},$$

where $\mathbf{U} \in \Re^{m \times r}$ and $\mathbf{V} \in \Re^{m \times r}$ contain the left and right singular vectors of $\mathbf{D}$ that constitute orthonormal bases of the range space of $\mathbf{D}$ and $\mathbf{D}^T$, respectively (if $\mathbf{D}$ is symmetric then $\mathbf{U} = \mathbf{V}$), and $\mathbf{\Sigma} \in \Re^{r \times r}$ is a diagonal matrix with singular values. Then the orthogonal projection onto the range space of $\mathbf{D}$ and $\mathbf{D}^T$ can be described as $\mathcal{P}_{\mathbf{U}} = \mathbf{U}\mathbf{U}^T$ and $\mathcal{P}_{\mathbf{V}} = \mathbf{V}\mathbf{V}^T$, resp. Using these, we define two parameters $\mu_1 > 0$ and $\mu_2 > 0$ such that

$$\max_{i,j} \max \left\{ \left| [\mathcal{P}_{\mathbf{U}}]_{ij} - \frac{r}{m}\chi_{i=j} \right|, \; \left| [\mathcal{P}_{\mathbf{V}}]_{ij} - \frac{r}{m}\chi_{i=j} \right| \right\} \leq \mu_1 \frac{\sqrt{r}}{m} \tag{6}$$

where $\chi_{i=j}$ is an indicator function returning 1 if $i = j$ and 0 otherwise, and,

$$\max_{i,j} |[\mathbf{U}\mathbf{V}^T]_{ij}| \leq \mu_2 \frac{\sqrt{r}}{m}. \tag{7}$$

Finally, the parameter $\mu > 0$ is defined simply as

$$\mu := \max\{\mu_1, \mu_2\},$$

and we say $\mathbf{D}$ satisfies the strong incoherence property when μ is small (i.e., $\mu = \mathcal{O}(\sqrt{\log m})$). Conceptually speaking, small values of μ indicate that observing an element in $\mathbf{D}$ provides good information about the other elements in the same row and column, since its information is "dissolved" into many components of singular vectors and so are the other elements.

The next theorem states that when the rank r or the incoherence parameter μ of a matrix $\mathbf{D}$ we want to recover is small, then exact recovery via matrix completion (4) is possible with high probability with only a small number of observed entries.

Theorem 2 (Candès and Plan [5]). *Let* $\mathbf{D} \in \Re^{m \times m}$ *a matrix with rank* $r \in (0, m]$ *and a strong incoherence parameter* $\mu > 0$. *If the number of observed entries from* $\mathbf{D}$ *satisfies*

$$|\Omega| \geq C\mu^2 mr \log^6 m$$

for some constants $C > 0$, *then the minimizer of the matrix completion problem* (4) *is unique and equal to the original full matrix* $\mathbf{D}$ *with probability at least* $1 - m^{-3}$.

4.2 Classification Error Bounds for Using Estimated Kernels

When $|\Omega|$ is sufficiently large, then two factors can contribute to differences between the original and the estimated kernel matrices: (i) the noise in observed entries corresponding to Ω (note that Theorem 2 is for noise-free cases) and (ii) the gap induced by a projection onto the cone of symmetric positive semidefinite matrices (Lemma 1). Let us define the recovery error ϵ due to noise for elements corresponding to Ω as follows,

$$\epsilon := \sum_{(i,j)\in\Omega_n} (\mathbf{L}_{i\cdot}^*(\mathbf{R}_{j\cdot}^*)^T - \mathbf{D}_{ij})^2.$$

Using this, the gap between an original kernel matrix $\mathbf{K}$ and an estimated one $\widetilde{\mathbf{K}}$ can be stated as follows (for the case of the MKL kernel),

$$\Delta := \|\mathbf{K} - \widetilde{\mathbf{K}}\|_F \leq \underbrace{\|\mathbf{K} - \mathbf{L}^*(\mathbf{R}^*)^T\|_F}_{=\delta_1} + \underbrace{\|\mathbf{L}^*(\mathbf{R}^*)^T - \mathbf{Z}^*(\mathbf{Z}^*)^T\|_F}_{=\delta_2}, \qquad (8)$$

where the two terms on the right are bounded respectively as follows,

$$\delta_1 \leq 2\epsilon \left[2\sqrt{(1 + 2m^2/|\Omega|)\, m} + 1 \right], \qquad \text{(due to [5, Theorem 7])}$$

$$\delta_2 \leq \|\mathbf{L}^* - \mathbf{R}^*\|_F/4 + \|\mathbf{L}^*(\mathbf{R}^*)^T - \mathbf{R}^*(\mathbf{L}^*)^T\|_F/2 \qquad \text{(due to Lemma 1).}$$

For simplicity we do not discuss the case of Schur kernels here since the first term in the right-hand side of (8) becomes quite complicated.

To show a classification error bound using an estimate kernel matrix $\widetilde{\mathbf{K}}$, we use a result from Nguyen et al. [15] that if a kernel $\mathbf{K}$ is universal on the input space [19] as the Gaussian kernels, the *error coefficient* $\mathcal{E}_{\mathbf{K}}$ approaches the Bayes error in probability, that is,

$$\mathcal{E}_{\mathbf{K}} = \frac{1}{m}\mathbf{1}^T\boldsymbol{\alpha}_{\mathbf{K}}^* \;\rightarrow\; \inf_{f\in\mathcal{F}} \mathbb{P}(y \neq f(\mathbf{x})),$$

where $\boldsymbol{\alpha}_{\mathbf{K}}^*$ is the optimal solution of an SVM (1) using $\mathbf{K}$, and $\mathcal{F}$ is an arbitrary family of measurable functions that contains the optimal Bayes classifier. This implies that the classification error of using an estimate kernel matrix $\widetilde{\mathbf{K}}$ can be characterized by the distance between $\mathcal{E}_{\widetilde{\mathbf{K}}}$ and $\mathcal{E}_{\mathbf{K}}$, as we state in the next theorem (the proof is in Appendix):

Theorem 3. *Suppose that $\widetilde{\mathbf{K}}$ is an MKL (3) kernel matrix constructed with remote kernel information recovered by matrix completion (4) with a sufficiently large sample index set Ω satisfying the condition of Theorem 2 and projection (5), where errors identified by δ_1 and δ_2 as above. Then, with very high probability, the classification error of using $\widetilde{\mathbf{K}}$ instead of the full-information MKL kernel $\mathbf{K}$ is bounded by*

$$|\mathcal{E}_{\widetilde{\mathbf{K}}} - \mathcal{E}_{\mathbf{K}}| \leq \frac{C}{\lambda_1(\mathbf{Q})}(\delta_1 + \delta_2),$$

where C is a parameter for the SVM, and $\lambda_1(\mathbf{Q})$ is the smallest eigenvalue of $\mathbf{Q} = \mathbf{YKY}$, $\mathbf{Y} = diag(\mathbf{y})$.

Note that when observations are made without noise, then $\delta_1 = 0$ with probability at least $1 - m^{-3}$ by Theorem 2. Also, δ_2 is expected to be small due to our symmetry construction of Ω, in which case $\mathbf{L}^* \approx \mathbf{R}^*$. Moreover, we often specify $C = C'/m$ for some $C' > 0$, and therefore if $1/\lambda_1(Q) = o(m)$, that is, $\lim_{m \to \infty} \frac{o(m)}{m} = 0$, then the term $C/\lambda_1(\mathbf{Q})$ becomes small as m increases.

5   Experiments

For experiments, we used five benchmark data sets from the UCI machine learning repository [1], summarized in Table 1, and also their subset composed of 5000 training and 5000 test examples (denoted by 5k/5k) to study characteristics of algorithms under various circumstances.

We have implemented our algorithm as open-source in C++, based on the JELLYFISH code[3] [17] for matrix completion, and SVMLIGHT[4] [8] for solving SVMs. Our implementation makes use of the union SVs set theorem (Theorem 1) for the MKL approach to reduce kernel completion time, but not for Schur since the theorem does not apply for this case.

Table 1. Data sets and their training parameters. Different values of C were used for the full data sets (column C) and smaller $5k/5k$ sets (column C ($5k/5k$)).

Name	m (train)	Test	p	C	C ($5k/5k$)	γ
ADULT	40701	8141	124	10	10	0.001
MNIST	58100	11900	784	0.1	1162	0.01
CCAT	89702	11574	47237	100	156	1.0
IJCNN	113352	28339	22	1	2200	1.0
COVTYPE	464809	116203	54	10	10	1.0

For all experiments, we split the original input feature vectors into subvectors of almost equal lengths, one for each node of $N = 3$ nodes (for 5k/5k sets) and $N = 10$ nodes (for full data sets). The tuning parameters C and γ were determined by cross validation for the full sets, and the C values for the 5k/5k subsets were determined by independent validation subsets, both with SVMLIGHT. The results of SVMLIGHT were included for a comparison to a non-distributed SVM training. Following [12], the local Gaussian kernel parameters for MKL were adjusted to $\gamma_n = \frac{p}{p_n} \approx N\gamma$ for a given γ, so that $\gamma_n \|\mathbf{x}_i[n] - \mathbf{x}_j[n]\|$ will have the same order of magnitude $\mathcal{O}(\gamma p)$ as $\gamma \|\mathbf{x}_i - \mathbf{x}_j\|$.

---

[3] Available for download at http://hazy.cs.wisc.edu/hazy/victor/jellyfish/.
[4] Available at http://svmlight.joachims.org/.

5.1   Characteristics of Kernel Matrices

The first set of experiments is to verify that how well kernel matrices fit for matrix completion. For this, we computed the two types of full kernel matrices, Schur (2) and MKL (3), accessing all features of the small 5k/5k subsets of the five UCI data sets (in this case the Schur kernel is simply the Gaussian kernel).

The important characteristics of the kernel matrices with respect to matrix completion are its rank (r) and coherence parameters μ_1 and μ_2 defined in (6) and (7). When these values are small, then Theorem 2 tells that we only need small number of observations for perfect matrix completion with high probability.

Table 2 summarizes these characteristics. Clearly, the rank (numerically effective rank, with eigenvalues larger than a threshold of 0.01) and coherence values were small in cases of ADULT, IJCNN, and COVTYPE compared to the other data sets, and the conditions were improved when the MKL kernel was used compared to the Schur kernel. So it was expected that matrix completion would perform better for these three data sets compared to the rest, MNIST and CCAT, for a fixed size of the sample index set $|\Omega|$.

Table 2. The density, rank r, and coherence parameters μ_1 and μ_2 defined in (6) and (7) of kernel matrices. Effective numbers of ranks are shown, which correspond to eigenvalues larger than a threshold (0.01). Smaller rank and coherence parameters are better for matrix completion.

	Schur				MKL			
	Density	r	μ_1	μ_2	Density	r	μ_2	μ_2
ADULT	1.0	789	25.7	24.4	1.0	222	12.0	5.5
MNIST	1.0	4782	68.1	68.5	1.0	4568	66.6	66.2
CCAT	1.0	4984	69.6	70.6	1.0	4982	69.6	70.6
IJCNN	1.0	1516	37.2	37.8	1.0	698	25.1	6.6
COVTYPE	1.0	1423	35.9	35.8	1.0	424	19.3	2.7

5.2   The Effect of Sampling Size

Next, we have used the 5k/5k data sets to investigate how the prediction performance of SVMs changed over several difference sizes of the sample index set Ω. We define the sampling ratio as

$$\text{Sampling Ratio} := |\Omega|/(m^2),$$

where the value of m is 5000 in this experiment. We compared the prediction performance of using Schur and MKL to that of SVMLIGHT.

Figure 2 illustrates the test accuracy values for five sampling ratios in up to 10 %. The statistics are over $N = 3$ nodes and over random selections of Ω. The performance on ADULT, IJCNN, and COVTYPE was close to that of SVMLIGHT,

Table 3. Test prediction performance on full data sets (mean and standard deviation). Two sampling ratios (2 % and 10 %) are tried for our method. The SVMLIGHT results are from using the classical Gaussian kernels with matching parameters. $|\cup_n S_n^*|^2/m^2$ is the fraction of the reduced number of variables compared to m^2.

		MKL		ASSET	SVMLIGHT
	$\|\cup_n S_n^*\|^2/m^2$	2 %	10 %		
ADULT	0.37	81.4±1.00	84.2±0.18	80.0±0.02	84.9
MNIST	0.98	78.9±1.69	87.0±0.20	88.9±0.39	98.9
CCAT	0.71	87.2±1.00	92.0±0.35	73.7±1.00	95.8
IJCNN	0.31	96.0±0.35	96.5±0.23	90.9±0.88	99.3

and it kept increasing with the growth of $|\Omega|$. This behavior was expected in the previous section as their kernel matrices had good conditions for matrix completion. On the other hand, the performance on MNIST and CCAT was far inferior to that of SVMLIGHT, as also expected.

The bottom-right corner of Fig. 2 shows the concentration of eigenvalue spectrum in the five kernel matrices. The height of each box represents the magnitude of the corresponding normalized eigenvalue, so that the height a stack of boxes represents the proportion of entire spectrum concentrated in the top 10 eigenvalues. The plot shows that 90 % of the spectrum in ADULT is concentrated in the top 10 eigenvalues, indicating that its kernel matrix has a very small numerically effective rank. This gives one explanation why our method performs as good as SVMLIGHT for the case of ADULT.

Comparing Schur to MKL, both showed similar prediction performance. However, higher concentration of the eigen spectrum of MKL indicated that it would make a good alternative to Schur, also considering the extra saving with MKL discussed in Sect. 3.3.

5.3 Performance on Full Data Sets

In the last experiment, we used the full data sets for comparing our method to one of the closely related approaches, ASSET [12]. Since ASSET admits only MKL-type kernels, we have omitted the Schur kernel in comparison. Among the several versions of ASSET in [12], we used the "Separate" version with central optimization. COVTYPE was excluded due to runtime issues of SVMLIGHT.

The results are in Table 3. The second column shows the ratio between a union SV set and an entire training set. The square of these numbers indicates the saving we have achieved by the union SVs trick, for example the size of matrix is reduced to 37 % of the original size for ADULT. The saving was substantial for ADULT and IJCNN. In terms of prediction performance, we have achieved test accuracy approaching to that of SVMLIGHT (within 1 % point (ADULT), 3.8 % points (CCAT), and 2.8 % points (IJCNN) on average) with 10 % sampling ratio, except for the case of MNIST where the gap was significantly larger (11.9 %): this

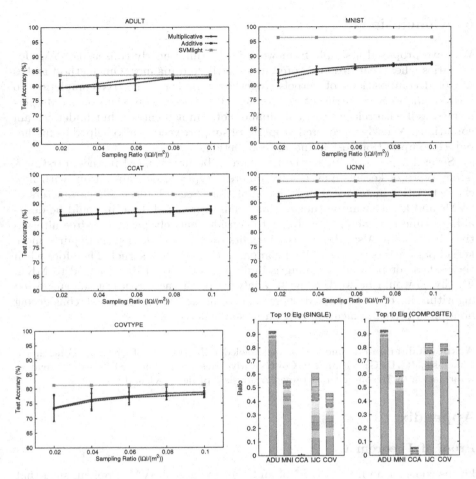

Fig. 2. Prediction accuracy on test sets for 5k/5k subsets of the five UCI data sets, over different sampling ratios in kernel completion. The average and standard deviation over multiple trials with random Ω and $N = 3$ nodes are shown. The bottom-right plot illustrates the proportion of the entire eigen-spectrum concentrated in the top ten eigenvalues.

result was consistent to the discussion in Sects. 5.1 and 5.2. Our method (with 10 % sampling) also outperformed ASSET (by 4.2 %, 18.3 %, and 5.6 % on average for ADULT, CCAT, and IJCNN respectively) except for the case of MNIST with a small but not negligible margin (1.9 %). We conjecture that the approximation of kernel mapping in ASSET have fitted particularly well for MNIST, but it remains to be investigated.

6   Conclusions

We have proposed a simple framework for learning nearly consensus SVMs for scenarios where features are stored in a distributed manner. Our method makes use of decompositions of kernels, together with kernel matrix completion to recover unobserved entries of remote kernel matrices. The resulting SVMs performed well with relatively small numbers of sampled entries, but under certain conditions. A newly discovered property of support vectors also helped us further reduce computation cost in matrix completion.

Several aspects of our method remain to be investigated further. First, different types of kernels may involve different types of decomposition, with new characteristics in terms of matrix completion. Second, although parameters of SVMs and kernels can be tuned using small aggregated data, it would be desirable to tune parameters locally, or to consider entirely parameter-free alternatives if possible. Also, despite the benefits of the MKL kernel, it requires more kernel parameters to be specified compared to the Schur kernel. Therefore when the budget for parameter tuning is limited, Schur would be preferred to MKL. Finally, it would be worthwhile to analyze the characteristics of the suggested algorithm in real communication systems to make it more practical, considering non-uniform communication cost on non-symmetric networks.

Acknowledgements. The authors acknowledge the support of Deutsche Forschungs-gemeinschaft (DFG) within the Collaborative Research Center SFB 876 "Providing Information by Resource-Constrained Analysis", projects A1 and C1.

Appendix

Proof of Theorem 1

Let us consider an index $i \in S^*$ of an SV of the global SVM problem, such that $[\boldsymbol{\alpha}^*]_i > 0$. Suppose that the ith component of the gradient of all local SVM problems at $\boldsymbol{\alpha}^*$ is strictly positive, that is,

$$[\mathbf{Y}\mathbf{M}_n\mathbf{Y}\boldsymbol{\alpha}^* - \mathbf{1}]_i > 0, \quad \forall n \in \{1, 2, \ldots, N\}. \tag{9}$$

Let us look into the optimality condition of the global SVM, regarding the ith component of the optimizer $\boldsymbol{\alpha}^*$. From the KKT conditions, we have

$$\frac{1}{N}\sum_{n=1}^{N}[\mathbf{Y}\mathbf{M}_n\mathbf{Y}\boldsymbol{\alpha}^* - \mathbf{1}]_i - [\mathbf{p}^*]_i + [\mathbf{q}^*]_i = 0, \quad [\mathbf{p}^*]_i[\boldsymbol{\alpha}^*]_i = 0, \quad [\mathbf{q}^*]_i[C\mathbf{1} - \boldsymbol{\alpha}^*]_i = 0,$$

where $\mathbf{p}^* \in \Re_+^m$ and $\mathbf{q}^* \in \Re_+^m$ are the Lagrange multipliers for the constraints $\boldsymbol{\alpha} \geq 0$ and $\boldsymbol{\alpha} \leq C\mathbf{1}$, respectively. Then $[\boldsymbol{\alpha}^*]_i > 0$ implies $[\mathbf{p}^*]_i = 0$, and therefore

$$\frac{1}{N}\sum_{n=1}^{N}[\mathbf{Y}\mathbf{M}_n\mathbf{Y}\boldsymbol{\alpha}^* - \mathbf{1}]_i + [\mathbf{q}^*]_i = 0.$$

If (9) is true, then we have a contradiction here since the first term above becomes strictly positive, where the second term satisfies $[\mathbf{q}^*]_i \geq 0$, and therefore the equality cannot hold. This implies that there exists at least one node n for which the condition in (9) is not satisfied, that is, $[\mathbf{YM}_n\mathbf{Y}\boldsymbol{\alpha}^* - \mathbf{1}]_i \leq 0$. This means that if we search for the local SVM solution at the node n starting from $\boldsymbol{\alpha}^*$, we must increase the value of the ith component from $[\boldsymbol{\alpha}^*]_i$ to reach the minimizer $[\boldsymbol{\alpha}_n^*]_i$ of this local SVM problem, since otherwise we will increase the objective function value. That is,

$$[\boldsymbol{\alpha}_n^*]_i \geq [\boldsymbol{\alpha}^*]_i > 0.$$

This implies that the index i also becomes an SV of at least one local SVM problem. Therefore, $i \in \cup_{n=1}^N S_n^*$, which implies the claim.

Proof of Theorem 3

From the definition of the error coefficient and the Cauchy-Schwarz inequality, we have

$$|\mathcal{E}_{\widetilde{\mathbf{K}}} - \mathcal{E}_{\mathbf{K}}| = \frac{1}{m}\mathbf{1}^T(\boldsymbol{\alpha}_{\widetilde{\mathbf{K}}}^* - \boldsymbol{\alpha}_{\mathbf{K}}^*) \leq \frac{1}{\sqrt{m}}\|\boldsymbol{\alpha}_{\widetilde{\mathbf{K}}}^* - \boldsymbol{\alpha}_{\mathbf{K}}^*\|_2. \tag{10}$$

On the other hand, from the optimality conditions of (1) with $\mathbf{K}$ and $\widetilde{\mathbf{K}}$,

$$(\boldsymbol{\alpha}_{\widetilde{\mathbf{K}}}^* - \boldsymbol{\alpha}_{\mathbf{K}}^*)^T(\mathbf{Q}\boldsymbol{\alpha}_{\mathbf{K}}^* - \mathbf{1}) \geq 0, \quad (\boldsymbol{\alpha}_{\mathbf{K}}^* - \boldsymbol{\alpha}_{\widetilde{\mathbf{K}}}^*)^T(\widetilde{\mathbf{Q}}\boldsymbol{\alpha}_{\widetilde{\mathbf{K}}}^* - \mathbf{1}) \geq 0,$$

and adding them up gives

$$(\boldsymbol{\alpha}_{\widetilde{\mathbf{K}}}^* - \boldsymbol{\alpha}_{\mathbf{K}}^*)^T(\mathbf{Q}\boldsymbol{\alpha}_{\mathbf{K}}^* - \widetilde{\mathbf{Q}}\boldsymbol{\alpha}_{\widetilde{\mathbf{K}}}^*) \geq 0.$$

This implies that (the idea is from [15]),

$$(\boldsymbol{\alpha}_{\widetilde{\mathbf{K}}}^* - \boldsymbol{\alpha}_{\mathbf{K}}^*)^T(\mathbf{Q} - \widetilde{\mathbf{Q}})\boldsymbol{\alpha}_{\widetilde{\mathbf{K}}}^* \geq (\boldsymbol{\alpha}_{\widetilde{\mathbf{K}}}^* - \boldsymbol{\alpha}_{\mathbf{K}}^*)^T\mathbf{Q}(\boldsymbol{\alpha}_{\widetilde{\mathbf{K}}}^* - \boldsymbol{\alpha}_{\mathbf{K}}^*).$$

From $(\boldsymbol{\alpha}_{\widetilde{\mathbf{K}}}^* - \boldsymbol{\alpha}_{\mathbf{K}}^*)^T(\mathbf{Q} - \widetilde{\mathbf{Q}})\boldsymbol{\alpha}_{\widetilde{\mathbf{K}}}^* \leq \|\boldsymbol{\alpha}_{\widetilde{\mathbf{K}}}^* - \boldsymbol{\alpha}_{\mathbf{K}}^*\|_2\|\mathbf{Q} - \widetilde{\mathbf{Q}}\|_2\|\boldsymbol{\alpha}_{\widetilde{\mathbf{K}}}^*\|_2$ (Cauchy-Schwarz and the definition of operator norms) and $(\boldsymbol{\alpha}_{\widetilde{\mathbf{K}}}^* - \boldsymbol{\alpha}_{\mathbf{K}}^*)^T\mathbf{Q}(\boldsymbol{\alpha}_{\widetilde{\mathbf{K}}}^* - \boldsymbol{\alpha}_{\mathbf{K}}^*) \geq \lambda_1(\mathbf{Q})\|\boldsymbol{\alpha}_{\widetilde{\mathbf{K}}}^* - \boldsymbol{\alpha}_{\mathbf{K}}^*\|_2^2$ (properties of the Rayleigh quotient; $\lambda_1(\mathbf{Q})$ is the smallest eigenvalue of $\mathbf{Q}$), the above inequality leads to (with $\|\boldsymbol{\alpha}_{\widetilde{\mathbf{K}}}^*\|_2 \leq C\sqrt{m}$),

$$\|\boldsymbol{\alpha}_{\widetilde{\mathbf{K}}}^* - \boldsymbol{\alpha}_{\mathbf{K}}^*\|_2 \leq \frac{C\sqrt{m}}{\lambda_1(\mathbf{Q})}\|\mathbf{Q} - \widetilde{\mathbf{Q}}\|_2.$$

From (10), the facts that $\mathbf{Q} = \mathbf{YKY}$ for $\mathbf{Y} = \text{diag}(\mathbf{y})$, $\|\mathbf{AB}\|_2 \leq \|\mathbf{A}\|_2\|\mathbf{B}\|_2$, and $\|\mathbf{A}\|_2 \leq \|\mathbf{A}\|_F$, and finally by (8), we have the claim,

$$|\mathcal{E}_{\widetilde{\mathbf{K}}} - \mathcal{E}_{\mathbf{K}}| \leq \frac{C}{\lambda_1(\mathbf{Q})}\|\mathbf{Q} - \widetilde{\mathbf{Q}}\|_2 = \frac{C}{\lambda_1(\mathbf{Q})}\|\mathbf{K} - \widetilde{\mathbf{K}}\|_F \leq \frac{C}{\lambda_1(\mathbf{Q})}(\delta_1 + \delta_2).$$

References

1. Bache, K., Lichman, M.: UCI machine learning repository (2013). http://archive. ics.uci.edu/ml
2. Boser, B.E., Guyon, I.M., Vapnik, V.N.: A training algorithm for optimal margin classifiers. In: Proceedings of the Fifth Annual Workshop on Computational Learning Theory, pp. 144–152 (1992)
3. Boyd, S., Parikh, N., Chu, E., Peleato, B., Eckstein, J.: Distributed optimization and statistical learning via the alternating direction method of multipliers. Found. Trends Mach. Learn. **3**(1), 1–122 (2011)
4. Candes, E.J., Tao, T.: The power of convex relaxation: near-optimal matrix completion. IEEE Trans. Inf. Theor. **56**(5), 2053–2080 (2010)
5. Candes, E., Plan, Y.: Matrix completion with noise. Proc. IEEE **98**(6), 925–936 (2010)
6. Crammer, K., Dredze, M., Pereira, F.: Confidence-weighted linear classification for natural language processing. J. Mach. Learn. Res. **13**, 1891–1926 (2012)
7. Forero, P.A., Cano, A., Giannakis, G.B.: Consensus-based distributed support vector machines. J. Mach. Learn. Res. **11**, 1663–1707 (2010)
8. Joachims, T.: Making large-scale support vector machine learning practical. In: Schölkopf, B., Burges, C., Smola, A. (eds.) Advances in Kernel Methods - Support Vector Learning, chap. 11, pp. 169–184. MIT Press, Cambridge (1999)
9. Lanckriet, G., Cristianini, N., Bartlett, P., E.G., L., Jordan, M.: Learning the kernel matrix with semidefinite programming. In: Proceedings of the 19th International Conference on Machine Learning (2002)
10. Lee, S., Bockermann, C.: Scalable stochastic gradient descent with improved confidence. In: Big Learning - Algorithms, Systems, and Tools for Learning at Scale, NIPS Workshop (2011)
11. Lee, S., Pölitz, C.: Kernel completion for learning consensus support vector machines in bandwidth-limited sensor networks. In: International Conference on Pattern Recognition Applications and Methods (2014)
12. Lee, S., Stolpe, M., Morik, K.: Separable approximate optimization of support vector machines for distributed sensing. In: De Bie, T., Cristianini, N., Flach, P.A. (eds.) ECML PKDD 2012, Part II. LNCS, vol. 7524, pp. 387–402. Springer, Heidelberg (2012)
13. Mangasarian, O.L., Musicant, D.R.: Lagrangian support vector machines. J. Mach. Learn. Res. **1**, 161–177 (2001)
14. Morik, K., Bhaduri, K., Kargupta, H.: Introduction to data mining for sustainability. Data Min. Knowl. Disc. **24**(2), 311–324 (2012)
15. Huang, L., Huang, L., Joseph, A.D., Joseph, A.D., Nguyen, X.L., Nguyen, X.L.: Support vector machines, data reduction, and approximate kernel matrices. In: Goethals, B., Goethals, B., Daelemans, W., Daelemans, W., Morik, K., Morik, K. (eds.) ECML PKDD 2008, Part II. LNCS (LNAI), vol. 5212, pp. 137–153. Springer, Heidelberg (2008)
16. Recht, B., Fazel, M., Parrilo, P.A.: Guaranteed minimum-rank solutions of linear matrix equations via nuclear norm minimization. SIAM Rev. **52**(3), 471–501 (2010)
17. Recht, B., Ré, C.: Parallel stochastic gradient algorithms for large-scale matrix completion. Technical report, University of Wisconsin-Madison, April 2011
18. Scholkopf, B., Smola, A.J.: Learning with Kernels: Support Vector Machines, Regularization, Optimization, and Beyond. MIT Press, Cambridge (2001)

19. Steinwart, I.: On the influence of the kernel on the consistency of support vector machines. J. Mach. Learn. Res. **2**, 67–93 (2002)
20. Stolpe, M., Bhaduri, K., Das, K., Morik, K.: Anomaly detection in vertically partitioned data by distributed core vector machines. In: Nijssen, S., Železný, F., Blockeel, H., Kersting, K. (eds.) ECML PKDD 2013, Part III. LNCS, vol. 8190, pp. 321–336. Springer, Heidelberg (2013)
21. Trefethen, L.N., Bau, D.: Numerical Linear Algebra. SIAM, Philadelphia (1997)

An Empirical Comparison of Support Vector Machines Versus Nearest Neighbour Methods for Machine Learning Applications

Mori Gamboni, Abhijai Garg, Oleg Grishin, Seung Man Oh, Francis Sowani,
Anthony Spalvieri-Kruse, Godfried T. Toussaint[✉], and Lingliang Zhang

Faculty of Science, New York University Abu Dhabi,
P.O. Box 129188, Saadiyat Island, Abu Dhabi, United Arab Emirates
{mg3794,ag3754,og402,smo304,fts215,ask417,gt42,lz781}@nyu.edu

Abstract. Support vector machines (SVMs) are traditionally considered to be the best classifiers in terms of minimizing the empirical probability of misclassification, although they can be slow when the training datasets are large. Here SVMs are compared to the classic k-Nearest Neighbour (k-NN) decision rule using seven large real-world datasets obtained from the University of California at Irvine (UCI) Machine Learning Repository. To counterbalance the slowness of SVMs on large datasets, three simple and fast methods for reducing the size of the training data, and thus speeding up the SVMs are incorporated. One is blind random sampling. The other two are new linear-time methods for guided random sampling which we call Gaussian Condensing and Gaussian Smoothing. In spite of the speedups of SVMs obtained by incorporating Gaussian Smoothing and Condensing, the results obtained show that k-NN methods are superior to SVMs on most of the seven data sets used, and cast doubt on the general superiority of SVMs. Furthermore, random sampling works surprisingly well and is robust, suggesting that it is a worthwhile preprocessing step to either SVMs or k-NN.

Keywords: Machine learning · Data mining · Support vector machines · SMO · Training data condensation · k-Nearest neighbour methods · Blind and guided random sampling · Wilson editing · Gaussian Condensing

1 Introduction

One of the most attractive learning machine models for pattern recognition applications, from the point of view of high classification accuracy, appears to be the support vector machine (SVM) [1]. There exists empirical evidence that SVMs yield lower rates of misclassification than even the classical k-Nearest Neighbour (k-NN) rule [2], in spite of the fact that (at least in theory) the latter is asymptotically Bayes optimal for all underlying probability distributions [3]. A major drawback of SVMs is their worst-case complexity, which is $O(N^3)$, where N is the

© Springer International Publishing Switzerland 2015
A. Fred et al. (Eds.): ICPRAM 2014, LNCS 9443, pp. 110–129, 2015.
DOI: 10.1007/978-3-319-25530-9_8

number of instances in the training set, so that for very large datasets the training time may become prohibitive [4]. Therefore much effort has been devoted to finding ways to speed up SVMs [5–12]. The simplest approach is to select a small random sample of the data for training [13]. This method is called *blind random sampling* because it uses no explicit information about the underlying structure of the data. This approach may be trivially implemented in $O(N)$ worst-case time. Non-blind random sampling techniques such as Progressive Sampling (PS) and Guided Progressive Sampling (GPS) have also been investigated with some success [14–16]. Non-random sampling methods attempt to use intelligent data analysis such as genetic algorithms [17] or proximity graphs [2,11] to preselect a supposedly better representative subset of the training data, which is then fed to the SVM, in lieu of the large original set of data. However, the use of guided data condensation methods usually incurs an additional worst-case cost of $O(N \log N)$ to $O(N^3)$. Since 1968, the literature contains a plethora of such algorithms and heuristics of varying degrees of computational complexity, for preselecting small subsets of the training data that will perform well under a variety of circumstances [18–20]. Although such techniques naturally speed up the training phase of the SVMs, by virtue of the smaller size of the training data, many studies primarily focus on (and report) only the number of support vectors retained, ignoring the additional time taken to perform the pre-selection. Indeed, it has been shown empirically, for methods that used proximity graphs for training data condensation, that if the additional time taken by the pre-selection step is taken into consideration, the overall training time is generally much worse than that of simple blind random sampling [11].

Some hybrid methods that combine blind random sampling with structured search for good representatives of datasets have also been tried. An original method combining blind random sampling with SVM and near neighbour search has recently been suggested by Li, Cervantes, and Yu [10]. Their approach first uses blind random sampling to select a small subset of the data, from which the support vectors are extracted using a preliminary SVM. These support vectors are then used to select points from the original training set (the data recovery step) that are near the preliminary support vectors, thus yielding the condensed training set on which the final SVM is applied.

In this paper several methods for speeding up the running time of SVMs are compared in terms of the speed-up factor and the classification accuracy, using seven large real world datasets taken from the University of California at Irvine (UCI) Machine Learning Repository [21]. All the methods are based on efficiently reducing the size of the training data that is subsequently fed to an SVM with sequential minimal optimization (SMO), whilst maintaining a high classification accuracy. Three probabilistic methods are investigated that run in $O(N)$ worst-case time. The first is blind random sampling and the second and third are new methods proposed here for guided random sampling (called Gaussian Condensing and Gaussian Smoothing), as well as an algorithm which can combine these classifiers that can sometimes yield better results. These are compared with nearest neighbour methods for reducing the size of the training set (k-NN condensation) and for smoothing the decision boundary (Wilson editing), both of which run in $O(N^2)$ worst-case time. One aim of this paper was also to compare SVMs with their leading competitor, the k-NN rule; thus, every method for reducing the size of the training data was also tested on a k-NN classifier.

2 The Classifiers Tested

2.1 Blind Random Sampling

Blind random sampling is the simplest method for reducing the size of the training set, both conceptually and computationally, with a running time of $O(N)$. Its possible drawback is that it is blind with respect to the quality of the resulting reduced training set, although this need not result in poor performance. In the experiments reported here the percentages of training data randomly selected for training the SVM were varied from 10 % to 90 % in increments of 10 %.

2.2 Wilson Editing (Smoothing)

Wilson's editing algorithm was used for smoothing the decision boundary [22]. Each instance in the training set is classified using the k-Nearest Neighbour rule by means of a *majority* vote. If the instance is misclassified it is *marked*. After all instances have been classified, all the marked points are deleted. This condensed set is then used as the testing set. Wilson editing was not designed to significantly reduce the size of the training set; its goal is rather to improve classification accuracy, and is used here as a pre-processing step before reducing the training set further with methods tailored for that purpose. Using packages available in the Weka Machine Learning Software [23], a value of $k = 3$ was chosen, and thus Wilson editing runs in $O(N^2)$ worst-case time with a straightforward, naïve implementation. Wilson editing was preceded by blind random sampling in a second set of experiments (see Gaussian Smoothing below).

2.3 k-Nearest-Neighbour Condensation

When all k nearest neighbours of a point X belong to the class of X, the k-NN rule makes a decision with very high confidence. In other words the point X is located relatively far from the decision boundary. This suggests that many points with this property could be safely deleted. Before classifying each testing set, the corresponding training set is condensed as follows. Each instance in the training set is classified using the k-NN rule. If the instance is correctly classified with *very high confidence* it is marked. After all instances are classified, all marked points are deleted. This condensed set is then used to classify the testing set. High confidence in the classification of X is measured by the proportion of the k nearest neighbours of X that belong to the class of X. The standard k-NN rule uses a *majority* vote as its measure of confidence. In our approach we use the *unanimity* vote (all the k nearest neighbours belong to the same class) and select a good value of k. This algorithm runs in $O(N^2)$ worst-case time using the naïve straightforward implementation and packages available in Weka. Note that when data of different classes are widely separated it may happen (at least in theory) that for every point X its k nearest neighbours all belong to the class of X. In such a situation unbridled k-NN condensation might discard the entire training set. For such an eventuality, if for some pattern class all training instances are marked for deletion, the mean of those instances is retained as the representative of that class. Experiments were also performed with k-NN condensation preceded by Wilson editing.

2.4   Gaussian Condensing

Gaussian Condensing is a novel heuristically guided random sampling algorithm introduced here. The heuristic implemented assumes that instances with feature values relatively close to the mean of their own class are likely to be furthest from the decision boundary, and therefore not expected to contain much discrimination information. Conversely, points relatively far from the mean are likely to be closer to the decision boundary, and expected to contain the most useful information. First, for each class, the mean value of each feature is calculated. Then, for each feature in every instance, the ratio between the Gaussian function of the feature value of that instance and the Gaussian function of the mean is computed. This determines a parameter termed the *partial discarding probability*. Finally, all instances are discarded probabilistically in parallel with a probability equal to the mean of the partial discarding probabilities of all their features. The main attractive attribute of this algorithm is that it runs in $O(N)$ worst-case time, where N is the number of training instances. It is therefore linear with respect to the size of the training data, and thus much faster than previous discarding methods that use proximity graphs, which are either quadratic or cubic in N. Indeed, the complexity of Gaussian Condensing is as low as that of blind random sampling.

The goal of Gaussian Condensing is to invert the probability distribution function of instances for all features of each class. Hence, points near the mean are certain to be thrown away, and points near the boundaries are almost never thrown away. If applied to data with a Gaussian distribution, the probability distribution function would result in an inverted bell curve, with the minimum point occurring at the center, and increasing towards the boundaries before decreasing again. A similar idea was introduced by Chen, Zhang, Xue, and Liu, [12], with strong results. However, their algorithm deletes a ratio of the total data closest to the mean. The approach proposed here is superior in two ways: (1) it does not require a method to decide the ratio of data that should be optimally kept, and (2) it does not create a hole? in the data, but rather preserves the entire distribution of points, by simply altering the density. Experiments were also performed with Gaussian Condensing preceded by Wilson editing.

2.5   Gaussian Smoothing

Gaussian Smoothing is another original method presented here that was added after experiments were done with Gaussian Condensing. This algorithm can be considered the opposite of Gaussian Condensing, in that it tries to discard points which are near the boundary. First, the mean feature value of every feature in every class is computed, then, lists of these values are grouped by feature and independently sorted (since the number of features is very small compared to the number of instances, sorting time does not dominate the total running time of the algorithm). For every feature of every instance, if the Gaussian function of the feature value of that instance is smaller than the Gaussian function of the closest mean feature value regardless of the class, the partial discarding probability is the ratio between the two values subtracted from one ($p = 1 - \frac{instance}{neighbor}$), otherwise, the partial discarding probability is zero. The mean of the partial discarding probabilities gives the final discarding probability and instances are discarded in parallel, exactly as in Gaussian Condensing. Thus, the partial discarding probability starts to increase for points when they are closer to the center of the neighboring class than the center of their own class (the partial discarding probability is zero where the two Gaussian functions intersect). Gaussian Smoothing thus runs in $O(N)$ worst-case time. k-NN condensation was not included in the second set of experiments as it performed poorly (see Sect. 4.2); Gaussian Smoothing was tested alone, in conjunction with Gaussian Condensing (in both possible orders), and preceded by blind random sampling.

2.6   Combined Gaussian Filter

In practice, using both Gaussian Condensing and Smoothing does not always give good accuracies (see Subsect. 4.2). However, a slight modification of the algorithm can sometimes give better results. To calculate the partial probabilities, the methods of condensing and smoothing are applied when the Gaussian function of the feature value of an instance is respectively higher than or lower than or equal to that of the closest mean feature value. In other words, if the Gaussian function of the feature value of the instance is lower than or equal to the closest mean feature value, then the partial discarding probability is given by the ratio of those two values subtracted by one (smoothing), if it is higher, the probability is given by the ratio between the Gaussian function of the feature value and the mean feature value of the class of the instance (condensing). The final discarding probability is calculated – and the points discarded – in the same manner as the other methods. This algorithm ideally causes the probability distribution to become inverted near the boundary of the classes, which is the case with perfect normal distributions, thus potentially keeping the best attributes of both the condensing and smoothing methods. The Combined Gaussian Filter runs in $O(N)$ for a constant number of classes, just as in Gaussian Smoothing.

3   The Datasets Tested

Year Prediction Million Song Data. The original dataset is extremely large, (515,345 instances) and therefore some of the data were randomly discarded. The pattern classes were converted from years to decades (1950 s through 2000 s) and then 3,000 instances of each class were chosen, comprising six classes with a total of 18,000 instances.

Letter Image Data. This dataset contains black-and-white rectangular pixel displays of the 26 upper-case letters in the English alphabet. The letter images were constructed from twenty different fonts. Each letter from the twenty fonts was randomly distorted to produce 20,000 unique instances. Each instance is described using 17 attributes: a letter category (A, B, C,..., Z) and 16 numeric features.

Wearable Computing Data. This dataset (PUC-Rio) contains information matching accelerometer readings from various parts of the human body, with the readings taken while the actions were performed. Accelerometers collected $x, y,$ and z axes data from the waist, left-thigh, right ankle, and right upper-arm of the subjects. In each instance, the subjects were either sitting-down, standing-up, walking, in the process of standing, or in the process of sitting. Metadata about the gender, age, height, weight and BMI of each subject are also provided. In total 165,632 instances of such data are included in this dataset.

MAGIC Gamma Telescope Data. This data consist of Monte-Carlo generated simulations of high-energy gamma particles. There are ten attributes, each continuous, and two classes ('g' and 'h'). The number of instances was 12,332 for 'g' and 6,688 for 'h'. For the purpose of this study approximately half of class 'g' was removed at random, since the goal of the present research is the improvement of the running time of SVMs, rather than the minimization of the probability of misclassification for this particular application.

Spambase Data. The Spambase dataset provides information about email spam. The emails are classified into two categories: spam and non-spam. The data labelled spam were collected from postmasters and individuals who had reported spam, and the non-spam data were collected from filed work and personal emails. The dataset was created with the goal of designing a personal spam filter. It contains 4601 instances, of which 1813 (39.4 %) are spam. These instances are characterized by 57 attributes (57 continuous features and one nominal class label). The class label is either 1 or 0, indicating that the email is either spam or non-spam, respectively.

Wine Quality Data. The white wine quality dataset includes over 2000 different *vinho verde* wines (instances). The dataset comprises twelve features that include acidity and sulphate content. There are ten classes defined in terms of quality ratings that vary between 1 and 10.

Handwritten Digits Data. This dataset contains 32 by 32 bitmaps that have been obtained by centering and normalising the input images from 43 different people. The training set consisting of 5,620 instances and has data from 30 people, while the test set comes from the 13 others, so as to prevent learning algorithms from classifying digits based on the writing style rather than features of the shape of the digits themselves. To decrease the dimensionality of the data, the bitmaps are divided into 4 by 4 blocks and the number of pixels in each block is counted. The total number of features is thus 64 and the number of classes is 10, the digits 0 through 9.

4   Results and Discussion

Note: Due to space restrictions, only particularly illustrative graphs are shown.

4.1   The Computation Platform

The timing experiments were performed on the fastest high-performance computer available in the United Arab Emirates (second fastest in the Gulf region): *BuTinah*, operated by New York University Abu Dhabi. The computer consists of 512 nodes, each one equipped with 12 Intel XeonX5675 CPU's clocked at 3.07 GHZ and 48 GB of RAM with 10 GB of swap memory. *BuTinah* operates at approximately 70 trillion floating-point operations per second (70 teraflops). The experiments utilized seven nodes, in total, consuming 9 h of computation time and 12 GB of memory. The testing environment was programmed in Java, using the Weka data Mining Package, produced by the University of Waikato.

4.2   Blind Random Sampling

The SMO (Sequential Minimization Optimization) version of SVM – invented by John Platt [24] and improved by Keerthi, Shevade, Bhattacharyya, and Murthy [25] – that is installed in the Weka machine learning package was compared to the classical k-NN decision rule when both are preceded by blind random removal of data before feeding the remaining data to each classifier. A typical result obtained with the Song dataset is shown in Fig. 1, for the classification accuracy (left vertical axis) and the total running time (right vertical axis). Total time refers to the sum of the times taken for training data condensation, training time, and testing time (results for the three individual timings will be presented in a following section). In this and all other experiments the classification accuracies and timings were obtained by the method of K-fold cross-validation (or Π method) with a value of $K = 10$ [26]. This means that for each of the classifiers and condensing methods tested the procedure for estimating the classification accuracy for each fold was the following. Let X denote the entire dataset. The i-th fold is obtained by taking the i-th 10 % of $\{X\}$ as the testing set (denoted by $\{X_{TS-i}\}$), and the remaining 90 % of the data as the training set (denoted by $\{X_{TR-i}\}$). Estimates of the misclassification accuracy of any classifier are then obtained by training the classifier on $\{X_{TR-i}\}$, and testing it on $\{X_{TS-i}\}$, for $i = 1, 2, \ldots, 10$ yielding a total of ten estimates. Similarly, when estimating the classification accuracy of an editing (or condensing) method, the editing (or condensing) is first applied to $\{X_{TR-i}\}$, and the resulting edited (condensed) set is used to classify $\{X_{TS-i}\}$. Finally, in all cases the average of the ten estimates obtained in this way is calculated. Thus the results shown in the figures are the mean values over the ten folds. This method also permits the computation of standard deviations (over the ten folds) to serve as indicators of statistically significant differences between the means. The error bars in the figures indicate ± one standard deviation.

All seven datasets exhibit similar behaviour to that depicted in Fig. 1 with respect to how the classification accuracy varies as a function of the percentage of training data removed. The classification accuracy results are not unanimous, but favour k-NN over SMO, the latter having significantly better accuracy than k-NN only for the Song data (Fig. 1). For the Letter Image, Wearable Computing, and MAGIC Gamma datasets k-NN did significantly better (e.g. Letter Image data set in Fig. 2). Furthermore, for the Wine and Handwritten Digits data no significant differences were observed between SMO and k-NN (Figure not shown). For the Spambase data SMO is significantly better only when 60 %, 70 % or 90 % of the data are discarded (Figure not shown).

With respect to the total time taken, for all the datasets, SMO takes considerably less running time than k-NN, and all show similar behaviour. For example,

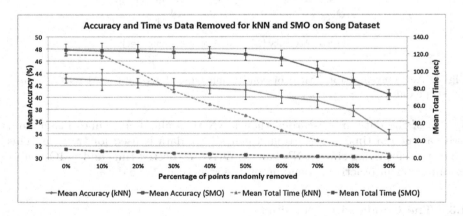

Fig. 1. Accuracy and total time vs. % of training data removed by blind random sampling for the Year Prediction Million Song data.

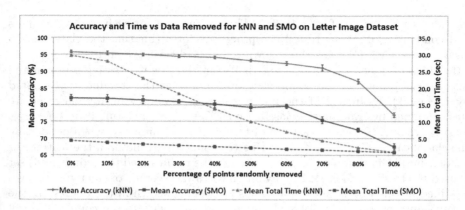

Fig. 2. Accuracy and total time vs. % of training data removed by blind random sampling for the Letter Image data.

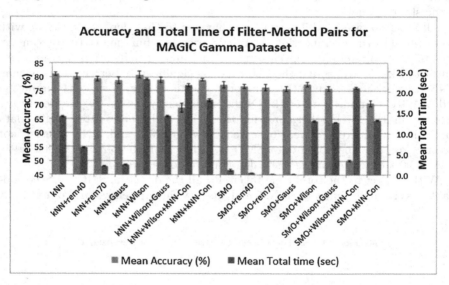

Fig. 3. Accuracy and total time of condensing algorithms for the MAGIC Gamma Telescope data.

if 70 % of the data are discarded then k-NN runs about five times faster (and SMO about ten times faster) than when all the data is used for training. This is not too surprising since k-NN runs in $O(N^2)$ expected time and SMO is able to run faster in practice depending on the structure of the data.

4.3 The Condensing Classifiers

Experiments were done applying various training data condensation classifiers to reduce the size of the training dataset that was fed to both the SMO and k-NN classifiers. The condensing classifiers tried in the first experiment were: (1) Wilson Editing, (2) Gaussian Condensation, (3) Wilson Editing + Gaussian Condensation, (4) Wilson Editing + k-NN Condensation, and (5) k-NN Condensation. In the second experiment, the classifiers tested were: (1) Gaussian Condensation + Gaussian Smoothing, (2) Gaussian Condensation + Wilson Editing, (3) Combined Gaussian Filters, (4) Gaussian Smoothing, (5) Gaussian Smoothing + Gaussian Condensation, and (6) Wilson Editing.

Figures 3–5 show the percent mean accuracy and mean total running times (in seconds) as well as the breakdown of the total time for all the five condensation classifiers in the first experiment, plus the results for blind random sampling obtained by discarding 40 % and 70 % of the training data. Figures 8–16 show the same variables for the six classifiers in the second set, as well as Gaussian Smoothing and Wilson Editing preceded by discarding 70 % of the data. In the figures, 'k-NN-Con' indicates k-NN condensation, 'Wilson' denotes Wilson Editing, 'Gauss' stands for Gaussian Condensation, 'GaussSmooth' for Gaussian Smoothing, 'GaussComb' for the Combined Gaussian Filters, and 'rem40' and 'rem70'/'%' are blind random removal of 40 % and 70 % of the data, respectively.

Experiment 1. Perusal of Figs. 3–5 reveals that none of the condensing methods improve the accuracy of the classifiers that do not use condensing. In fact, k-NN condensation performs the worst for all but two of the data sets – Song and Wine data – and for those two, other methods give poor results (e.g. see Fig. 4 for Wine data). With k-NN classification, random removal of 70 % of the training set and Gaussian Condensing were the fastest methods for all the datasets (e.g. MAGIC Gamma data in Fig. 3) other than the Handwritten Digits (see Fig. 5). Similar relative behavior was observed with SMO, however, in all cases the running times with SMO were much smaller than those with k-NN.

In Figs. 3–5 the times plotted are *total* times: condensing time + training time + testing time. One of the main goals of this research is to compare the testing times of the various classifiers, since these reflect the speed of the classifier on all future data. However, if the condensing times and training times dominate the testing time, then the total times listed in the figures may hide the testing times. Therefore a breakdown of the individual times was also plotted for all the experiments. Note from Fig. 5 that the classifiers with the smallest total times are: SMO, SMO+rem40, SMO+rem70, and SMO+Gauss. Furthermore their accuracies are not significantly different. Therefore the classifiers with the fastest testing times would be preferred in this case.

Figure 6 shows the breakdown of condensing, training, testing, and total time for the Song data on a linear scale in seconds. This figure clearly shows how large the testing time for k-NN is compared to all other classifiers, thus making it difficult to compare the four classifiers of main interest. To zoom in on their performance the data from Fig. 6 are shown on a logarithmic scale in Fig. 7, where it can be clearly seen that SMO has shorter testing times with rem40, rem70, and Gaussian Condensing, but is slower when no filter is used (although total time is similar). All data sets show similar behaviour, and thus are not shown here.

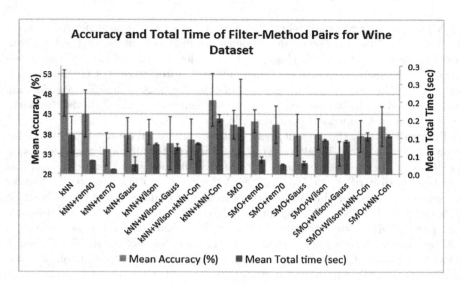

Fig. 4. Accuracy and total time of condensing algorithms for the Wine Quality data.

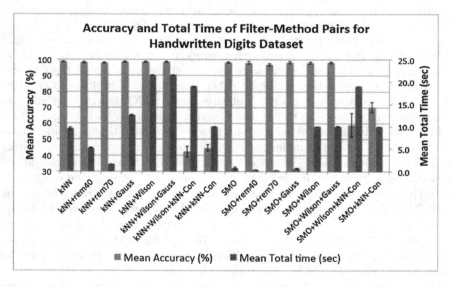

Fig. 5. Accuracy and total time of condensing algorithms for the Digits data.

Experiment 2. Figures 8–13 show that the classifiers tested give greatly varying results across data sets. Only the Letter Image and Wearable Computing data show very similar behaviour (Wearable Computing data shown in Fig. 8). In general, k-NN classification with Gaussian Smoothing and Condensing together, in both orders (but not with the combined algorithm), Gaussian Condensing and

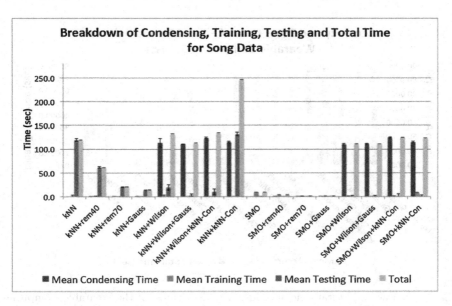

Fig. 6. Breakdown of condensing, training, testing, and total time for the Year Prediction Million Song data.

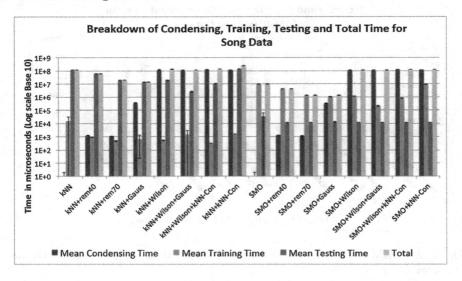

Fig. 7. The data of Fig. 8 on a logarithmic scale.

Wilson Editing, and random removal with Gaussian Smoothing or Wilson Editing are all fast, but give mixed results in terms of accuracy. In the Letter Image and Wearable Computing datasets, k-NN with Combined Gaussian Filters gives the greatest accuracy but is slightly slow, while Gaussian Smoothing and Con-

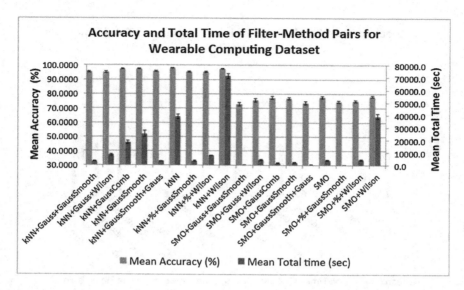

Fig. 8. Accuracy and total time of condensing algorithms for the Wearable Computing data.

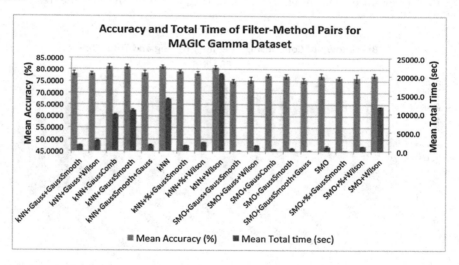

Fig. 9. Accuracy and total time of condensing algorithms for the MAGIC Gamma Telescope data.

densing together in either order, and random removal with Gaussian Smoothing both give good accuracy (see Fig. 8). The MAGIC Gamma data show similar behaviour, although the latter three classifiers are instead better with SMO (see Fig. 9). On average, k-NN classification gives better accuracy than SMO classification for these data sets, as well as the Song data (See Fig. 10). The latter is

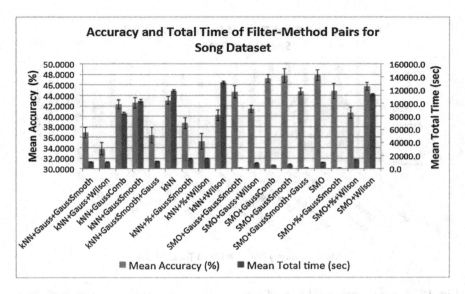

Fig. 10. Accuracy and total time of condensing algorithms for the Year Prediction Million Song data.

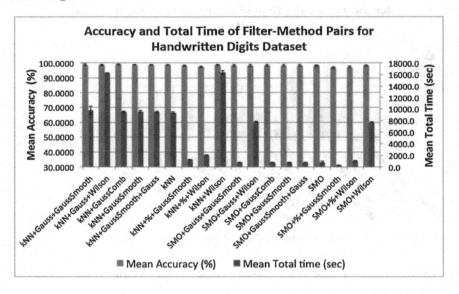

Fig. 11. Accuracy and total time of condensing algorithms for the Digits data.

also similar to MAGIC Gamma data since they both have the same three well performing classifiers, however in the Song data, SMO with Combined Gaussian Filters is overall the best method, SMO with Gaussian Condensation and Wilson Editing or random removal and Wilson Editing give poor accuracy, while

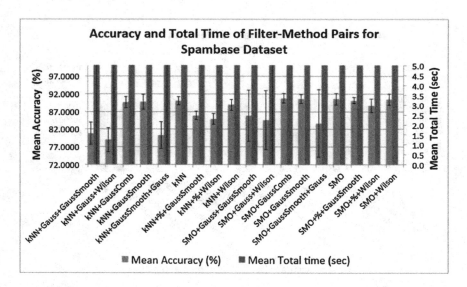

Fig. 12. Accuracy and total time of condensing algorithms for the Spambase data.

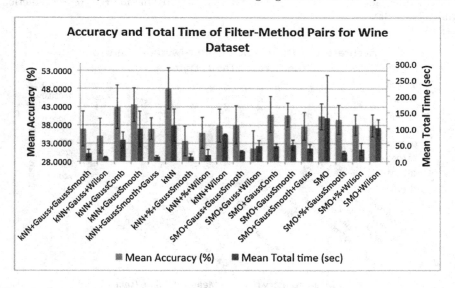

Fig. 13. Accuracy and total time of condensing algorithms for the Wine Quality data.

only Gaussian Condensing with Wilson Editing gives good accuracy and short running times out of all k-NN classifiers. The Handwritten Digit data gives generally excellent accuracy with all classifiers, but SMO with random removal and Gaussian Smoothing is the fastest (see Fig. 11). In the Spambase data (Fig. 12) SMO classification is much better than k-NN classification. Finally, the Wine

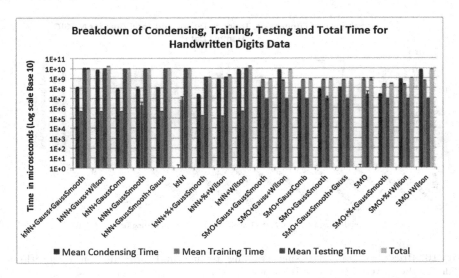

Fig. 14. Breakdown of condensing, training, testing, and total time on a logarithmic scale for the Digits data.

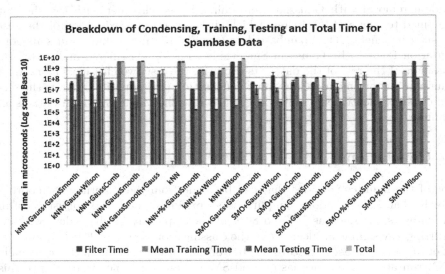

Fig. 15. Breakdown of condensing, training, testing, and total time on a logarithmic scale for the Spambase data.

data (Fig. 13) shows very little similarity with other data sets, and yields generally poor results for every classifier.

As in Experiment 1, a breakdown of the condensing, training, testing, and total time was also graphed (only the logarithmic versions are shown for the same reasons outlined above). The Handwritten Digits data (see Fig. 14) show

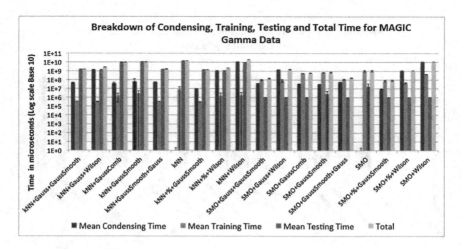

Fig. 16. Breakdown of condensing, training, testing, and total time on a logarithmic scale for the MAGIC Gamma Telescope data.

random removal with Gaussian Smoothing as the classifier with the shortest combined testing and total time, which gives a good approximation of the best classifier in general. The Spambase data (see Fig. 15) also include Gaussian Condensation followed by Gaussian Smoothing as a top classifier. Finally, the top classifiers of the rest of the data sets comprise the last two classifiers along with Gaussian Smoothing and Gaussian Condensation (the opposite order as the second classifier mentioned), shown in Fig. 16 using the Magic GAMMA data as an example.

5   Conclusions

One of the main conclusions that can be made from the experiments reported here is that blind random sampling is surprisingly good and robust. For all the datasets, as much as 70 % to 80 % of the data may be discarded, without incurring any significant decrease in the classification accuracy. Furthermore, for six of the seven datasets, discarding 70 % of the data at random in this way made k-NN run about five times faster, and SMO about ten times faster. Since this method is so simple and requires so little computation time we believe that it should play a role as a pre-processing step for speeding up SVMs.

Previous research has shown that SVMs perform better than k-NN. However, some of the comparisons have used synthetically generated datasets that do not resemble real world data. On the other hand, the results of the present study with seven real-world datasets tell a different story. SMO is significantly better in terms of accuracy only for the Song data, whereas k-NN does better for the Letter Images, Wearable Computing, and Magic Gamma datasets. For the other three datasets (Spam, Wine, and Handwritten Digits) there are no significant differences between SMO and k-NN. However, SMO has the advantage of running faster than k-NN, which may be more important in some applications. In future research we hope to discover structural features of the data that predict when SMO is expected to outperform k-NN.

One of the goals of this research was to test how much Wilson editing and k-NN condensation improve the overall performance of classifiers in practice. It was found that for all seven datasets using Wilson editing as a pre-processing step to either SVM or k-NN, yielded no statistically significant improvement in accuracy, and k-NN condensation led to poorer accuracy. Furthermore, both Wilson editing and k-NN condensation incur a considerable additional cost in the filter (editing+condensing) time. However, if used after Gaussian Condensing or blind random sampling, it runs quickly, and can sometimes give good levels of accuracy.

Another main goal of this research project was to compare the new proposed methods for condensing and smoothing training data in $O(N)$ worst-case time: Gaussian Condensing/Smoothing, along with the algorithm for combining both Gaussian filters. This probabilistic method falls in the category of guided (or intelligent) random sampling and is almost as fast as blind random sampling. The results of this study show that Gaussian Smoothing used alone or together with Gaussian Condensing can give surprisingly good results, and is especially fast with SMO classification. The Combined Gaussian Filters also performed quite well in general. Finding ways to predict which combination of random sampling methods will do best depending on the dataset is an area of research we would also like to investigate further.

Acknowledgements. This research was supported by a grant from the Provost's Office of New York University Abu Dhabi in the United Arab Emirates. The authors are grateful to the University of California at Irvine for making available their large collection of data at the Machine Learning Repository.

References

1. Vapnik, V.: The Nature of Statistical Learning Theory. Springer, New York (1995)
2. Toussaint, G.T., Berzan, C.: Proximity-graph instance-based learning, support vector machines, and high dimensionality: an empirical comparison. In: Perner, P. (ed.) MLDM 2012. LNCS (LNAI), vol. 7376, pp. 222–236. Springer, Heidelberg (2012)
3. Devroye, L.: On the inequality of cover and hart in nearest neighbour discrimination. IEEE Trans. Pattern Anal. Mach. Intell. **3**, 75–78 (1981)
4. Bordes, A., Ertekin, S., Weston, J., Bottou, L.: Fast kernel classifiers with online and active learning. J. Mach. Learn. Res. **6**, 1579–1619 (2005)
5. Almeida, M.B., Braga, A.P., Braga, J.P.: SVM-KM: speeding SVMs learning with a priori cluster selection and k-means. In: Proceedings of the 6th Brazilian Symposium on Neural Networks, pp. 162–167 (2000)
6. Chen, J., Chen, C.: Speeding up SVM decisions based on mirror points. In: Proceedings of the 6th International Conference Pattern Recognition, vol. 2, pp. 869–872 (2002)

7. Panda, N., Chang, E.Y., Wu, G.: Concept boundary detection for speeding up SVMs. In: Proceedings of the 23 International Conference on Machine Learning, Pittsburgh (2006)

8. Wang, Y., Zhou, C.G., Huang, Y.X., Liang, Y.C., Yang, X.W.: A boundary method to speed up training support vector machines. In: Liu, G.R., et al. (eds.) Computational Methods, pp. 1209–1213. Springer, Netherlands (2006)

9. Chen, J., Liu, C.-L.: Fast multi-class sample reduction for speeding up support vector machines. In: Proceedings of the IEEE International Workshop on Machine Learning for Signal Processing, Beijing, China, 18–21 September (2011)

10. Li, X., Cervantes, J., Yu, W.: Fast classification for large datasets via random selection clustering and support vector machines. Intell. Data Anal. **16**, 897–914 (2012)

11. Liu, X., Beltran, J.F., Mohanchandra, N., Toussaint, G.T.: On speeding up support vector machines: proximity graphs versus random sampling for pre-selection condensation. In: Proceedings of the International Conference Computer Science and Mathematics, Dubai, United Arab Emirates, 30–31 January, vol. 73, pp. 1037–1044 (2013)

12. Chen, J., Zhang, C., Xue, X., Liu, C.-H.: Fast instance selection for speeding up support vector machines. Knowl. Based Syst. **45**, 1–7 (2013)

13. Lee, Y.L., Mangasarian, O.L.: RSVM: reduced support vector machines. In: Proceedings of the First SIAM International Conference on Data Mining, 5–7 April (CD-ROM). SIAM, Chicago (2001)

14. Provost, F., Jensen, D., Oates, T.: Efficient progressive sampling. In: Fifth ACM SIGKDD International Conference on Knowledge Discovery and Data Mining, San Diego, USA (1999)

15. Ng, W.Q., Dash, M.: An evaluation of progressive sampling for imbalanced datasets. In: Sixth IEEE International Conference on Data Mining Workshops, Hong Kong, China (2006)

16. Portet, F., Gao, F., Hunter, J., Quiniou, R.: Reduction of large training set by guided progressive sampling: application to neonatal intensive care data. In: Proceedings of Intelligent Data Analysis in Biomedicine and Pharmacology, Amsterdam, pp. 43–44 (2007)

17. Kawulok, M., Nalepa, J.: Support vector machines training data selection using a genetic algorithm. In: Gimel'farb, G.L., et al. (eds.) SSPR & SPR 2012. LNCS, vol. 7626, pp. 557–565. Springer, Heidelberg (2012)

18. Hart, P.E.: The condensed nearest neighbour rule. IEEE Trans. Inf. Theory **14**, 515–516 (1968)

19. Sriperumbudur, B.K., Lanckriet, G.: Nearest neighbour prototyping for sparse and scalable support vector machines. Technical report No. CAL-2007-02, University of California San Diego (2007)

20. Toussaint, G.T.: Geometric proximity graphs for improving nearest neighbour methods in instance-based learning and data mining. Int. J. Comput. Geom. Appl. **15**, 101–150 (2005)

21. Bache, K., Lichman, M.: UCI Machine Learning Repository (2013). http://archive.ics.uci.edu/ml

22. Wilson, D.L.: Asymptotic properties of nearest neighbour rules using edited-data. IEEE Trans. Syst. Man Cybern. **2**, 408–421 (1973)

23. Witten, I., Frank, E.: WEKA: machine learning algorithms in java. In: Data Mining: Practical Machine Learning Tools and Techniques with Java Implementations, pp. 265–320. MorganKaufmann (2000)

24. Platt, J.C.: Fast training of support vector machines using sequential minimial optimization. In: Scholkopf, B., Burges, C., Smola, A. (eds.) Advances in Kernel Methods: Support Vector Machines. MIT Press, Cambridge (1998)
25. Keerthi, S.S., Shevade, S.K., Bhattacharyya, C., Murthy, K.R.K.: Improvements to Platt's SMO algorithm for SVM classifier design. Neural Comput. **13**, 637–649 (2001)
26. Toussaint, G.T.: Bibliography on estimation of misclassification. IEEE Trans. Inf. Theory **20**, 472–479 (1974)

Applications

Improving the Detection of Relations Between Objects in an Image Using Textual Semantics

Dennis Medved[1], Fangyuan Jiang[2], Peter Exner[1], Magnus Oskarsson[2], Pierre Nugues[1(✉)], and Kalle Åström[2]

[1] Department of Computer Science, Lund University, Lund, Sweden
{dennis.medved,peter.exner,pierre.nugues}@cs.lth.se
[2] Department of Mathematics, Lund University, Lund, Sweden
{fangyuan,magnuso,kalle}@maths.lth.se

Abstract. In this article, we describe a system that classifies relations between entities extracted from an image. We started from the idea that we could utilize lexical and semantic information from text associated with the image, such as captions or surrounding text, rather than just the geometric and visual characteristics of the entities found in the image.

We collected a corpus of images from Wikipedia together with their corresponding articles. In our experimental setup, we extracted two kinds of entities from the images, human beings and horses, and we defined three relations that could exist between them: *Ride*, *Lead*, or *None*. We used geometric features as a baseline to identify the relations between the entities and we describe the improvements brought by the addition of bag-of-word features and predicate–argument structures that we extracted from the text. The best semantic model resulted in a relative error reduction of more than 18 % over the baseline.

Keywords: Semantic parsing · Relation extraction from images · Machine learning

1 Introduction

A large percentage of queries to retrieve images relate to people and objects [12,20] as well as relations between them: the 'story' within the image [8]. Although the automatic recognition, detection and segmentation of objects in images has reached relatively high levels of accuracy, reflected by the Pascal VOC Challenge evaluation [1,6,10], the identification of relations is still a territory that is yet largely unexplored. Notable exceptions include [2,15]. The identification of these relations would result in a richer model of the image content and would enable users to search images illustrating two or more objects more accurately.

Relations between objects within images are often ambiguous and captions are intended to help us in their interpretation. As human beings, we often have to read the caption or the surrounding text to understand what happened in a

© Springer International Publishing Switzerland 2015
A. Fred et al. (Eds.): ICPRAM 2014, LNCS 9443, pp. 133–145, 2015.
DOI: 10.1007/978-3-319-25530-9_9

scene and the nature of the relations between the entities. This combined use of text and images has been explored in automatic interpretation mostly in the form of bag of words, see Sect. 2. This approach might be inadequate however, as bags of words do not take the word or sentence context into account. This model inadequacy formed the starting idea of this project: As we focused on relations in images, we tried to model their counterparts in the text and reflect them not only with bags of words but also in the form of predicate–argument structures.

2   Related Work

To the best of our knowledge, no work has been done to identify relations in images using a combined analysis of image and text data. There are related works however:

Reference [16] combined image segmentation with a text-based classifier using image captions as input. They used bags of words and applied a $TF \cdot IDF$ weighting on the text. The goal was to label the images as either taken indoor or outdoor. They improved the results by using both text and image information together, compared to using only one of the classifiers.

Reference [3] used a set of 100 image-text pairs from *Yahoo! News* and automatically annotated the images utilizing the associated text. The goal was to detect the presence of specific humans, but also more general objects. They analyzed the image captions to find named entities. They also derived information from discourse segmentation, which was used to determine the saliency of entities.

Reference [14] used a large corpus of French news articles, composed of a text, images, and image captions. They combined an image detector to recognize human faces and logos, with a named entity detection in the text. The goal was to correctly annotate the faces and logos found in the images. The images were not annotated by humans, instead named entities in the captions were used as the ground truth, and the classification was based on the articles.

Reference [19] used a large collection of images from *Flickr* that users had annotated by supplying keywords and short descriptions. The goal was to categorize the images, utilizing a combination of features derived from image analysis, together with relevant image labels extracted from the text associated with the images.

Reference [13] used a semantic network and image labels to integrate prior knowledge of inter-class relationships in the learning step of a classifier to achieve better classification results. All of these works combined text and image analysis for classification purposes, but they did not identify relations in the images. Another area of related work is the generation of natural language descriptions of an image scene, see [7,9].

3   Data Set and Experimental Setup

The internet provides plenty of combined sources of images and text including news articles, blogs, and social media. Wikipedia is one of such sources that, in addition to a large number of articles, is a substantial repository of images

Fig. 1. The upper row shows: a ford mustang, the 3rd light horse regiment hat badge, and a snuff bottle. The lower row shows: a human riding a horse, one human leading the horse and one bystander, and seven riders and two bystanders. Bounding boxes are displayed.

illustrating the articles. As of today, the English version has over 4 million articles and about 2 million images [21]. It is not unusual for editors to use an image for more than one article, and an image can therefore have more than one article or caption associated with it. The images used in the articles are stored in Wikimedia Commons, which is a database of freely reusable media files.

We gathered a subset of images and articles from Wikipedia restricted to two object categories: *Horse* and *Human*. We extracted the articles containing the keywords *Horse* or *Pony* and we selected their associated images. This resulted in 901 images, where 788 could be used. Some images were duplicates and some did not have a valid article associated with them.

An image connected to the articles with the words *Horse* or *Pony* does not necessarily depict a real horse. It can show something associated with the words for example: a car, a statue, or a painting. Some of the images also include humans, either interacting with the horse or just being part of the background, see Fig. 1 for examples. An image can therefore have none or multiple horses, and none or multiple humans.

We manually annotated the horses and humans in the images with a set of possible relations: *Ride*, *Lead*, and *None*. *Ride* and *Lead* are when a human is riding or leading a horse and *None* is an action that is not *Ride* or *Lead* including no action at all. The annotation gave us the number of respective humans and horses, their sizes and their locations in the image.

We processed the articles with a semantic parser [4], where the output for each word is its lemma and part of speech, and for each sentence, the dependency graph and predicate-argument structures it contains. We finally applied a coreference solver to each article.

Table 1. The number of different objects in the source material.

Item	Count
Extracted images	901
Usable images	788
Human-horse pairs	2,235
Relation: *None*	1,935
Relation: *Ride*	233
Relation: *Lead*	67

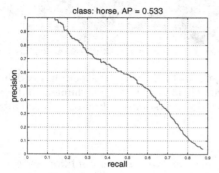 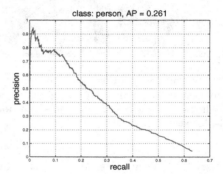

Fig. 2. The precision-recall curves on our image test set using [6]'s detector and generically trained models for the horse (left) and person (right) categories.

4   Visual Parsing

As our focus was to investigate to what extent the use of combinations of text and visual cues could improve the interpretation or categorization precision, we set aside the automatic detection of objects in the images. We manually identified the objects within the images by creating bounding boxes around horses and humans. We then labeled the interaction between the human-horse pair if the interaction corresponded to *Lead* or *Ride*. The *None* relationships were left implicit. It resulted in 2,235 possible human-horse pairs in the images, but the distribution of relations was quite heavily skewed towards the None relation. The Lead relation had significantly fewer examples; see Table 1.

The generation of the bounding boxes could be produced automatically by an object detection algorithm trained on the relevant categories (in our case people and horses) such as e.g. the deformable part-based model described in [6]. Figure 2 shows the precision-recall curve using this detector with generically trained models for the horse (left) and person (right) categories. Such a detection step would have enabled us to skip the manual annotation. Nonetheless, in the experiment we report here, we focused on the semantic aspects and we used manually created bounding boxes.

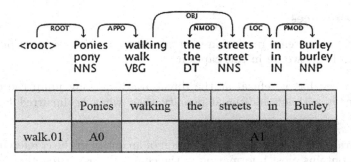

Fig. 3. A representation of a parsed sentence: the upper part shows the syntactic dependency graph and the lower part shows the predicate, *walk*, and its two arguments the parser has extracted: *Ponies* and *the streets in Burley*.

5 Semantic Parsing

We used the Athena parsing framework [4] in conjunction with a coreference solver [18] to parse the Wikipedia articles. For each word, the parser outputs its lemma and part of speech (POS). The lemma is the dictionary form of the word, for example the lemma of *running* is *run*. The POS is the word category. We used the Penn Treebank tag set [11], where, for example, JJ denotes an adjective and NNS, a plural noun. In addition, the parser produces a dependency graph with labeled edges for each sentence, corresponding to grammatical functions, as well as the predicates the sentence contains and their arguments. For each Wikipedia article, we also identify the words or phrases that refer to a same entity i.e. words or phrases that are coreferent.

Figure 3 shows the dependency graph and the predicate–argument structure of the caption: *Ponies walking the streets in Burley*[1].

5.1 Predicates

The predicates correspond to actions or relations such as *jump, walk,* or *own.* Each predicate can have one or more senses, where each sense will correspond to a distinct predicate–argument structure. The semantic parser uses the PropBank [17] nomenclature, where the predicate sense is explicitly shown as a number added after the word. The sentence in Fig. 3 contains one predicate: walk.01 with its two arguments A0 and A1, where A0 corresponds to the walker and A1, the path walked.

PropBank predicates can also have modifying arguments denoted with the prefix "AM-". There exist 14 different types of modifiers in PropBank such as:

AM-DIR: shows motion along some path,
AM-LOC: indicates where the action took place, and
AM-TMP: shows when the action took place.

---

[1] http://en.wikipedia.org/wiki/New_Forest, retrieved November 9, 2012.

5.2   Coreferences

We applied a coreference resolution to create sets of coreferring mentions as with *the rider* and the two *he* in this caption:

> If *the rider* has a refusal at the direct route *he* may jump the other
> B element without additional penalty than what *he* incurred for the
> refusal.[2]

The phrase *the rider* is the first mention of an entity in the coreference chain. It usually contains most information in the chain. We use it together with part-of-speech information and we substitute coreferent words with this mention in a document, although this is mostly useful with pronouns. The modified documents can thereafter be used with different lexical features.

6   Feature Extraction

We used classifiers with visual and semantic features to identify the relations. The visual features formed a baseline system. We then added semantic features to investigate the improvement over the baseline.

6.1   Visual Features

The visual parsing annotation provided us with a set of objects within the images and their bounding boxes defined by the coordinates of the center of each box, its width, and height.

To implement the baseline, we derived a larger set of visual features from the bounding boxes, such as the overlapping area, the relative positions, etc., and combinations of them. We ran an automatic generation of feature combinations and we applied a feature selection process to derive our visual feature set. We evaluated the results using cross-validation. However, as the possible number of combinations was very large, we had to discard manually a large part of them. Once stabilized, the baseline feature set remained unchanged while developing and testing lexical features. It contains the following features:

F_Overlap Boolean feature describing whether the two bounding boxes overlap or not.

F_Distance numerical feature containing the normalized length between the centers of the bounding boxes.

F_Direction(8) nominal feature containing the direction of the human relative the horse, discretized into eight directions.

F_Angle numerical feature containing the angle between the centers of the boxes.

F_OverlapArea numerical feature containing the size of the overlapping area of the boxes.

---

[2] http://en.wikipedia.org/wiki/Eventing, retrieved November 9, 2012.

Table 2. Precision, recall and F_1 for visual features.

	Precision	Recall	F_1
None	0.9472	0.9648	0.9559
Ride	0.7685	0.7553	0.7619
Lead	0.4285	0.2239	0.2941
Mean			0.6706

Table 3. The confusion matrix for visual features.

		Predicted class		
		None	*Ride*	*Lead*
Actual class	*None*	1867	49	19
	Ride	56	176	1
	Lead	48	4	15

F_MinDistanceSide numerical feature containing the minimum distance between the sides of the boxes.

F_AreaDifference numerical feature containing the quotient of the areas.

We used logistic regression and to cope with nonlinearities, we used pairs of features to emulate a quadratic function. The three following features are pairs involving a numerical and a Boolean features, creating a numerical feature. The Boolean feature is used as a step function: if it is false, the output is a constant; if it is true, the output is the value of the numeric feature.

F_Distance+F_LowAngle(7) numerical feature, F_LowAngle is true if the difference in angle is less than $7°$.

F_Angle+F_LowAngle(7) numerical feature.

F_Angle+F_BelowDistance(100) numerical feature, F_BelowDistance(100) is true if the distance is less than 100.

Without these feature pairs, the classifier could not correctly identify the *Lead* relation and the F_1 value for it was 0. With these features, F_1 increased to 0.29. Table 2 shows the recall, precision, and F_1 for the three relations using visual features. Table 3, shows the corresponding confusion matrix.

6.2 Semantic Features

We extracted the semantic features from the Wikipedia articles. We implemented a selector to choose the size of the input between: complete articles, partial articles (the paragraph that is the closest to an image), captions, and file names. The most specific information pertaining to an image is found in the caption and the file name, followed by the partial article, and finally, the whole article.

Bag-of-Words Features. A bag-of-word (BoW) feature was created for each of the four different inputs. A BoW feature is represented by a vector of weighted word frequencies. The different versions have separate settings and dictionaries. We also used a combined bag-of-word feature vector consisting of the concatenation of the partial article, caption, and filename feature vectors.

The features have a filter that can exclude words that are either too common, or not common enough, based on their frequency, controlled by a threshold. We used a $TF \cdot IDF$ weighting on the included words.

We used file names as one of the inputs, as it is common to have a long descriptive names of the images in Wikipedia. However, they are not as standardized as the captions. Some images have very long descriptive titles; others were less informative, for example: "DMZ1.jpg". The file names were not semantically parsed, but we defined a heuristic algorithm, which was used to break down the file name strings into individual words.

Predicate Features. Instead of using all of the words in a document, we used information derived from the predicate–argument structure to filter out more relevant terms. We created a feature that only used the predicate names and their arguments as input. The words that are not predicates, or arguments to the predicates, are removed as input to the feature. The arguments can be filtered depending on their type, for example A0, A1, or AM-TMP. We can either consider all of the words of the arguments, or only the heads.

As for the BoW, we created predicate features with articles, partial articles, and captions as input. We never used the file names, because we could not carry a semantic analysis on them. We also created a version of the predicate-based features that we could filter further on the basis of a list of predicate names, including only predicates present in a predefined list, specified by regular expressions.

7   Classification

To classify the relations, we used the LIBLINEAR [5] package and the output probabilities over all the classes. The easiest way to classify a horse-human pair is to take the corresponding probability vector and pick the class with the highest probability. But sometimes the probabilities are almost equal and there is no clear class to chose. We selected a threshold using cross-validation. If the maximum probability in the vector is not higher than the threshold, the pair is classified as *None*. We observed that because *None* represents a collection of actions and nonaction, it is more likely to be the true class when *Ride* and *Lead* have low probabilities.

Even with the threshold, this scheme can classify two or more humans as riding or leading the same horse. Although possible, it is more likely that only one person is riding or leading the horse at a time. Therefore we added constraints to the classification: a horse can only have zero or one rider, and zero or one leader. For each class, only the most probable human is chosen, and only if it is higher than the threshold.

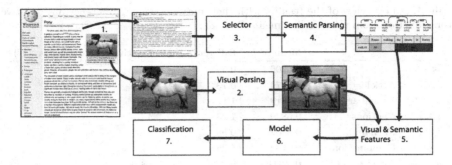

Fig. 4. An overview of the system design, see Sect. 8 for description.

For each human-horse pair, the predicted class is compared to the actual class. The information derived from this can be used to calculate the precision, recall, and F_1 for each class. The arithmetic mean of the three F_1 values is calculated, and can be used as a comparison value. We also computed the number of correct classifications and a confusion matrix.

8   System Architecture

Figure 4 summarizes the architecture of the whole system:

1. Wikipedia is the source of the images and the articles. The text annotation uses the Wiki markup language.
2. Image analysis: placement of bounding boxes, classification of objects and actions. This was done manually, but could be replaced by an automatic system.
3. Text selector between: the whole articles, paragraphs that are the closest to the images, filenames, or captions.
4. Semantic parsing of the text, see Sect. 5.
5. Extraction of feature vectors based on the bounding boxes and the semantic information.
6. Model training using logistic regression from the LIBLINEAR package. This enables us to predict probabilities for the different relations.
7. Relation classification using probabilities and constraints.

9   Results

We used the L2-regularized logistic regression (primal) solver from the LIBLIN-EAR package and we evaluated the results of the classification with the different feature sets starting from the baseline geometric features and adding lexical features of increasing complexity. We carried out a 5-fold crossvalidation.

We evaluated permutations of features and settings and we report the set of combined BoW features that yielded the best result. Table 4 shows an overview of the results:

Table 4. An overview of the results, with their mean F_1-value, difference and relative error reduction from the baseline mean F_1-value.

		Mean of F_1	Difference (pp)	Relative error reduction (%)
Baseline		0.6706	0.00	0.00
BoW	Articles	0.6779	0.73	2.22
	Partial articles	0.6818	1.12	3.40
	Captions	0.6829	1.23	3.73
	Filenames	0.6802	0.96	2.91
	Combination	0.7132	4.26	12.9
Predicate	Articles	**0.7318**	**6.12**	**18.6**
	Partial articles	0.6933	2.27	6.89
	Captions	0.6791	0.85	2.58
	Articles + Words	0.6830	1.24	3.76
	Articles + Coref	0.7280	5.74	17.4

- The baseline corresponds to the geometrical features; we obtained a mean F_1 of 0.67 with them;
- BoW corresponds to the baseline features and the bag-of-word features described in Sect. 6.2; whatever the type of text we used as input, we observed an improvement. We obtained the best results with a concatenation of the partial article, caption, and filename (combination, $F_1 = 0.71$);
- predicate corresponds to the baseline features and the predicate feature vector described in Sect. 6.2. Predicate features using only one lexical feature vector from the article text gave better results than combining different portions of the text ($F_1 = 0.73$).

Our best feature set is the predicate features utilizing whole articles as input. It achieves a relative error reduction of 18.6 percent compared to baseline.

Tables 2 and 3 show the detailed results of the baseline with the geometric features only. Tables 5 and 6 show the results of the best BoW feature combination: a concatenation of the feature vectors from the inputs: partial articles, captions, and filenames. Tables 7 and 8 show the result of the best predicate features.

Table 5. Precision, recall, and F_1 for the concatenation of BoW features with the inputs: partial articles, captions and filenames.

	Precision	Recall	F_1
None	0.9638	0.9638	0.9638
Ride	0.7642	0.8626	0.8104
Lead	0.5135	0.2835	0.3653
Mean			0.7132

Table 6. The confusion matrix for BoW for the concatenation of BoW features with the inputs: partial articles, captions and filenames.

		Predicted class		
		None	*Ride*	*Lead*
Actual class	*None*	1865	57	13
	Ride	27	201	5
	Lead	43	5	19

Table 7. Precision, recall and F_1 for predicate feature on articles.

	Precision	Recall	F_1
None	0.9745	0.9498	0.9620
Ride	0.7301	0.9055	0.8084
Lead	0.4500	0.4029	0.4251
Mean			0.7318

Table 8. The confusion matrix for predicate feature on articles.

		Predicted class		
		None	*Ride*	*Lead*
Actual class	*None*	1838	70	27
	Ride	16	211	6
	Lead	32	8	27

10   Discussion

Classifying the *Lead* relation with geometric features with only bounding boxes as the input revealed quite difficult. There is indeed very little visual difference between standing next to a horse and leading it. We were not able to classify any *Lead* correctly until we added the combination features.

For single BoW features, the captions gave the best result, followed by partial articles, filenames, and lastly articles. The order of the results was what we expected, based on how specific information the features had about the images. But for the predicate features, the order was reversed: articles produced the best result, followed by partial articles, and captions.

Using a specific list of predicates did not produce good results although, depending on the list, results vary greatly. Using a list with the words: *ride, lead, pull,* and *race*, with articles as input, gave the best result, but Table 4 shows a relative drop of 4.88 compared to no filtering. The negative results could possibly be explained by the fact that it is not common to explicitly describe the relations in the images, and only utilizing keywords such as *ride* is of little help.

Applying coreference resolution on the documents lowered the results. Table 4 shows a relative drop of 0.38 if applied on the predicate feature based on articles.

Despite these negative results, we still believe that solving coreferences could improve the results. The solver was designed to be used with another set of semantic information. To be able to use the solver, we altered its source code and possibly made it less accurate. We checked manually coreference chains and we could observe a significant number of faulty examples, leading us to believe that the output quality of the solver left to be desired.

11   Conclusions and Future Work

We designed a supervised classifier to identify relations between pairs of objects in an image. As input to the classifier, we used geometric, bag-of-words, and semantic features. The results we obtained show that semantic information, in combination with geometric features, proved useful to improve the classification of relations in the images. Table 4 shows that the relative error reduction is 12.9 percent by utilizing a combination of bag-of-words features. An even greater improvement is made using predicate information with an relative error reduction of 18.6 percent compared to baseline.

Coreference resolution lowered the performance, but the interface between the semantic parser and the coreference solver was less than optimal. There is room for improvement regarding this solver, either with the interface to the semantic parser or with to another solver. It could also be interesting to try other types of classifiers, not just logistic regression, and see how they perform.

Using automatically annotated images as input to the program could be relatively easily implemented and would automate all the steps in the system. A natural continuation of the work is to expand the number of objects and relations. [6], for example, use 20 different classifiers for common objects: cars, bottles, birds, etc. All, or a subset of it, could be chosen as the objects, together with some common predicates between the objects as the relations.

It would also be interesting to try other sources of images and text than Wikipedia: either using other resources available online or creating a new database with images captioned with text descriptions. Another interesting expansion of the work would be to map entities found in the text with objects found in the image. For example, if a caption includes the name of a person, one could create a link between the image and information about the entity.

Acknowledgements. This research was supported by Vetenskapsrådet, the Swedish research council, under grant 621-2010-4800 and *Det digitaliserade samhället* thematic grant, and the Swedish e-science program: eSSENCE.

References

1. Carreira, J., Sminchisescu, C.: Constrained parametric min-cuts for automatic object segmentation. In: IEEE International Conference on Computer Vision and Pattern Recognition (2010)
2. Chen, N., Zhou, Q.-Y., and Prasanna, V.: Understanding web images by object relation network. In: Proceedings of the 21st International Conference on World Wide Web, WWW 2012, pp. 291–300. ACM, New York (2012)

3. Deschacht, K., Moens, M.-F.: Text analysis for automatic image annotation. In: Proceedings of the 45th Annual Meeting of the Association of Computational Linguistics, pp. 1000–1007, Prague (2007)
4. Exner, P., Nugues, P.: Constructing large proposition databases. In: Proceedings of the Eight International Conference on Language Resources and Evaluation (LREC 2012), Istanbul (2012)
5. Fan, R.-E., Chang, K.-W., Hsieh, C.-J., Wang, X.-R., Lin, C.-J.: LIBLINEAR: a library for large linear classification. J. Mach. Learn. Res. **9**, 1871–1874 (2008)
6. Felzenszwalb, P.F., Girshick, R.B., McAllester, D., Ramanan, D.: Object detection with discriminatively trained part based models. IEEE Trans. Pattern Anal. Mach. Intell. **32**(9), 1627–1645 (2010)
7. Gupta, A., Verma, Y., Jawahar, C.: Choosing linguistics over vision to describe images. In : Proceeding of the Twenty-Sixth AAAI Conference on Artificial Intelligence (2012)
8. Jörgensen, C.: Attributes of images in describing tasks. Inf. Process. Manage. **34**(2–3), 161–174 (1998)
9. Kulkarni, G., Premraj, V., Dhar, S., Siming, L., Choi, Y., Berg, A., Berg, T.: Baby talk: understanding and generating image descriptions. In: Proceedings of Conference Computer Vision and Pattern Recognition (2011)
10. Torr, P.H.S., Torr, P.H.S., Ladicky, L., Ladicky, L., Kohli, P., Kohli, P., Russell, C., Russell, C.: Graph cut based inference with co-occurrence statistics. In: Maragos, P., Maragos, P., Daniilidis, K., Daniilidis, K., Paragios, N., Paragios, N. (eds.) ECCV 2010, Part V. LNCS, vol. 6315, pp. 239–253. Springer, Heidelberg (2010)
11. Marcus, M., Marcinkiewicz, M.A., Santorini, B.: Building a large annotated corpus of english: the penn treebank. Comput. Linguis. **19**(2), 313–330 (1993)
12. Markkula, M., Sormunen, E.: End-user searching challenges indexing practices in the digital newspaper photo archive. Inf. Retrieval **1**(4), 259–285 (2000)
13. Marszalek, M., Schmid, C.: Semantic hierarchies for visual object recognition. In: Proceeding of Conference Computer Vision and Pattern Recognition (2007)
14. Moscato, V., Picariello, A., Persia, F., Penta, A.: A system for automatic image categorization. In: IEEE International Conference on Semantic Computing, ICSC 2009, pp. 624–629. IEEE (2009)
15. Myeong, H., Chang, J.Y., Lee, K.M.: Learning object relationships via graph-based context model. In: CVPR, pp. 2727–2734 (2012)
16. Paek, S., Sable, C., Hatzivassiloglou, V., Jaimes, A., Schiffman, B., Chang, S., Mckeown, K.: Integration of visual and text-based approaches for the content labeling and classification of photographs. In: ACM SIGIR, vol. 99 (1999)
17. Palmer, M., Gildea, D., Kingsbury, P.: The proposition bank: an annotated corpus of semantic roles. Comput. Linguis. **31**(1), 71–105 (2005)
18. Stamborg, M., Medved, D., Exner, P., Nugues, P.: Using syntactic dependencies to solve coreferences. In: Joint Conference on EMNLP and CoNLL - Shared Task, pp. 64–70. Association for Computational Linguistics, Jeju Island, Korea (2012)
19. Tirilly, P., Claveau, V., Gros, P., et al.: News image annotation on a large parallel text-image corpus. In: 7th Language Resources and Evaluation Conference, LREC 2010 (2010)
20. Westman, S., Oittinen, P.: Image retrieval by end-users and intermediaries in a journalistic work context. In: Proceedings of the 1st International Conference on Information Interaction in context, pp. 102–110. ACM (2006)
21. Wikipedia. Wikipedia statistics English (2012). http://stats.wikimedia.org/EN/TablesWikipediaEN.htm

A ToF-Aided Approach to 3D Mesh-Based Reconstruction of Isometric Surfaces

S. Jafar Hosseini[✉] and Helder Araujo

Institute of Systems and Robotics,
Department of Electrical and Computer Engineering,
University of Coimbra, Coimbra, Portugal
{jafar,helder}@isr.uc.pt

Abstract. In this paper, we investigate structure-from-motion (SfM) for surfaces that deform isometrically. Our SfM framework is intended for the estimation of both the 3D surface and the camera motion at one time through a template-based approach founded on the combination of a ToF sensor and a conventional RGB camera. The objective is to take advantage of depth maps acquired by the ToF sensor so that a considerable enhancement can be achieved in the reconstruction of the non-rigid structure using the high-resolution images captured by means of the RGB camera. A triangular mesh is adopted to represent isometric surfaces. The depth of a sparse set of 3D feature points spread all over the surface will be obtained with the help of the ToF camera, thereby enabling the recovery of the depth of the mesh vertices using a multivariate linear system. Subsequently, a non-linear constraint is formed based on the projected length of each edge of the mesh. A second non-linear constraint is then used for minimizing re-projection errors. These constraints are finally incorporated into an optimization scheme to solve for structure and motion. Experimental results show that the proposed approach has good performance even if only a low-resolution depth image is used.

Keywords: Structure from motion · Isometric surface · ToF camera · 3D reconstruction

1 Introduction

Structure-from-motion can be defined as the problem of simultaneous inference of the motion of a camera and the 3D geometry of the scene solely from a sequence of images. SfM was also extended to the case of deformable objects. Non-rigid SfM is under-constrained, which means that the recovery of non-rigid 3D shape is an inherently ambiguous problem [23,24]. Given a specific configuration of points on the image plane, different 3D non-rigid shapes and camera motions can be found that fit the measurements. To solve this ambiguity, prior knowledge on the shape and motion should be used to constrain the solution. For example, Aanaes et al. [1] impose the prior knowledge that the reconstructed shape does not vary much from frame to frame while Del Bue et al. [2] impose the constraint

© Springer International Publishing Switzerland 2015
A. Fred et al. (Eds.): ICPRAM 2014, LNCS 9443, pp. 146–161, 2015.
DOI: 10.1007/978-3-319-25530-9_10

that some of the points on the object are rigid. The priors can be divided in two main categories: the statistical and the physical priors. For instance, the methods relying on the low-rank factorization paradigm [1, 2] can be classified as statistical approaches. Learning approaches such as [3, 21, 22] also belong to the statistical approaches. Physical constraints include spatial and temporal priors on the surface to be reconstructed [6, 7]. A physical prior of particular interest is the hypothesis of having an inextensible (i.e. isometric) surface [8–10]. In this paper, we consider this type of surface. This hypothesis means that the length of the geodesics between every two points on the surface should not change across time, which makes sense for many types of material such as paper and some types of fabric.

3D reconstruction of non-rigid surfaces from images is an under-constrained problem and many different kinds of priors have been introduced to restrict the space of possible shapes to a manageable size. Based on the type of the surface model (or representation) used, we can classify the algorithms for reconstruction of deformable surfaces. The point-wise methods only reconstruct the 3D position of a relatively small number of feature points resulting in a sparse reconstruction of the 3D surface [9]. Physics-based models such as superquadrics [11], triangular meshes [10], Thin-Plate Splines (TPS) [9], or tensor product B-splines [18] have been also utilized in other algorithms. In TPS, the 3D surface is represented as a parametric 2D-3D map between the template image space and the 3D space. Then, a parametric model is fit to a sparse set of reconstructed 3D points in order to obtain a smooth surface which is not actually used in the 3D reconstruction process. There has been increasing interest in learning techniques that build surface deformation models from training data. More recently, linear models have been learned for SfM applications [12, 13]. There have also been a number of attempts at performing 3D surface reconstruction without using a deformation model. One approach is to use lighting information in addition to texture clues to constrain the reconstruction process [14], which has only been demonstrated under very restrictive assumptions on lighting conditions and is therefore not generally applicable. A common assumption in deformable surface reconstruction is to consider that the surface is inextensible. In [9], the authors propose a dedicated algorithm that enforces the inextensibility constraints. However, the inextensibility constraint alone is not sufficient to reconstruct the surface. Another sort of implementation is given by [4, 10]. In these papers, a convex cost function combining the depth of the reconstructed points and the negative of the reprojection error is maximized while enforcing the inequality constraints arising from the surface inextensibility. The resulting formulation can be easily turned into a SOCP problem. A similar approach is explored in [8]. The approach of [9] is a point-wise method. The approaches of [4, 8, 10] use a triangular mesh as surface model, and the inextensibility constraints are applied to the vertices of the mesh.

1.1 Model and Approach

In this work, we aim at the combined inference of the 3D surface and the camera motion while preserving the geodesics by using a RGB camera aided by a ToF range sensor. Usually, RGB cameras have high image resolutions. With these cameras, one can use efficient algorithms to calculate the depth of the scene, recover object shape or reveal structure, but at a high computational cost. ToF cameras deliver a depth map of the scene in real-time but with insufficient resolution for some applications. So, a combination of a common camera and a ToF sensor can exploit the capabilities of both. We assume that the fields of view of both the RGB and the ToF cameras mostly overlap. The surface is represented as a triangular 3D mesh and a set of correspondences between 3D feature points and 2D locations in an input image is available. In practice, they are obtained by matching SIFT features between the input image and a reference image in which the surface shape is known. The 2D points in the reference image correspond to 3D feature points on the mesh. The goal of the algorithm is to allow the 3D reconstruction of the surface mesh when matching is difficult and depth estimates are available for a limited number of points on the surface. Our approach performs SfM under the constraint that the deformation be isometric.

1.2   Outline of the Paper

This paper is organized as follows: to represent an isometric surface, a triangular mesh as well as a planar reference configuration is used. In Sect. 3, the matching between data from the range and the RGB cameras is described. Next, the estimation of the depth of the mesh vertices based on the depth of the feature points is described. The entire approach for the estimation of the 3D shape and motion is based on minimizing the sum of both the re-projection errors and the errors on the projected length of the mesh edges. Experimental results and quantitative evaluation are presented in the last section. We show that our approach is able to handle isometry indirectly without having to directly apply this constraint. In addition, it obviates the need for a dense set of 3D points lying on the surface by effective use of a ToF sensor.

2   Notation and Background

2.1   Notation

Matrices are represented as bold capital letters ($\mathbf{A} \in \mathbb{R}^{n \times m}$, n rows and m columns). Vectors are represented as bold small letters ($\mathbf{a} \in \mathbb{R}^n$, n elements). By default, a vector is considered a column. Small letters (a) represent one dimensional elements. By default, the jth column vector of $\mathbf{A}$ is specified as $\mathbf{a}_j$. The jth element of a vector $\mathbf{a}$ is written as a_j. The element of $\mathbf{A}$ in the row i and column j is represented as $A_{i,j}$. $\mathbf{A}^{(1:2)}$ and $\mathbf{a}^{(1:2)}$ indicate the first 2 rows of $\mathbf{A}$ and $\mathbf{a}$. $\mathbf{A}^{(3)}$ and $\mathbf{a}^{(3)}$ denote the third row of $\mathbf{A}$ and $\mathbf{a}$, respectively. Regular capital letters (A) indicate one dimensional constants. We use $\mathbb{R}$ after a vector or matrix to denote that it is represented up to a scale factor. There might be few cases opposed to this notation, however, the aim is to comply with it as closely as possible.

2.2   Barycentric Coordinates

In geometry, the barycentric coordinate system is a coordinate system in which the location of a point of a simplex (a triangle, tetrahedron, etc.) is specified as the center of mass, or barycenter, of masses placed at its vertices.

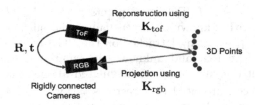

Fig. 1. RGB/ToF camera setup.

3   Combining Depth and RGB Images

3.1   Mapping Between Depth and RGB Images

The resolutions of the depth and the RGB images are different. A major issue that directly arises from the difference in resolution is that a pixel-to-pixel correspondence between the two images can not be established even if the FOVs fully overlap. Therefore the two images have to be registered so that the mapping between the pixels in the ToF image and in the RGB image can be established. The depth map provided by the ToF camera is sparse and affected by errors. Several methods can be used to improve the resolution of the depth images [15–17,25] allowing the estimation of a dense depth image. We will use a simple approach based on linear interpolation.

To estimate depth for all the pixels of the RGB image, based on the depth provided by the ToF camera, a simple linear approach is used. We assume that the relative pose between both cameras, specified by the rotation matrix $\mathbf{R}'$ and translation vector $\mathbf{t}'$ has been estimated. We also assume that both cameras are internally calibrated, i.e., their intrinsic parameters are known. Let $\mathbf{p}_{tof}$ and $\mathbf{p}_{rgb}$ represent the 3D coordinates of a 3D point in the coordinate system of the Tof and the RGB cameras, respectively.

We use a pinhole camera model for both the RGB and the ToF cameras. Assume that the relative pose of the RGB camera and ToF sensor is fixed with a rotation $\mathbf{R}'$ and a translation $\mathbf{t}'$: $\mathbf{p}_{rgb} = \mathbf{R}' \, \mathbf{p}_{tof} + \mathbf{t}'$ as shown in Fig. 1. The point cloud $\mathbf{p}_{tof}$ is obtained directly from the calibrated ToF camera. Since the relative pose is known as well as the intrinsic parameters for both cameras, $\mathbf{p}_{rgb}$ can be obtained from $\mathbf{p}_{tof}$. To estimate depth for all points of the RGB image, a simple linear interpolation procedure is used. For each 2D point of the RGB image, we select the 4 closest neighbors whose depth was obtained from the depth image. Then, a bilinear interpolation is performed. Another possibility would be to select the 3 closest neighboring points (therefore, defining a triangle) and assume that the corresponding 3D points define a plane. An estimate for the depth of the point could then be obtained by intersecting its projecting ray with the 3D plane defined by the three 3D points.

3.2   Recovery of the Mesh Depth

Assume that a sparse set of 3D feature points $\mathbf{p}^{ref} = \{\mathbf{p}_1^{ref}, \cdots, \mathbf{p}_N^{ref}\}$ on a reference template with a known shape (usually a flat surface), and a set of 2D image points $\mathbf{q} = \{\mathbf{q}_1, \cdots, \mathbf{q}_N\}$ tracked on the RGB input image of the same surface but with a different and unknown deformation are given. As already stated, we represent the surface as a triangulated 3D mesh with n_v vertices $\mathbf{v}_i$ (and n_{tr} triangles) concatenated in a vector $\mathbf{s} = [\mathbf{v}_1^T, \cdots, \mathbf{v}_{n_v}^T]^T$, and denote by $\mathbf{s}^{ref}$ the reference mesh, and $\mathbf{s}$ the mesh we seek to recover. Let $\mathbf{p}_i$ be a feature point on the mesh $\mathbf{s}$ corresponding to the point $\mathbf{p}_i^{ref}$ in the reference configuration. We can express $\mathbf{p}_i$ in terms of the barycentric coordinates of the triangle it belongs to:

$$\mathbf{p}_i = \sum_{j=1}^{3} a_{ij} \mathbf{v}_j^{[i]} \tag{1}$$

where the a_{ij} are the barycentric coordinates and $\mathbf{v}_j^{[i]}$ are the vertices of the triangle containing the point $\mathbf{p}_i$. Since we are dealing with rigid triangles, these barycentric coordinates remain constant for each point and can be easily computed from points $\mathbf{p}_i^{ref}$ and the mesh $\mathbf{s}^{ref}$. Let us denote by $\mathbf{A} = \{\mathbf{a}_1, \cdots, \mathbf{a}_N\}$ the set of barycentric coordinates associated to the 3D feature points, where $\mathbf{a}_i = [a_{i1}, a_{i2}, a_{i3}]$. The rigidity of a triangle enforces that the sum of the relative depths around a closed triangle be zero. Assuming that the depth of the vertices of a triangle is denoted as $v_{z,1}$, $v_{z,2}$ and $v_{z,3}$, we have: $(v_{z,1}-v_{z,2})+(v_{z,2}-v_{z,3})+(v_{z,3}-v_{z,1}) = 0$. Substituting $(v_{z,1}-v_{z,2})$, $(v_{z,2}-v_{z,3})$ and $(v_{z,3}-v_{z,1})$ for rz_1, rz_2 and rz_3, respectively, which denote the relative depth of the edges of the triangle, we can represent the above equation differently as: $rz_1 + rz_2 + rz_3 = 0$ where $rz_1 = v_{z,1}-v_{z,2}$, $rz_2 = v_{z,2}-v_{z,3}$, and $rz_3 = v_{z,3}-v_{z,1}$. Having the above equations for any triangle of the mesh makes a total of $n_{tr}+n_e$ (the number of triangles + the number of edges) linear equations which can be jointly expressed as $\mathbf{M}_{1(n_{tr}+n_e) \times (n_v+n_e)} \mathbf{x}_{1(n_v+n_e) \times 1} = 0$. This homogeneous system of equations must be satisfied at each time instant (i.e. for any deformation). However, finding a unique solution is not possible. More specifically, $\mathbf{M}_1$ is rank-deficient by n_v, that is, it does not have $n_v + n_e$ linearly independent columns ($\mathrm{rank}(\mathbf{M}_1) = n_e$). So, there will be a n_v-dimensional basis for the solution space to $\mathbf{M}_1\mathbf{x}_1 = 0$. Any solution is a linear combination of basis vectors. In order to constrain the solution space and determine just one solution out of the infinite possibilities, in a way that this linear system matches only one particular deformation, it is necessary to add n_v independent equations. To add additional constraints, we augment this system with the z coordinate of few properly distributed feature points in this arrangement: using the method described in the previous section, we can obtain an estimate for the depth of a feature point i, indicated by $p_{z,j}$. From the Eq. 1, we can derive $p_{z,i} = a_{i1}v_{z,1}^{[i]} + a_{i2}v_{z,2}^{[i]} + a_{i3}v_{z,3}^{[i]}$. This non-homogeneous system of equations can be represented as $\mathbf{M}_{2(N \times n_v)}\mathbf{x}_{2(n_v \times 1)} = \mathbf{p}_z$. It can be verified that $\mathbf{x}_1 = \begin{bmatrix} \mathbf{rz} \\ \mathbf{x}_2 \end{bmatrix}$. $\mathbf{rz}$ is a n_e-vector of the relative depth of the edges. Having the above equation for any feature point results in N linear independent equations. Putting together both sets of equations just explained, we end up with $n_{tot} = n_{tr} + n_e + N$ linear equations ($\mathbf{Mx}_1 = \begin{bmatrix} \mathbf{0} \\ \mathbf{p}_z \end{bmatrix}$) where the only unknowns are the depth of the vertices and of the edges (i.e. $n_v + n_e$ unknowns), which means that the resulting linear system is overdetermined. In fact, we obtain $n_e + N$ independent equations out of n_{tot} equations. Yet, this is not enough to find the right single solution because there are still an infinitude of further solutions that minimize $\left\| \mathbf{Mx}_1 - \begin{bmatrix} \mathbf{0} \\ \mathbf{p}_z \end{bmatrix} \right\|$ in the least-squares sense. One possible approach after the 3D coordinates are estimated is to fit an initial surface using polynomial interpolation, to the data which consists in xy-coordinates of the feature points on the reference configuration as input and their z-coordinates on the input deformation as output. Once the parameters of the interpolant have been found, we can obtain initial estimates of depth for the vertices, with their xy-coordinates on the reference configuration as input. The interpolated depth has proved to be very close to the correct one. Then, we add an equality constraint for each vertex as $\mathbf{I}_{n_v \times n_v}\mathbf{x}_2 = \mathbf{v}_z'$ ($\mathbf{v}_z'$ is the interpolated depth of the vertices). The new linear system $\mathbf{M}_{new}\mathbf{x}_1 = \mathbf{b}$ has most likely full column-rank. So, the number of independent equations out of $n_{tot} + n_v$ equations would be $n_e + n_v$. Since the number of independent equations is equal to the number of unknowns, there must be a unique solution, which can be computed via the normal equations. In principle, finding the least-sqaures estimate is recommended.

4 Global Metric Estimation of Structure and Motion

Next we describe two non-linear constraints applied to the estimation problem. These two constraints are used to solve for SfM so that metric reconstruction of the shape is achieved and the motion matrices lie on the appropriate motion manifold. Furthermore, when there are too few correspondences without additional knowledge (as is the case here), shape recovery would not be effective. So, we need to limit the space of possible shapes by applying a deformation model. This model adequately fills in the missing information while being flexible enough to allow reconstruction of complex deformations [3]. We assume we can model the mesh deformation as a linear combination of a mean shape $\mathbf{s}_0$ and n_m basis shapes (deformation modes) $\mathbf{S} = [\mathbf{s}_1, ..., \mathbf{s}_{n_m}]$:

$$\mathbf{s} = \mathbf{s}_0 + \sum_{k=1}^{n_m} w_k \mathbf{s}_k = \mathbf{s}_0 + \mathbf{Sw} \tag{2}$$

4.1   Constraint 1: Projected Length

Assume that the RGB camera motion relative to the world coordinate system is expressed as a rotation matrix $\mathbf{R}$ and a translation vector $\mathbf{t}$. A common approach to solve for the camera motion and surface structure is to minimize the image re-projection error, namely by bundle adjustment. The cost function being minimized is the geometric distance between the image points and the re-projected points. However, we are going to adapt bundle adjustment to our own problem rather than use it directly, as follows: the errors to be minimized will be the difference between the observed and the predicted projected lengths of an edge.

Orthographic Camera. Under orthographic projection, if we assume that the mesh vertices are registered with respect to the image centroid, we can drop the translation vector. The modified formulation of bundle adjustment can be specified as the following non-linear constraint:

$$e_{pl} = \sum_{i=1}^{n_e} \left(l_i - \left\| \mathbf{R}^{(1:2)} \left[\mathbf{s}_1^{[i]} - \mathbf{s}_2^{[i]} \right] \right\| \right)^2 \tag{3}$$

where the leftmost term is the measurement (observation) of the projected length of an edge (the computation of l_i is trivial with the help of estimated mesh depth). n_e is the number of edges. $\mathbf{s}_1^{[i]}$ and $\mathbf{s}_2^{[i]}$ denote 2 entries of the mesh, which account for the ending vertices of the edge i. e_{pl} can be also expressed as a quadratic function.

Perspective Camera. In this case, we formulate a non-linear constraint based on what we call "unnormalized projected length", as:

$$e_{pl} = \sum_{i=1}^{n_e} \left(l_i - \left\| \mathbf{K}_{rgb}^{\circ} \left[\mathbf{R} | \mathbf{t} \right] \left[\begin{bmatrix} \mathbf{s}_1^{[i]} \\ 1 \end{bmatrix} - \begin{bmatrix} \mathbf{s}_2^{[i]} \\ 1 \end{bmatrix} \right] \right\| \right)^2 \tag{4}$$

where $\mathbf{K}_{rgb}^{\circ}$ is a known calibration matrix equivalent to $\begin{bmatrix} f & 0 & 0 \\ 0 & f & 0 \\ 0 & 0 & 1 \end{bmatrix}$. From the esti-mated mesh depth, l_i can be easily measured using simple mathematical manipulation. Since there is a subtraction in the above cost function, the translation vector $\mathbf{t}$ can be removed. Also, note that the 2-norm is applied to the first 2 entries of a 3-vector to estimate the square of unnormalized projected length. So, only the 2 first rows of the product of $\mathbf{K}_{rgb}^{\circ}.\mathbf{R}$ are involved in the constraint:

$$e_{pl} = \sum_{i=1}^{n_e} \left(l_i - \left\| \mathbf{f}^{[i]}(\mathbf{R}^{(1:2)}, w) \right\| \right)^2 \tag{5}$$

4.2   Constraint 2: Reprojection Error

Several difficulties may affect the estimation of the depths namely:

Errors due to the depth interpolation;
Irregular distribution of the feature points over the object surface.

As a result of these factors, the depth estimate for the mesh vertices may be significantly inaccurate. In addition, there are also reprojection errors, that is, errors on the image positions of the 3D feature points. We should thus account for the reprojection error by adding a term to the function to be optimized. By combining Eqs. 1 and 2, we'll have:

$$\mathbf{p}_i = \sum_{j=1}^{3} a_{ij} \mathbf{s}_j^{[i]} \tag{6}$$

where $\mathbf{s}_{0j}^{[i]}$ and $\mathbf{S}_j^{[i]}$ are the subvector of $\mathbf{s}_0$ and the submatrix of $\mathbf{S}$ (respectively), corresponding to the vertex j of the triangle in which the feature point i resides.

The term corresponding to the reprojection error can be obtained as indicated below.

Orthographic Camera

$$e_{re} = \sum_{i=1}^{N} \left\| \mathbf{q}_i - \mathbf{R}^{(1:2)} \mathbf{p}_i \right\|^2 \tag{7}$$

Perspective Camera

$$e_{re} = \sum_{i=1}^{N} \left\| \lambda_i \begin{bmatrix} \mathbf{q}_i \\ 1 \end{bmatrix} - \left[\mathbf{K}_{rgb}^{\circ} \left[\mathbf{R} | \mathbf{t} \right] \begin{bmatrix} \mathbf{p}_i \\ 1 \end{bmatrix} \right] \right\|^2 \tag{8}$$

The projective depths λ_i can be determined using the estimated depth for feature's image points on the RGB image. Subsequently, errors in λ_i (induced by the first condition mentioned above) would introduce false search directions in the e_{re}-based minimization problem. Therefore, it is advantageous to reformulate the above equations so that λ_i is removed from them. So, we take into account the equation below:

$$\lambda_i \begin{bmatrix} \mathbf{q}_i \\ 1 \end{bmatrix} = \mathbf{K}_{rgb}^{\circ} \left[\sum_{j=1}^{3} a_{ij} \mathbf{R}.\mathbf{s}_j^{[i]} \right] + \mathbf{K}_{rgb}^{\circ}.\mathbf{t} \tag{9}$$

After some simple algebraic manipulation, we obtain:

$$\left[a_{i1} \mathbf{A}_i \ \ a_{i2} \mathbf{A}_i \ \ a_{i3} \mathbf{A}_i \right]_{2 \times 9} \begin{bmatrix} \mathbf{R}.\mathbf{s}_1^{[i]} \\ \mathbf{R}.\mathbf{s}_2^{[i]} \\ \mathbf{R}.\mathbf{s}_3^{[i]} \end{bmatrix}_{9 \times 1} + \mathbf{A}_i.\mathbf{t} =$$

$$\begin{bmatrix} e_1^{[i]}(\mathbf{R}, \mathbf{w}, \mathbf{t}) \\ e_2^{[i]}(\mathbf{R}, \mathbf{w}, \mathbf{t}) \end{bmatrix}_{2 \times 1} = 0 \quad \text{where } \mathbf{A}_i = \mathbf{K}_{rgb}^{\circ(1:2)} - \mathbf{q}_i . \mathbf{K}_{rgb}^{\circ(3)} \tag{10}$$

This equation provides 2 linear constraints as: $e_1^{[i]}(.) = 0$ and $e_2^{[i]}(.) = 0$. Thus, the modified e_{re} takes a form free of λ_i as follows: $e_{mre} = \sum_{i=1}^{N} \left(e_1^{[i]}(.)^2 + e_2^{[i]}(.)^2 \right)$, where e_{mre} denotes the modified e_{re}. e_{pl} is a function of $\mathbf{R}^{(1:2)}$ and $\mathbf{w}$ whereas e_{mre} (or e_{re}) is a function of $\mathbf{R}$, $\mathbf{w}$ and $\mathbf{t}$. In order to simplify e_{pl}, we modify it by considering that: 1- the translation vector $\mathbf{t}$ is fixed and the camera setup has only rotational movement relative to the world coordinate system. 2- adding the following function to $\mathbf{f}^{[i]}(\mathbf{R}^{(1:2)}, \mathbf{w})$ in the first constraint, we are able to solve for the full matrix $\mathbf{R}$:

$$f_{rz}^{[i]}(\mathbf{R}^{(3)}, \mathbf{w}) = \left(\mathbf{R}^{(3)} \left[\mathbf{s}_1^{[i]} - \mathbf{s}_2^{[i]} \right] \right) \tag{11}$$

$$e_{rz} = rz_i - f_{rz}^{[i]}(\mathbf{R}^{(3)}, \mathbf{w}) \tag{12}$$

where $rz_i = v_{z,1}^{[i]} - v_{z,2}^{[i]}$. e_{rz} is actually the difference between the observed and the predicted relative depths of edge i. Combining $\mathbf{f}^{[i]}(.)$ and $f_{rz}^{[i]}(.)$, it yields:

$$e_{mpl} = \sum_{i=1}^{n_e} \left(\sqrt{(l_i^2 + rz_i^2)} - \left\| \begin{bmatrix} \mathbf{f}^{[i]}(\mathbf{R}^{(1:2)}, \mathbf{w}) \\ f_{rz}^{[i]}(\mathbf{R}^{(3)}, \mathbf{w}) \end{bmatrix} \right\| \right)^2 \tag{13}$$

where e_{mpl} represents a modified version of e_{pl}. As a result, we brought e_{mpl} and e_{mre} into a common form where both are functions of $\mathbf{R}$ and $\mathbf{w}$.

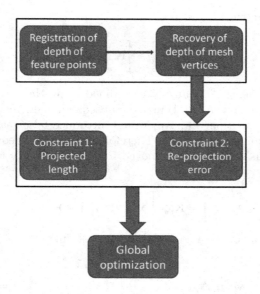

Fig. 2. Representation of the approach via block diagram.

4.3   Objective Function

So far, we have derived two constraints expressed as two separate non-linear problems. However, we intend to integrate both constraints into one single objective function so that they are taken into account at once, when estimating all the parameters. To do so, we minimize the weighted summation of them in such a way that the reprojection error term is assigned a weight m that accounts for its relative influence within the combined objective function. A block diagram of the overall structure of the approach is demonstrated in Fig. 2. In our global optimization, we first consider a simplified formulation of the objective function by excluding the camera motion $[\mathbf{R}|\mathbf{t}]$. We include it back in the second case.

Estimation of Structure only. The constraints are simplified so that the only unknown parameter is the structure (we assume that the camera motion is set to $[\mathbf{I}|0]$).

    Orthographic Camera: $\min_{\mathbf{w}} e_{tot} = (e_{pl} + m.e_{re})$
Perspective Camera: $\min_{\mathbf{w}} e_{tot} = (e_{mpl} + m.e_{mre})$

Estimation of both Structure and Camera Motion. We consider now the full optimization by including the camera motion.

    Orthographic Camera: $\min_{\mathbf{R}^{(1:2)},\mathbf{w}} e_{tot} = (e_{pl} + m.e_{re})$
Perspective Camera: $\min_{\mathbf{R},\mathbf{w}} e_{tot} = (e_{mpl} + m.e_{mre})$

The above optimization problems can be solved using a non-linear minimization algorithm such as Levenberg-Marquardt (LMA). The rotation estimates obtained from this optimization may not satisfy the orthormality constraints. So, the optimization algorithm must be fed with a good initialization. To provide initial estimates relatively close to the true ones, we do the following: if initial guesses for $R^{(1:2)}$ and R are not given, they can be initialized using well-known methods that attempt to solve for SfM through non-rigid factorization of $\{q_{ij}\}$ and $\{\lambda_{ij}q_{ij}\}$ from all frames, for instance, as in [13]. In these methods, the factorization is followed by a refinement step to upgrade the reconstruction to metric. The deformation coefficients $\mathbf{w}_k$ are initialised to random small values. One possible solution to further meet the rotation constraints is to subsequently apply Procrustes [19,20].

4.4   Additional Constraint

Non-linear optimization may converge to local minima. The probability of such occurrence can be reduced by adding a new regularization term that requires the estimated depth data to be as close to the measured one as possible. So, we would have:

$$e_z = \sum_{i=1}^{n_v} \left(v_z^{[i]} - \left(\mathbf{R}^{(3)}\mathbf{s}^{[i]} + \mathbf{t}^{(3)} \right) \right)^2 \tag{14}$$

where $v_z^{[i]}$ is the depth of the vertex i, already recovered and $\mathbf{s}^{[i]}$ is the 3D position corresponding to the vertex i. Notice that this regularization is very dependent on the accuracy of $\mathbf{v}_z^{[i]}$.

5   Experiments

5.1   Synthetic Data

Next, we evaluate the methods described above using synthetic data. We synthetized a number of frames of a deforming circle-like paper (radius = 20 cm) approximated by a 9×9 mesh such as the one shown in Fig. 2. The reason to use a circular mesh is that it is uniform and has a symmetric shape. Therefore, it has similar shapes (up to a rotation) for a number of different deformations, which, in fact, brings more complexity to the reconstruction of the right deformation. The inextensible meshes used for training have been built using Blender and PCA was then applied to estimate the deformation model. In order to generate the input data, we get a sparse set of 3D feature points ($N = 32$) well-distributed on the surface of a reference planar mesh. The experiments are repeated equally for both the orthographic and the perspective cameras. For the perspective case, the camera model is defined such that the focal length is $f = 500$ pixels. The model assumes that the surface is located 50 cm in front of the cameras (along the optical axis). The 3D feature points across the surface are then projected onto the 2D camera and a zero-mean Gaussian noise with 1-pixel standard deviation (Std) was then added to these projections. The depth data of feature points is also generated by adding a zero-mean Gaussian noise with $0.1 - cm$ Std. The results of the quantitative assessment represent an average obtained from five deformations randomly selected. By performing 50 trials for each deformation, each average value was acquired from 250 trials. Two of the estimated deformations and their equivalent ground-truth are qualitatively illustrated in Fig. 3.

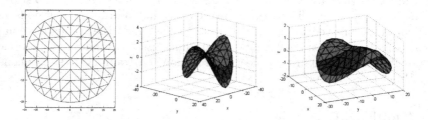

Fig. 3. Left: a 9×9 template mesh with sparse feature points - radius = 20 cm. Right: metric coordinates in cm - overlap between the ground-truth shapes (blue) and the recovered ones (red) (Color figure online).

Reconstruction Error. The accuracy of the method is reported in terms of reconstruction errors. The reconstruction errors are computed with respect to two measures as:

Table 1. Preliminary results.

Reconstruction error	PRE	MRE	RotationAccuracy
Our approach - orthographic	0.0608	0.0755	0.002
Our approach - perspective	0.0603	0.0751	$< 1°$

1- Point reconstruction error (PRE): The normalized Euclidean distance between the observed ($\hat{\mathbf{p}}_i$) and the estimated ($\mathbf{p}_i$) world points according to $PRE = \frac{1}{N} \sum_{i=1}^{N} \left[\|\mathbf{p}_i - \hat{\mathbf{p}}_i\|^2 / \|\hat{\mathbf{p}}_i\|^2 \right]$.

2- Mesh reconstruction error (MRE): The normalized Euclidean distance between the observed ($\hat{\mathbf{v}}_i$) and the estimated ($\mathbf{v}_i$) mesh vertices, which is computed as $MRE = \frac{1}{n_v} \sum_{i=1}^{n_v} \left[\|\mathbf{v}_i - \hat{\mathbf{v}}_i\|^2 / \|\hat{\mathbf{v}}_i\|^2 \right]$. The reprojection error of the feature points can be also regarded as another measure of precision. The accuracy of the Stiefel rotation matrix is evaluated based on the orthonormality constraint as $RotationAccuracy = \left\| \mathbf{R}^{(2 \times 3)} \mathbf{R}^{(2 \times 3)T} - \mathbf{I}^{(2 \times 2)} \right\|_F^2$. In case of the perspective camera, we compare the axis-angle of the recovered and ground-truth rotations as $RotationAccuracy = \left| angle - \hat{angle} \right|^2$.

The quantitative output can be seen in Table 1. Our approach takes into consideration just few feature points, though we take advantage of the ToF sensor to get the depth of them.

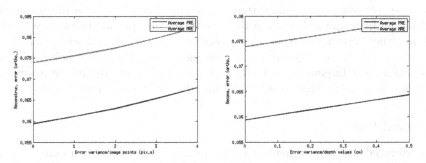

Fig. 4. Orthographic camera - left: average PRE and average MRE with respect to the increasing noise in image points. Right: average PRE and average MRE with respect to the increasing noise in depth data.

Length of the Edges. When a 3D surface is reconstructed in a truly inextensible way, the length of the recovered edges must be the same as that of the template edges. So, in order to see to what extent the lengths remain the same along the deformation path, we specify a metric to figure out the discrepancy between the initial and recovered lengths as: $IsometryExtent = \left(1 - \left(\frac{1}{n_e} \sum_{i=1}^{n_e} \left(\left|L_i - \hat{L}_i\right| / \hat{L}_i\right)\right)\right) \times 100\%$ which has been found to be 95.77 % for the proposed method, which indicates that it preserves the length of the edges greatly, confirming that isometry constraint is satisfied to a large degree.

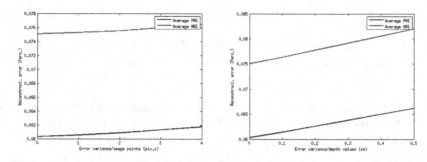

Fig. 5. Perspective camera - left: average PRE and average MRE with respect to the increasing noise in image points. Right: average PRE and average MRE with respect to the increasing noise in depth data.

The Impact of Noise. Different levels of noise (whether in image points or in depth data) have been simulated to demonstrate how robustly the approach reacts to the noise. Each of these 2 types of noise has been investigated separately. Figures 4 and 5 illustrate results for increasing levels of Gaussian noise in feature's image points, where the Std varied from 0 to 4 pixels with 1-pixel increments, together with the reconstruction error for various levels of Gaussian noise in depth of feature points, with $0.1 - cm$ increments of Std, which was computed following the remark that, since the depth variation of the surface itself is small, the deviations from the true depth of every 3D point may be very close together, varying at each trial according to a Gaussian distribution. From the Figs. 4 and 5, we may draw the conclusion that the white noise does not make a dramatic impact on the output, ensuring that the performance remains pretty stable and the algorithm carries on efficiently in the face of noise.

5.2   Real Data

We performed also experiments with real data recorded using a camera setup comprising a ToF camera and a RGB camera. The camera configuration is set up in a way that makes the FOV of the ToF camera be part of the FOV of the 2D camera and the camera setup was calibrated both internally and externally. Bilinear interpolation was applied to estimate the depth of each 2D point track. We used a piece of cardboard to make real inextensible deformations and proceeded with the tracking and matching of few feature points with respect to the reference template using SIFT local feature descriptor. The same deformation model as the one acquired in synthetic experiments was employed. Some deformations and their recovered shapes are shown in Fig. 7. Although it was not possible to quantitatively assess the results and do benchmarking, the efficiency of the approach was visible from the 3D reconstruction output.

5.3   Comparative Evaluations Using Motion Capture Data

Rather than generate the training data synthetically using Blender, we take advantage of datasets recorded using Vicon which is able to capture real deformations accurately. Since the synthetically deformed meshes might not exactly overlap the real deformations, we rebuilt the deformation model based on this real data and redid the experiments. The template configuration is now composed of equal triangles and covers a 20×20-cm square-like area. As an example, the reconstructed surfaces in Fig. 7 look better than the ones in Fig. 6. Consequently, when learned with real data, the deformation model would be more robust to the deformations.

As a general rule, two different entities can be compared only when they meet identical conditions which characterize them. To this end we analyzed the state-of-art literature and selected the approach described in [3]. In particular this approach also uses a triangular mesh and can use the same types of data sets required by our approach. As a result, to show how the real training data will influence the 3D reconstruction, we performed a set of simulations as we already did with Blender data and we compare the performance of our SfM framework to this approach (where the authors use a second-order cone program (SOCP) to accomplish the 3D reconstruction of inextensible surfaces). Their approach is known to be very robust and efficient, where a linear local deformation model integrates local patches into a global surface and requires many feature points distributed throughout the surface. To account for noise in our approach, like before, a Gaussian noise with 1-pixel Std was added to the image points and a Gaussian noise with $0.1 - cm$ Std to the depth data. The SOCP-based approach was evaluated without noise. We obtained the results for 5 deformations after having done 50 trials for each one. From the Table 2, it can be seen that the result of our approach is comparable to that of the SOCP-based method. The reconstruction errors are considerably lower than those in Table 1, which may imply that the use of good-quality real data for training might improve significantly the results.

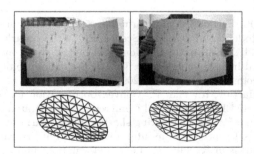

Fig. 6. Real deformations; A 20×20-cm square was selected from the intermediate part of the cardboard and the corresponding circle was reconstructed.

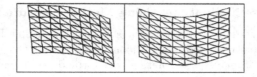

Fig. 7. The reconstructed shape of the corresponding squares in Fig. 6.

Table 2. Comparison between the proposed approach and the SOCP-based one.

Reconstruction error	PRE	MRE
Our approach	0.0120	0.0185
SOCP-based approach	0.0162	0.0217

6   Conclusions

In this paper, we have proposed a SfM framework combining a monocular camera and a ToF sensor to reconstruct surfaces which deform isometrically. The ToF camera was used to provide us with the depth of a sparse set of feature points, from which we can recover the depth of the mesh using a multivariate linear system. The key advantage of the RGB/ToF system is to benefit from the high-resolution RGB data in combination with the low-resolution depth information. As a result, our approach to inextensible surface reconstruction is formulated as an optimization problem on the basis of two non-linear constraints. Finally, we carried out a set of experiments showing that the approach generates good results. As next objective, we'll extend the approach to deal with non-rigid surfaces which are not isometric e.g. conformal surfaces and etc.

References

1. Aans, H., Kahl, F.: Estimation of deformable structure and motion. In: Workshop on Vision and Modelling of Dynamic Scenes, ECCV, Denmark (2002)
2. Del-Bue. A., Llad, X., Agapito, L.: Non-rigid metric shape and motion recovery from uncalibrated images using priors. In: IEEE Conference on Computer Vision and Pattern Recognition, New York (2006)
3. Salzmann, M., Hartley, R., Fua, P.: Convex optimization for deformable surface 3-D tracking. In: IEEE International Conference on Computer Vision (2007)
4. Salzmann, M., Fua, P.: Reconstructing sharply folding surfaces: a convex formulation. In: IEEE Conference on Computer Vision and Pattern Recognition (2007)
5. Salzmann, M., Urtasun R., Fua, P.: Local deformation models for monocular 3D shape recovery. In: IEEE Conference on Computer Vision and Pattern Recognition, pp. 1054–1061 (2008)
6. Gumerov, N., Zandifar, A., Duraiswami, R., Davis, L.S.: Structure of applicable surfaces from single views. In: Pajdla, T., Matas, J.G. (eds.) ECCV 2004. LNCS, vol. 3023, pp. 482–496. Springer, Heidelberg (2004)

7. Prasad, M., Zisserman, A., Fitzgibbon, A.W.: Single view reconstruction of curved surfaces. In: IEEE Conference on Computer Vision and Pattern Recognition, pp. 1345–1354 (2006)

8. Shen, S., Shi, W., Liu, Y.: Monocular 3-D tracking of inextensible deformable surfaces under L2-norm. IEEE Trans. Image Process. **19**, 512–521 (2010)

9. Perriollat, M., Hartley, R., Bartoli, A.: Monocular template-based reconstruction of inextensible surfaces. Int. J. Comput. Vis. **95**(2), 124–137 (2010)

10. Salzmann, M., Moreno-Noguer, F., Lepetit, V., Fua, P.: Closed-form solution to non-rigid 3D surface registration. In: Forsyth, D., Torr, P., Zisserman, A. (eds.) ECCV 2008, Part IV. LNCS, vol. 5305, pp. 581–594. Springer, Heidelberg (2008)

11. Metaxas, D., Terzopoulos, D.: Constrained deformable superquadrics and nonrigid motion tracking. PAMI **15**, 580–591 (1993)

12. Torresani, L., Hertzmann A., Bregler C.: Learning non-rigid 3d shape from 2d motion. In: NIPS, pp. 580–591 (2003)

13. Llado, X., Bue A.D., Agapito, L.: Non-rigid 3D factorization for projective reconstruction. In: BMVC (2005)

14. White, R., Forsyth, D.: Combining cues: shape from shading and texture. In: CVPR (2006)

15. Diebel, J., Thrun, S.: An application of markov random fields to range sensing. In: Proceedings of NIPS (2005)

16. Yang, R., Davis, J., Nister, D.: Spatial-Depth Super Resolution for Range Images. In: IEEE Conference on Computer Vision and Pattern Recognition, CVPR 2007, Minneapolis, pp. 1–8 (2007)

17. Kim, H., Tai, Y.W., Brown, M.S.: High quality depth map upsampling for 3D-TOF cameras. In: IEEE International Conference on Inso Kweon Computer Vision (ICCV), Barcelona, pp. 1623–1630 (2011)

18. Brunet, F., Hartley, R., Bartoli, A., Navab, N., Malgouyres, R.: Monocular template-based reconstruction of smooth and inextensible surfaces. In: Kimmel, R., Klette, R., Sugimoto, A. (eds.) ACCV 2010, Part III. LNCS, vol. 6494, pp. 52–66. Springer, Heidelberg (2011)

19. Akhter, I., Sheikh, Y., Khan, S.: In defense of orthonormality constraints for non-rigid structure from motion. In: CVPR, pp. 1534–1541 (2009)

20. Xiao, J., Chai, J., Kanade, T.: A closed-form solution to non-rigid shape and motion recovery. In: Pajdla, T., Matas, J.G. (eds.) ECCV 2004. LNCS, vol. 3024, pp. 573–587. Springer, Heidelberg (2004)

21. Zhou, H., Li, X., Sadka, A.H.: Nonrigid structure-from-motion from 2-D images using markov chain monte carlo. MultMed **14**(1), 168–177 (2012)

22. Srivastava, S., Saxena, A., Theobalt, C., Thrun, S.: Rapid interactive 3D reconstruction from a single image. In: VMV, pp. 9–28 (2009)

23. Paladini, M., Bue, A.D., Stosic, M., Dodig, M., Xavier, J., Agapito, L.: Factorization for non-rigid and articulated structure using metric projections. In: Proceedings of IEEE Conference on Computer Vision and Pattern Recognition, pp. 2898–2905 (2009)

24. Dai, Y., Li, H., He, M.: A simple prior-free method for non-rigid structure-from-motion factorization. In: CVPR, pp. 2018–2025 (2012)

25. Kim, Y.M., Theobalt, C., Diebel, J., Kosecka, J., Miscusik, B., Thrun, S.: Multi-view image and tof sensor fusion for dense 3D reconstruction. In: Computer Vision Workshops (ICCV Workshops), pp. 1542–1549 (2009)

Segmentation of Tomatoes in Open Field Images with Shape and Temporal Constraints

Ujjwal Verma[1]([✉]), Florence Rossant[2], Isabelle Bloch[3], Julien Orensanz[4], and Denis Boisgontier[4]

[1] Department of Electronics and Communication Engg, Manipal Institute of Technology, Manipal University, Manipal, Karnataka, India
ujjwal.verma@isep.fr, ujjwal.verma@manipal.edu
[2] Institut Superieur d'Electronique de Paris (ISEP), Paris, France
florence.rossant@isep.fr
[3] LTCI, CNRS, Télécom ParisTech, Université Paris-Saclay, Paris, France
isabelle.bloch@telecom-paristech.fr
[4] Cap2020, Gironville sur Essonne, France

Abstract. With the aim of estimating the growth of tomatoes during the agricultural season, we propose to segment tomatoes in images acquired in open field, and to derive their size from the segmentation results obtained in pairs of images acquired each day. To cope with difficult conditions such as occlusion, poor contrast and movement of tomatoes and leaves, we propose to base the segmentation of an image on the result obtained on the image of the previous day, guaranteeing temporal consistency, and to incorporate a shape constraint in the segmentation procedure, assuming that the image of a tomato is approximately an ellipse, guaranteeing spatial consistency. This is achieved with a parametric deformable model with shape constraint. Results obtained over three agricultural seasons are very good for images with limited occlusion, with an average relative distance between the automatic and manual segmentations of 6.46 % (expressed as percentage of the size of tomato).

Keywords: Image segmentation · Parametric active contours · Shape constraint · Precision agriculture

1 Introduction

Optimal harvesting date and predicted yield are valuable information when farming open field tomatoes, making harvest planning and work at the processing plant much easier. Monitoring tomatoes during their early stages of growth is also interesting to assess plant stress or abnormal development. Satellite data and crop growth modeling are generally used for estimating the yield of a large region [10,13]. However, satellite data are affected by adverse climatic conditions (clouds, etc.) resulting in inaccurate predictions [10]. Crop growth modeling, which integrates information regarding the cultivated plant, soil and

© Springer International Publishing Switzerland 2015
A. Fred et al. (Eds.): ICPRAM 2014, LNCS 9443, pp. 162–178, 2015.
DOI: 10.1007/978-3-319-25530-9_11

weather conditions, considers the ideal case with no infected plant. Recent studies have concentrated on combining these two approaches [19]. Nevertheless these methods depend on the quality of the different parameters involved (vegetation indices, soil and weather information) and they are not accurate enough to detect abnormal development.

In this work, we present a different approach where we intend to monitor the growth of tomatoes and measure their size in an open field. For this purpose, two cameras are installed in the field and two images are captured at regular intervals. In order to avoid a complete 3D reconstruction, we assume that a tomato can be approximated by a sphere in the 3D space, which projects into an ellipse in the image plane. Hence, the first part of our system aims at detecting and segmenting the tomatoes in both images, using elliptic approximations. Then, the second part aims at estimating the sphere radius, using the camera parameters. An estimate of the yield is obtained from this information. In this paper, we focus on the segmentation procedure only.

Computer vision algorithms have been applied in the agricultural domain in order to replace human operators with an automated system. They have been used to grade and sort agricultural products [4,7,11], to detect weeds in a field [2,9,18], and to model the growth of fruits and then predict the yield [1,14]. In [1], the yield of an apple orchard is estimated using only the density of flowers. In [14], only 5 images captured at different stages of the apple maturation are studied in order to predict the yield. These methods [1,14] are limited to a controlled environment (apple orchards) where complex scenarios such as occlusions are not considered. Moreover, in [1] the observed scene is modified by placing a black cloth behind the tree in order to simplify the image processing tasks. However, to the best of our knowledge, there has not been any related work where the growth of a fruit or vegetable cultivated in open fields is studied based on the images captured during the entire agricultural season.

Since there is little growth of a tomato during a given day, only one image per day is analyzed in this work, thus creating a series of approximately 20–30 images. One of the difficulties of the segmentation part is occlusion: most of the tomatoes are partially hidden by other tomatoes and/or leaves (Fig. 1). Moreover, color information is not of much use as tomatoes are red only at the end of the ripening. Also, another difficulty is a very low contrast in some cases due to shadows.

In this work, the segmentation should be as automatic as possible. However, we assume that an operator validates each obtained segmentation. If the result is poor, the operator rejects it. Indeed, given the difficulties, the segmentation is a very challenging task, and a manual validation is preferable. This approach enables us to use the segmentation done in the ith image (if validated) as a reference for the segmentation of the same tomato in the $(i + 1)^{th}$ image.

In order to segment the tomatoes, we use a parametric active contour model, which allows us to introduce a priori knowledge on the shape of the object to be segmented, thus making the segmentation more robust to noise and occlusion. Using an elliptic shape constraint is consistent with our prior assumption.

(a) 6th image.   (b) 7th image.

Fig. 1. Two successive images of the tomato $S = 7$.

The main steps of the segmentation algorithm are as follows: first, gradient information is used in order to find the candidate contour points and propose several elliptic approximations using the RANSAC algorithm. Secondly, region information is added, enabling us to select the best ellipse for the initialization of the active contour and finding the regions of potential occlusions. Thirdly, the active contour with elliptic constraint is applied. Finally, four ellipse estimates are computed. The operator has only to select the best one as the final segmentation.

The original features of the proposed algorithm include the approximation of the tomatoes as ellipses and the conditioning of the computation of the image energy by the non-occluded regions. These features allow coping with occlusions and local loss of contour and edges.

We present the active contour model with shape constraint in Sect. 2 and the different steps of the segmentation algorithm in Sect. 3. Section 4 discusses the experimental results. A brief discussion on the second part of the system which aims at estimating the radius of the tomatoes is presented in Sect. 5. This paper extends our premilinary work in [15].

2   Active Contour with an Elliptical Shape Prior

Parametric active contour model or snake was originally introduced in [8] in order to detect a boundary of an object in an image. This algorithm deforms the contour iteratively from its initial position towards the edges of an object by minimizing an energy functional. The energy functional associated with the contour v is usually composed of three terms:

$$\mathbf{E}_T(v) = \mathbf{E}_{Int}(v) + \mathbf{E}_{Im}(v) + \mathbf{E}_{Ext}(v), \tag{1}$$

where $\mathbf{E}_{Int}(v)$ is the internal energy controlling the smoothness of the curve and $\mathbf{E}_{Im}(v)$ is the energy derived from image data. The external energy $\mathbf{E}_{Ext}(v)$ can express contextual information, such as shape information. The authors in [6] used Legendre moments to define an affine invariant shape prior in a region based active contour. In our case, the region information is significant but not stable enough due to the presence of leaves (occlusions) and other tomatoes of similar intensity profile. In [3], Fourier descriptors are used in order to align the active contour with the reference curve of suitable shape and orientation. In our work, the tomato in each image is assumed to have an ellipse shape, which is included as a constraint in a parametric active contour model.

Let us define the reference ellipse as z_e. This ellipse is estimated from the evolving contour z. Both curves are expressed in polar coordinates with the origin at the center of z_e:

$$z_e(\theta) = r_e(\theta)e^{j\theta}, z(\theta) = r(\theta)e^{j\theta}, \theta \in [0, 2\pi]. \tag{2}$$

Our energy functional with an elliptic shape regularization is defined as:

$$\mathbf{E}_T(r, r_e) = \int_0^{2\pi} \frac{\alpha}{2}|r'(\theta)|^2 d\theta + \int_0^{2\pi} E_{Im}(r(\theta)e^{j\theta})d\theta + \frac{\psi}{2}\int_0^{2\pi} |r(\theta) - r_e(\theta)|^2 d\theta. \tag{3}$$

In the above equation, the first term represents the internal energy which controls the variations of r and makes it regular. The second term is a classical image energy calculated from the gradient vector flow [17]. The last term restricts the evolving contour to be close to the reference ellipse. The parameter α controls the smoothness of the curve, and ψ controls the influence of the shape prior on the total energy. Note that instead of modifying a 2-D vector $v(s) = (x(s), y(s))$ as in the classical active contour model, only a 1-D vector $r(\theta)$ is modified for each value of the parameter θ. Moreover, the shape constraint makes the usual second derivative term in the internal energy useless, and is therefore not included in the proposed energy functional.

The minimum of $\mathbf{E}_T$ is obtained in two steps: first, a least square estimate of the ellipse z_e is computed from the initial contour z_0. Then, the evolving contour z is computed by minimizing $\mathbf{E}_T$ while assuming z_e fixed. From the evolving contour z so obtained, the parameters of the least square estimate of the ellipse z_e are regularly updated. This two-step iterative process is repeated, in order to obtain the minimum of $\mathbf{E}_T$.

The minimization of $\mathbf{E}_T$ with respect to r is equivalent to solving the following Euler equation:

$$-\alpha r''(\theta) + \nabla E_{Im}(\theta) \cdot n(\theta) + \psi(r(\theta) - r_e(\theta)) = 0, \tag{4}$$

where $n(\theta) = [\cos\theta, \sin\theta]^T$.

To find iteratively a solution of this equation, we introduce a time variable, and the resulting equation is discretized using finite differences, as in the case of the classical active contours.

3   Detailed Algorithm

In this section, we present an algorithm which allows us to follow the growth of a tomato, which has been manually segmented in the first image ($i = 1$).

Let us denote by im^{i+1} the $(i+1)^{th}$ image of the tomato S. In the rest of this paper, an ellipse centered at $[xc, yc]$, whose semi major and minor axes lengths are a and b, respectively, and which has a rotation angle of φ, is represented as $Ell = [xc, yc, a, b, \varphi]$. The tomato approximated by an ellipse in im^i is represented as $Ell^i = [xc^i, yc^i, a^i, b^i, \varphi^i]$. In our sequential approach, the computation of the contour in the $(i+1)^{th}$ image is based on both the information in im^{i+1} and the contour of the tomato in the ith image. The temporal regularization (assuming little growth and movement of the tomato during a day) and the spatial regularization (tomato modeled as a sphere in the 3D space) are used throughout the segmentation procedure.

3.1 Pre-processing

As mentioned above, the color information is not of much use. However, the edges of tomatoes are more prominent in the red component of the image, and hence only this component is considered. The original image is cropped around the position (xc^i, yc^i), resulting in a smaller image (imS_c^{i+1}). The contrast is enhanced by a contrast stretching transformation.

3.2 Updating the Tomato Position

Due to its increasing weight, the tomato tends to fall towards the ground (Fig. 2). Its position in imS_c^{i+1} is calculated using pattern matching. The bright areas, that may correspond to the tomato, are extracted by convolving the cropped image with a binary mask representing a white disk of radius χr^i where $r^i = \frac{a^i + b^i}{2}$ and χ is a constant determined empirically ($\chi = 1.25$ in our experiments). The local maxima $C_c^{i+1} = \{(x_k, y_k), k = 1, ..., k_n\}$ are then extracted. From these k_n points, the one $C_m = (x_m, y_m)$, which is the closest to (xc^i, yc^i), is selected as the new location of the tomato center (Fig. 2(b)).

A new cropped image imS^{i+1} is then extracted from im^{i+1}, centered at $C_m = (x_m, y_m)$. The size of this new image is adapted to the size of the tomato (derived from a^i and b^i) so that we restrict the region to be analyzed as much as possible, thus reducing the computation cost of the next steps. The contrast stretching transformation is applied to imS^{i+1}.

3.3 Elliptic Approximations

In order to obtain an initial contour for the active contour model, we first compute l_n points which may lie on the boundary of the tomato. From these l_n points, a RANSAC estimate is used to obtain several candidate ellipses. Finally, one of these ellipses is selected as the initial contour based on additional region information and size regularization.

Let us take C_m as the origin of the polar coordinate system. Then we select l_n points $P_l = p_l e^{j\theta_l}$, where $l = 1, ..., l_n$, $0 < \theta_l < 2\pi$, that satisfy the following

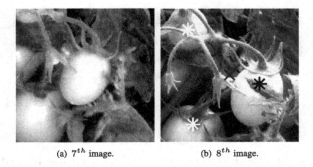

(a) 7th image.   (b) 8th image.

Fig. 2. Updating the position of the tomato: previous position (xc^i, yc^i) in red, candidate positions C_c^{i+1} in magenta and blue, and new position C_m in blue (Color figure online).

three conditions:

$$0.5r^i < p_l < 1.5r^i \tag{5}$$

$$|\arg(\nabla imS^{i+1}(P_l)) - \theta_l| \leq \tfrac{\pi}{8} \tag{6}$$

$$|\nabla imS^{i+1}(P_l)| > \eta \tag{7}$$

where $\nabla imS^{i+1}(P_l)$ is the gradient at P_l in imS^{i+1} and η is a constant whose value is determined experimentally ($\eta = 0.2$). The above conditions select the points of strong gradient whose direction is within an acceptable limit with respect to the vector normal to the circle with radius r^i. The threshold values have been set experimentally. As shown in Fig. 4(a), most points lying on the boundary of the tomato have been correctly detected along with some additional points lying on the leaves.

A least square estimate of an ellipse calculated from *all* l_n points might result in a contour far away from the actual boundary because of the detection of irrelevant points. Therefore, we use a RANSAC [5] estimate based on an elliptic model in order to compute several candidate ellipses. Note that the spatial (tomato modeled as an ellipse) and temporal (parameters of the model) regularization has been used in this step to increase the robustness of the segmentation procedure.

Under normal circumstances, the size and the orientation of the tomato in imS^{i+1} are supposed to be close to the ones in imS^i. This information is incorporated in the RANSAC estimation and only the ellipses whose parameters satisfy the following conditions are considered:

$$-0.1 < \tfrac{a^{i+1}-a^i}{a^i} < 0.2, -0.1 < \tfrac{b^{i+1}-b^i}{b^i} < 0.2 \tag{8}$$

$$-0.1 < \tfrac{SA^{i+1}-SA^i}{SA^i} < 0.25 \tag{9}$$

$$|\tfrac{Ecc^{i+1}-Ecc^i}{Ecc^i}| < 0.1 \tag{10}$$

$$|\tfrac{\varphi^{i+1}-\varphi^i}{\varphi^i}| < 0.2 \tag{11}$$

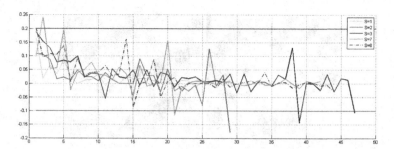

Fig. 3. Evolution of a. The abscissa represents the image number (i), and the ordinate represents $\frac{a^{i+1}-a^i}{a^i}$. The solid horizontal red lines show the selected threshold values (Color figure online).

where SA^{i+1} and SA^i represent the surface of the ellipses in imS^{i+1} and imS^i respectively. The eccentricity ($Ecc = \frac{a}{b}$) for the two ellipses is denoted by Ecc^{i+1} and Ecc^i respectively.

Negative variations for a and b (Eq. 8) are possible because of the movement of the tomato with respect to the camera or because of the variation in the orientation, as tomatoes are actually not perfect spherical objects. Equation 9 restricts the apparent size of the tomato while Eq. 10 restricts the admissible values for eccentricity, thus controlling the apparent shape of the tomato.

The threshold values in Eqs. 8–11 have been determined after studying the parameters of the ellipses obtained from the manual segmentation of five tomatoes. For example, Fig. 3 shows the relative evolution of the length of semi-major axis a of the ellipses. Most of the measurements are situated within the limits defined above. Note that the dissymmetry in the lower and upper bounds in Eqs. 8–9 is due to the fact that tomatoes are supposed to grow during the agricultural season.

From the N ellipses computed using the RANSAC algorithm, a total of N_a ellipses, with $N_a < N$, are retained, corresponding to the N_a ellipses with the largest number of inliers (Fig. 4(b)).

3.4   Adding Region Information

A region growing algorithm is applied in order to add region information and determine the best initialization for the active contour among the N_a ellipses Ell_u^{i+1}, where $u = 1, ..., N_a$. Moreover, potential occlusions are also derived from this information.

Let us denote by ω_u the binary image representing the region inside the ellipse Ell_u^{i+1}. We apply a classical region growing algorithm starting from ω_{seed} and limiting the growing to ω_{limit}, where:

$$\omega_{seed} = \bigcap_{u=1}^{N_a} \omega_u, \quad \omega_{limit} = \bigcup_{u=1}^{N_a} \omega_u. \tag{12}$$

The final region is denoted by ω_t (Fig. 4(c)).

We define τ_m as:

$$\tau_m = \min_{u=1,2,\dots N_a} \tau(u), \tag{13}$$

with

$$\tau(u) = \frac{|\omega_u \cap (1 - \omega_t)| + |(1 - \omega_u) \cap \omega_t|}{|\omega_u \cap \omega_t|}, \tag{14}$$

where $|A|$ represents the cardinality of a set A. The ratio $\tau(u)$ measures the consistency between the segmentation obtained through the contour analysis ω_u and the region analysis ω_t. It reaches a minimum (zero) when ω_u and ω_t match perfectly.

Let us denote by a_u^{i+1} and b_u^{i+1} the semi-axis lengths of the candidate ellipse $Ell_u^{i+1}, u \in [1, N_a]$. We select the ellipse v (Fig. 4(d)) that minimizes $\left[\left(a_u^{i+1} - a^i \right)^2 + \left(b_u^{i+1} - b^i \right)^2 \right]$ under the condition $\tau(v) \leq 1.1 \, \tau_m$. Thus, we have obtained the initial contour by combining the results obtained using two different segmentation methods, one based on boundary information and the other based on region information. The selected ellipse Ell_v^{i+1} is chosen among the ones for which both results are consistent, leading to a better robustness with respect to occlusions. Moreover, another regularization condition is added, which imposes that the size and shape of the ellipse in imS^{i+1} are close to the ones in imS^i.

The next step aims at finding the regions where occlusions could disturb the behavior of the active contour. For example, the region in which the tomato is attached to the plant has a different intensity from the one of the tomato.

Let Ell_{te} denote the ellipse which covers the convex hull of ω_t and which minimizes the number of pixels inside the ellipse Ell_{te} and not belonging to the region ω_t (Fig. 4(e)). Let ω_{te} be the region inside Ell_{te}. Then, the region of occlusion ω_{oc} can be computed as $\omega_{oc} = \omega_{te} \cap \omega_t^c$.

Using morphological operations (erosion followed by reconstruction by dilation), small regions are removed from ω_{oc}, so that the resulting ω_{oc} corresponds to actual leaves causing the occlusions (Fig. 4(f)). Apart from detecting the "head" of the tomato, any other additional occlusion (mostly due to leaves) can also be detected using this approach (Fig. 4(f)).

3.5   Applying Active Contours

The active contour (Sect. 2) is applied with the following initialization $Ell_{vc}^{i+1} = [xc_v^{i+1}, yc_v^{i+1}, 0.95a_v^{i+1}, 0.95b_v^{i+1}, \varphi_v^{i+1}]$. Indeed, the movement of the curve z is smoother and faster if initialized inside the tomato. For the first n_{start} iterations, the parameter ψ is set to zero, so that z moves towards the most prominent contours. Then the shape constraint is introduced for $n_{ellipse}$ iterations ($\psi \neq 0$) in order to guarantee robustness with respect to occlusion. Finally, the shape constraint is relaxed ($\psi = 0$) for a few n_{end} iterations, which guarantees reaching the boundary more accurately, as a tomato is not a perfect ellipse.

Note that the image forces are not considered in the region of occlusion ω_{oc}, in every step of this process.

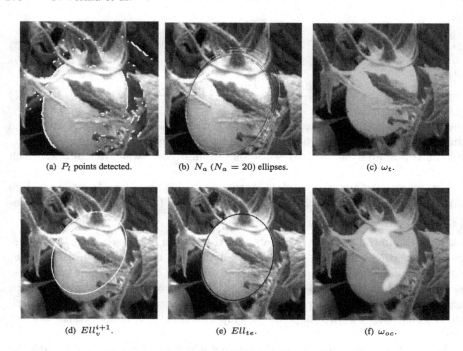

(a) P_l points detected.   (b) N_a ($N_a = 20$) ellipses.   (c) ω_t.

(d) Ell_v^{i+1}.   (e) Ell_{te}.   (f) ω_{oc}.

Fig. 4. (a,b) Points of strong gradient and ellipses detected using the RANSAC estimate. (c,d) ω_t: region representing tomato, Ell_v^{i+1}: selected initial ellipse. (e,f) Ell_{te}:
convex hull of ω_t and region of potential occlusion ω_{oc}.

As explained in Sect. 2, the reference ellipse z_e is regularly updated, every
n_{shape} iterations. A least square estimate calculated from all the points of the
curve z is not relevant, because some of them may lie on false contours (e.g.
leaves). So, the following algorithm aims at selecting a subset of points that
actually lie on the boundary of the tomato.

We use a polar coordinate system with the origin at the center of the current
reference ellipse z_e. As in Sect. 2, let us denote by $z(\theta) = r(\theta)e^{j\theta}$ a point of
the evolving curve, $z_e(\theta) = r_e(\theta)e^{j\theta}$ the corresponding point on the reference
ellipse, $n_e(\theta)$ the vector normal to the ellipse z_e, and $z_q(\theta) = r_q(\theta)e^{j\theta}$ the point
that maximizes the gradient module for $0 < r_q(\theta) < 1.1r_e(\theta)$. The point $z(\theta)$ is
selected as a point lying on the boundary of the tomato if it satisfies the following
conditions:

$$|\nabla imS^{i+1}(z(\theta)) \cdot n_e(\theta)| > \Gamma \tag{15}$$

$$\frac{|\nabla imS^{i+1}(z(\theta)) \cdot n_e(\theta)|}{|\nabla imS^{i+1}(z(\theta))|} > 0.75 \tag{16}$$

$$d(z_q(\theta), z(\theta)) < d_{max} \tag{17}$$

where · represents the vector dot product.

The first condition ensures that the magnitude of the gradient vector projected onto the normal of the ellipse is strong. The threshold Γ is determined

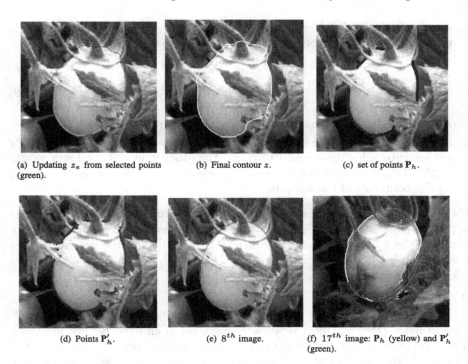

(a) Updating z_e from selected points (green).

(b) Final contour z.

(c) set of points $\mathbf{P}_h$.

(d) Points $\mathbf{P}'_h$.

(e) 8th image.

(f) 17th image: $\mathbf{P}_h$ (yellow) and $\mathbf{P}'_h$ (green).

Fig. 5. (a,b) Active contour with shape constraint. (c,d) Two different sets of points $\mathbf{P}_h$ and $\mathbf{P}'_h$. (e,f) Final ellipse estimates for two different images (Color figure online).

automatically [12]. The second condition ensures that the direction of the gradient is close to the vector normal to the ellipse. The last condition ($d_{max} = 2$ in our experiments) imposes that the considered point is a meaningful local maximum of the gradient. Finally, the parameters of the reference ellipse are updated by calculating a least square approximation from the subset of points lying on the evolving contour z selected using the above conditions (Fig. 5(a)).

3.6 Refining the Results

A least square estimate of an ellipse from z (Fig. 5(b)) is generally not relevant as outliers may be present due to occlusion. So, again, a selection procedure is applied. A first subset of points $\mathbf{P}_h$ (Fig. 5(c)) is obtained by using criteria similar to the ones described in Sect. 3.5 (Eqs. 15–17). Then, another subset $\mathbf{P}'_h$ is computed by relaxing the condition related to the gradient direction (Fig. 5(d)).

Then four ellipses are computed as follows:

1. A least square approximation $Ell_{f1}^{i+1} = [xc_{f1}^{i+1}, yc_{f1}^{i+1}, a_{f1}^{i+1}, b_{f1}^{i+1}, \varphi_{f1}^{i+1}]$ is computed from all the points of z.

2. Another estimate $Ell_{f2}^{i+1} = [xc_{f2}^{i+1}, yc_{f2}^{i+1}, a_{f2}^{i+1}, b_{f2}^{i+1}, \varphi_{f2}^{i+1}]$ is obtained from $\mathbf{P}'_h$ using the RANSAC algorithm with the following conditions:

$$0.9a_{f1}^{i+1} < a_{f2}^{i+1} < 1.1a_{f1}^{i+1} \tag{18}$$

$$0.9b_{f1}^{i+1} < b_{f2}^{i+1} < 1.1b_{f1}^{i+1} \tag{19}$$

3. A least square approximation Ell_{f3}^{i+1} is obtained from the subset $\mathbf{P}_h$.
4. A weighted least square estimate Ell_{f4}^{i+1} is obtained where the points of $\mathbf{P}_h$ are assigned a higher weight (0.75) and the other points of z a lower weight (0.25). This is done in order to give importance to the points that are surely on the boundary of the tomato.

If the images have a good contrast, and little or no occlusion, all the four ellipses will be almost identical (Fig. 5(e)). However, in case of occlusions and poor contrast, the four ellipses may be different (Fig. 5(f)), and the user selects the best one.

4　Results

Two cameras (Pentax Optio W80) were installed in an open field of tomatoes. The same setup was used for three agricultural seasons (April-August, 2011, 2012 and 2013). We have identified 21 tomatoes, covering different sites and different seasons, thus ensuring variability (614 images in total). The tomatoes were identified manually by observing the images of the entire agricultural season. Due to the severe occlusions, only a limited number of tomatoes were visible in most of the images of a given season. Therefore, only the tomatoes which were visible in more than 10 consecutive images were studied.

As discussed earlier, one of the main challenges of the segmentation is the occlusion and the poor quality of the images due to the poor illumination and/or shadow. Moreover, for the images acquired in the 2013 agricultural season ($S = 12, .., 21$) the size of the tomatoes was significantly smaller as compared to the one observed during the agricultural seasons in 2011 and 2012 ($S = 1, ..., 11$). This is due to the variation in the external climatic conditions. Also, in some images, a shadow created by the leaves (or the tomato itself) can be observed ($S = 8, 12$). As a result, a portion of the contour is not clear and distinct. This results in an ambiguity on the position of the contour. Given this ambiguity, even a manual segmentation is a challenging task on this portion of the contour. Moreover, a blurred contour was observed in some images of some sequences ($S = 3, 13, 18, 19, 20$), due to the presence of additional neighboring tomatoes in the background.

The data set contains images of varying contrast and degree of occlusions. Obviously, it is impossible to obtain a reliable segmentation, even manually, in case of severe occlusion. Consequently we studied experimentally the effect of the percentage of occlusion on the final estimation of the radius of a spherical object (considering the complete system, segmentation and partial 3D reconstruction).

In our experiments, the percentage P of occlusion corresponds to the occlusion of an arc with subtended angle equal to $\frac{2\pi P}{100}$. For less than 30 % occlusion, the variation in the estimated radius was very small, and for more than 30 % occlusion, significant change in the values of the estimated radius was observed. Thus, we identified three different categories:

– Category 1, containing images with an amount of occlusion P less than 30 % for which the estimation is very robust with respect to segmentation impreci-sion,
– Category 2, with 30 % $< P <$ 50 % which is more prone to segmentation error,
– Category 3, with $P >$ 50 % for which it is impossible to perform a reliable segmentation.

The percentage of occlusion was determined manually by selecting the end points of the occluded elliptic arc. Note that the percentage of occlusion was computed only to evaluate the segmentation procedure, and this is not a part of the algo-rithm.

The obtained segmentations A were compared with the manual segmenta-tions M (approximated by ellipses) by computing the average D^i_{mean} and maxi-mal D^i_{max} distances between A and M for the ith image (expressed in pixels). In order to better interpret the results, the maximum and mean distances between two contours are normalized by the size of the tomato as:

$$D^i_{meanR} = \frac{D^i_{mean}}{r^i}100, \; D^i_{maxR} = \frac{D^i_{max}}{r^i}100. \tag{20}$$

For this project, $D_{meanR} <$ 10 % is considered as the acceptable limit of error in order to follow the growth of tomatoes.

For the images of category 1, good results (Table 1) were obtained even in the presence of occlusion by nearby leaves/branches and tomatoes (Figs. 6(a) and 6(b)). In some images captured at the beginning of the season, when the size is very small, the occlusion due to leaves present on the "head" of the tomato results in an ambiguity on the position of the actual contour (Fig. 6(c)). Also, in some images (Fig. 6(d)), due to a shadow effect on a portion of the contour, the intensity profiles of the tomato and the adjacent leaves are nearly identical, resulting in a very low contrast. Such cases may result in comparatively high distance measures even in the absence of any occlusion.

Due to the smaller size of tomatoes in sequences $S = 12, ..., 21$, higher dis-tances were observed in these sequences (since the distances D^i_{meanR} and D^i_{maxR} are normalized by the size r^i of the tomato). For example, Fig. 6(e) shows the obtained segmentation Ell_{f4} on the 3rd image of sequence $S = 17$. The distances normalized by the size of the tomato are $D_{meanR} = 9.89$ % and $D_{maxR} = 26.60$ %. However, the distances expressed in pixels are significantly lower ($D_{mean} = 2$, $D_{max} = 5.38$ pixels). For most of the sequences a low $\mu_{D_{meanR}}$ along with lower $\sigma_{D_{meanR}}$ demonstrates the robustness of the proposed method. However, for some sequences ($S = 13, 17, 18$) higher values of $\mu_{D_{meanR}}$ and $\sigma_{D_{meanR}}$ were observed mainly due to the false detection of the position of the tomato (Sect. 3.2), or

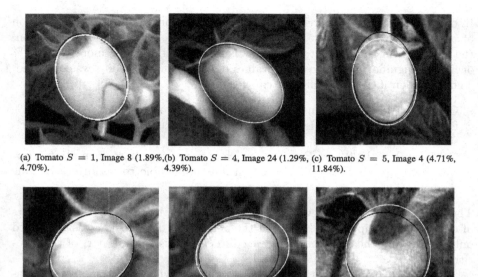

(a) Tomato $S = 1$, Image 8 (1.89%,(b) Tomato $S = 4$, Image 24 (1.29%, (c) Tomato $S = 5$, Image 4 (4.71%, 4.70%).   4.39%).   11.84%).

(d) Tomato $S = 8$, Image 5 (4.77%, (e) Tomato $S = 17$, Image 3 (9.89%, (f) Tomato $S = 12$, Image 7 (8.13%, 16.04%).   26.61%).   19.80%).

Fig. 6. Final segmentation Ell_{f4} (red) obtained on images of category 1, and values of (D_{meanR}, D_{maxR}) with respect to the manual segmentation (in cyan) (Color figure online).

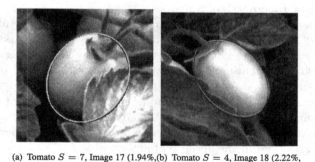

(a) Tomato $S = 7$, Image 17 (1.94%,(b) Tomato $S = 4$, Image 18 (2.22%, 5.45%).   6.16%).

Fig. 7. Final segmentation Ell_{f4} (red) obtained on images of category 2 (Color figure online).

due to the small size of the tomato, as discussed previously. For the sequence $S = 11$, all the images suffer from poor contrast and noise due to the shadow created by leaves. As a result, even a manual segmentation is a challenging task in this sequence.

Table 1. Mean (μ) and standard deviation (σ) of D_{meanR} and D_{maxR} by comparing ellipse Ell_{f4} and Ell_{opt} with the manual segmentation M. Only the images belonging to category 1 (i.e. with a low amount of occlusion) have been considered.

	N_S^1	Ell_{f4}				Ell_{opt}			
		$\mu_{D_{meanR}}$	$\sigma_{D_{meanR}}$	$\mu_{D_{maxR}}$	$\sigma_{D_{maxR}}$	$\mu_{D_{meanR}}$	$\sigma_{D_{meanR}}$	$\mu_{D_{maxR}}$	$\sigma_{D_{maxR}}$
$S=1$	26	1.72	0.77	5.06	2.76	1.34	0.68	3.76	2.21
$S=2$	4	1.85	0.46	5.45	1.91	1.57	0.40	4.43	0.88
$S=3$	21	3.4	2.24	9.79	6.88	2.87	2.05	8.40	6.58
$S=4$	14	2.73	1.92	7.81	5.71	2.20	1.77	6.28	5.34
$S=5$	5	4.81	1.3	13.05	3.56	4.54	1.14	12.44	3.46
$S=6$	0	-	-	-	-	-	-	-	-
$S=7$	25	1.88	0.65	4.81	1.97	1.7	0.49	4.61	1.62
$S=8$	20	6.07	5.75	15.41	10.61	5.4	4.88	14.9	10.37
$S=9$	1	5.26	0	11.86	0	5.24	0.00	11.86	0.00
$S=10$	5	2.25	0.56	6.59	2.25	1.75	0.36	4.40	1.25
$S=11$	4	11.81	4.99	32.66	11.39	10.18	5.38	29.21	14.74
$S=12$	19	4.74	1.33	11.69	3.23	4.25	1.34	11.1	3.44
$S=13$	5	41.5	16.55	84.98	27.57	40.48	16.18	83.51	26.85
$S=14$	4	9.57	2.35	29.48	7.3	9.18	2.57	28.06	7.84
$S=15$	0	-	-	-	-	-	-	-	-
$S=16$	21	4.68	1.08	10.22	2.74	4.46	1.18	10.12	2.68
$S=17$	20	11.78	2.44	26.75	4.21	11.56	2.48	26.85	4.04
$S=18$	23	14.18	20.06	35.76	33.83	13.94	20.09	35.15	33.42
$S=19$	0	-	-	-	-	-	-	-	-
$S=20$	5	8.76	5.38	20.05	14.93	6.88	2.3	13.65	2.98
$S=21$	25	7.34	3.18	16.56	8.93	7.12	3.15	15.76	8.94

For the images of category 2 containing a significant amount of occlusion, D_{meanR} is significantly higher than for images of category 1. This is because of heavy occlusions along with the poor quality of images (effects due to shadow and/or presence of other tomatoes). However, in some sequences ($S = 1, 2, 5, 6, 8, 9, 15, 18, 20, 21$) an average D_{meanR} of less than 10 % was observed. This is because of the good contrast on the non-occluded arc in these images, which results in a good segmentation. Finally, good results were obtained on 73 % of the images (Fig. 7), where $D_{meanR} < 10\%$ even in the presence of severe occlusions.

In the results presented so far, Ell_{f4} was compared with the manual segmentation. However Ell_{f4} is not necessarily the best ellipse, and was selected here for illustrative purpose only. Due to the variation in the contrast and occlusion, there is not a single ellipse (among the four ellipse estimates) which represents a good segmentation for *all* the images. Let us denote by Ell_{opt} the ellipse, among the four ellipse estimates (Ell_{f1}, Ell_{f2}, Ell_{f3} and Ell_{f4}), for which $\mu_{D_{meanR}}$ is minimum. Table 1 shows the distribution of D_{meanR} and D_{maxR} for ellipse Ell_{opt}. It can be observed that the values of D_{meanR} and D_{maxR} for Ell_{opt} are lower than for those of Ell_{f4}. The operator selects Ell_{opt} as the final segmentation.

5   Estimating the Size of the Tomato

From the obtained segmentation in both images and the camera parameters, we then estimate the size of the tomatoes. However, determining the image point pairs which correspond to the same point in the 3D space is a challenging task given the complexity of the scene. Instead we simplify the size estimation procedure by exploiting the spherical hypothesis.

The contour of the tomato is approximated by an ellipse whose parameters are calculated using the procedure presented above. Then, the sphere center in the 3D space is computed using triangulation from the centers of the ellipse calculated in both images. Next, the 3D space points situated on the contour of the tomato are computed using properties of projective geometry, independently from each image. Finally, a joint optimization procedure enables us to estimate the sphere radius.

In order to evaluate the size estimation procedure, the size of tomatoes observed in laboratory was measured. Since a tomato is not a perfect sphere, two reference values were measured manually and compared with the estimated radius of the sphere. For the manually segmented tomatoes observed in laboratory, we found that the relative percentage error between the largest of the reference value and the estimated radius was less than 5 % in 91 % of the cases. For the tomatoes cultivated in the open field, the relative percentage error was less than 10 % in 80 % of the cases [16]. The errors are mainly caused by the imperfect segmentation, due to shadowing effect and the poor quality of the images.

6   Conclusions

We presented a segmentation procedure used to monitor the growth of tomatoes from images acquired in an open field. Starting from an approximate computation of the position of the center of the tomato, segmentation algorithms based on contour and region information are proposed and combined, in order to determine a first estimate of the contour. Then, a parametric active contour with shape constraint is applied and four ellipse estimates representing the tomatoes are obtained. In all the steps of this process, a priori knowledge about the shape and the size of the tomatoes is modeled and incorporated as regularization terms, leading to a better robustness. It is supposed that the operator selects, at the end of the process for each image, the ellipse corresponding to the best elliptic estimation of the actual contour.

The segmentation of tomatoes is a challenging task due to the presence of occlusion and variation in contrast. In order to evaluate the robustness of the proposed algorithm, the entire image set was divided into three categories based on the amount of occlusion. For the images with an acceptable level of occlusion, good results were obtained on most (87 %) of images where D_{meanR} was less than 10 %. Also, the low standard deviation for D_{meanR} indicates the robustness of the proposed algorithm. Good results with $D_{meanR} < 10$ % were obtained on 73 % of the images that contain a significant amount of occlusion.

For the moment, it has been assumed that an operator manually selects one ellipse as the final segmentation. In future work, we wish to provide automatically the best representation of the tomato. Also, in some images, the position of the tomato is not detected correctly due to the presence of other tomatoes nearby. This could be improved by updating the position of the tomato globally by considering also the movement of adjacent tomatoes. One possible improvement for the active contour model is to restrict the size of the reference ellipse, as there is little growth between two consecutive images.

Acknowledgements. This work was partly supported by the MCUBE project (European Regional Development Fund (ERDF)), which aims at integrating multimedia processing capabilities in a classical Machine to Machine (M2M) framework, thus allowing the user to remotely monitor an agricultural field. The authors would like to thank Jérôme Grangier, for his participation in this project.

References

1. Aggelopoulou, A., Bochtis, D., Fountas, S., Swain, K., Gemtos, T., Nanos, G.: Yield prediction in apple orchards based on image processing. J. Precis. Agric. **12**, 448–456 (2011)
2. Aitkenhead, M., Dalgetty, I., Mullins, C., McDonald, A., Strachan, N.: Weed and crop discrimination using image analysis and artificial intelligence methods. Comput. Electron. Agric. **39**(3), 157–171 (2003)
3. Charmi, M., Ghorbel, F., Derrode, S.: Using Fourier-based shape alignment to add geometric prior to snakes. In: ICASSP, pp. 1209–1212 (2009)
4. Du, C., Sun, D.: Learning techniques used in computer vision for food quality evaluation: a review. J. Food Eng. **72**(1), 39–55 (2006)
5. Fischler, A., Bolles, C.: Random sample consensus: a paradigm for model fitting with applications to image analysis and automated cartography. Commun. ACM **24**(6), 381–395 (1981)
6. Foulonneau, A., Charbonnier, P., Heitz, F.: Affine-invariant geometric shape priors for region-based active contours. IEEE Trans. Pattern Anal. Mach. Intell. **28**(8), 1352–1357 (2006)
7. Jayas, D., Paliwal, J., Visen, N.: Review paper (automation and emerging technologies): multi-layer neural networks for image analysis of agricultural products. J. Agric. Eng. Res. **77**(2), 119–128 (2000)
8. Kass, M., Witkin, A., Terzopoulos, D.: Snakes: active contour models. Int. J. Comput. Vis. **1**(4), 321–331 (1988)
9. Lee, W.S., Slaughter, D.C., Giles, D.K.: Robotic weed control system for tomatoes. Precis. Agric. **1**, 95–113 (1999)
10. Mkhabela, M., Bullock, P., Raj, S., Wang, S., Yang, Y.: Crop yield forecasting on the Canadian prairies using MODIS NDVI data. Agric. Forest Meteorol. **151**(3), 385–393 (2011)
11. Narendra, V.G., Hareesh, K.S.: Prospects of computer vision automated grading and sorting systems in agricultural and food products for quality evaluation. Int. J. Comput. Appl. **1**(4), 1–9 (2010)
12. Otsu, N.: A threshold selection method from gray-level histograms. Automatica **11**(285–296), 23–27 (1975)

13. Prasad, A., Chai, L., Singh, R., Kafatos, M.: Crop yield estimation model for Iowa using remote sensing and surface parameters. Int. J. Appl. Earth Obs. Geoinf. **8**(1), 26–33 (2006)
14. Stajnko, D., Cmelik, Z.: Modelling of apple fruit growth by application of image analysis. Agric. Conspec. Sci. **70**, 59–64 (2005)
15. Verma, U., Rossant, F., Bloch, I., Orensanz, J., Boisgontier, D.: Shape-based segmentation of tomatoes for agriculture monitoring. In: International Conference on Pattern Recognition Applications and Methods (ICPRAM), Angers, France, pp. 402–411 (2014)
16. Verma, U., Rossant, F., Bloch, I., Orensanz, J., Boisgontier, D.: Tomato development monitoring in an open field, using a two-camera acquisition system. In: 12th International Conference on Precision Agriculture (2014)
17. Xu, C., Prince, J.: Snakes, shapes, and gradient vector flow. IEEE Trans. Image Process. **7**(3), 359–369 (1998)
18. Yang, C., Prasher, S., Landry, J., Ramaswamy, H., Ditommaso, A.: Application of artificial neural networks in image recognition and classification of crop and weeds. Can. Agric. Eng. **42**(3), 147–152 (2000)
19. Zhao, H., Pei, Z.: Crop growth monitoring by integration of time series remote sensing imagery and the WOFOST model. In: 2013 Second International Conference on Agro-Geoinformatics (Agro-Geoinformatics), pp. 568–571 (2013)

Fast and Accurate Pedestrian Detection in a Truck's Blind Spot Camera

Kristof Van Beeck[1,2]([envelope]) and Toon Goedemé[1,2]

[1] EAVISE, KU Leuven - Campus De Nayer, J. De Nayerlaan 5, 2860
Sint-katelijne-waver, Belgium
[2] ESAT/PSI - VISICS, KU Leuven, Kasteelpark Arenberg 10, 3001 Leuven, Belgium
{kristof.vanbeeck,toon.goedeme}@kuleuven.be

Abstract. We propose a multi-pedestrian detection and tracking framework targeting a specific application: detecting vulnerable road users in a truck's blind spot zone. Existing safety solutions are not able to handle this problem completely. Therefore we aim to develop an active safety system which warns the truck driver if pedestrians are present in the truck's blind spot zone, using solely the vision input from the truck's blind spot camera. This is not a trivial task, since—aside from the large distortion induced by such cameras—the application inherently requires real-time operation while at the same time attaining very high accuracy. To achieve this, we propose a fast and efficient pedestrian detection and tracking framework based on our novel *perspective warping window* approach. Experiments on real-life data show that our approach achieves excellent accuracy results at real-time performance, using a single core CPU implementation only.

Keywords: Computer vision · Pedestrian tracking · Real-time · Active safety systems

1 Introduction

Fast and meanwhile accurate pedestrian detection is necessary for many applications. Unfortunately these two demands are contradictory, and thus very difficult to unite. Even with today's cheaply available computational power it remains very challenging to achieve both goals. Indeed, recent state-of-the-art pedestrian detectors achieving real-time performance heavily rely on the use of parallel computing devices (e.g. multicore CPUs or GPUs) to perform this task. This often makes it unfeasible to use these algorithms in real-life applications, especially if these applications rely on embedded systems to perform their tasks.

In this paper we propose an efficient multi-pedestrian detection and tracking framework for a specific application: detection of pedestrians in a truck's blind spot zone. Statistics indicate that in the European Union alone, these blindspot accidents cause each year an estimated 1300 casualties [12]. Several commercial systems were developed to cope with this problem, both *active* and

© Springer International Publishing Switzerland 2015
A. Fred et al. (Eds.): ICPRAM 2014, LNCS 9443, pp. 179–195, 2015.
DOI: 10.1007/978-3-319-25530-9_12

Fig. 1. Example frame from our blind spot camera.

passive systems. Active safety systems automatically generate an alarm if pedestrians enter dangerous zones around the truck (e.g. ultrasonic distance sensors), whereas passive safety systems still rely on the focus of the truck driver (e.g. blind spot mirrors). However, none of these systems seem to adequately cope with this problem since each of these systems have their specific disadvantages. Active safety systems are unable to interpret the scene and are thus not able to distinguish static objects from actual pedestrians. Therefore they tend to generate many false alarms (e.g. with traffic signs). In practice the truck driver will find this annoying and often disables these type of systems. Existing passive safety systems are far from the perfect solution either. In fact, although blind spot mirrors are obliged by law in the European Union since 2003, the number of casualties did not decrease [18]. This is mainly due to the fact that these mirrors are not adjusted correctly; research indicates that truck drivers often use these mirrors to facilitate maneuvering. A passive blind-spot camera system with a monitor in the truck's cabin is always adjusted correctly, however it still relies on the attentiveness of the driver.

To overcome these problems we aim to develop an active safety system based on the truck's blind spot camera. Our final goal is to automatically detect vulnerable road users in the blind spot camera images, and warn the truck driver about their presence. Such an active safety system has multiple advantages over existing systems: it is independent of the truck driver, it is always adjusted correctly and it is easily implemented in existing passive blind spot camera systems. Due to the specific nature of this problem, this is a challenging task. Vulnerable road users are a very diverse class: besides pedestrians also bicyclists, mopeds, children and wheelchair users are included. Furthermore the specific position and type of the blind spot camera induces several constraints on the captured images. These wide-angle blind spot cameras introduce severe distortion while the sideway-looking view implies a highly dynamical background. See Fig. 1 for an example frame from our blind spot dataset.

However, the most challenging part is undoubtly the hard real-time constraint, combined with the need for high accuracy. In this paper we present part of such a total safety solution: we propose an efficient multi-pedestrian

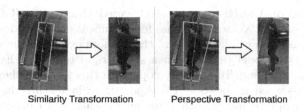

Similarity Transformation   Perspective Transformation

Fig. 2. Similarity vs perspective transformation model.

tracking- and detection framework based on blind spot camera images. Our algorithm achieves both high accuracy and high detection speeds. Using a single-core CPU implementation we reach an average of 13 FPS on our datasets.

In previous work [24, 25] we proposed our initial *warping window approach*. However, this initial approach was based solely on a naive similarity warp, running up against its limit (e.g. w.r.t. accuracy for our application). In this paper we propose our *perspective warping window approach*: we extensively redesigned and improved our previous work making it more elegant and accurate, without significantly increasing the algorithmic complexity. Moreover, we even obtain higher computation speeds. Figure 2 concisely compares our previous and our improved novel approach presented here.

Our proposed algorithm briefly works as follows. Traditional state-of-the-art pedestrian detectors use a *sliding window* paradigm: each possible position and scale in the image is evaluated. This however is unfeasible in real-time applications. Instead, we proposed our *warping window* approach: we eliminate the need to perform a full scale-space search using the exploitation of scene constraints. That is, at each position in the input image we locally model the transformation induced by the specific camera viewpoint and the lens distortion. During detection, we can then warp the regions of interest (ROIs) in the image and use a standard pedestrian detector at a single scale on each ROI. This approach is integrated in a tracking-by-detection framework and combined with temporal information, making it more robust while reducing the detection time. We performed extensive experiments to evaluate our algorithm concerning both speed and accuracy. For this we recorded several realistically simulated dangerous blind spot scenarios with a real truck.

The remainder of this paper is organised as follows. In the next section we describe related work concerning this topic. Section 3 describes our algorithm in more detail, while in Sect. 4 we propose our experiments and evaluation results. We then conclude our work in Sect. 5.

2   Related Work

In the past few years the accuracy of pedestrian detectors has been significantly improved. Currently, even on challenging datasets excellent accuracy results are presented [9]. Initially, Dalal and Triggs [5] proposed a pedestrian detection framework based on the Histograms of Oriented Gradients (HOG) combined with an SVM (Support Vector Machine) for classification. This idea was further refined in Felzenszwalb et al. [14] where the authors extended the concept with a part-based HOG model rather than a single rigid template. Evidently, this increases calculation time. To partially cope with this problem they proposed a more efficient cascaded framework [13]. Apart from increasing the model complexity, one can opt to increase the number of features to improve detection accuracy. Indeed, such a detector is presented in [7], called *Integral Channel Features*. However, each of these detectors still uses a sliding window approach. Across the entire image the features are calculated at all scales. To avoid such an exhaustive full scale-space search several optimisation techniques were proposed; e.g. Lampert et al. [17] proposed an efficient subwindow search. Dollár et al. [6] introduced the *Fastest Pedestrian Detector in the West (FPDW)* approach, in which they approximate feature responses from scales nearby thus eliminating the need to fully construct the scale-space pyramid. Extensive comparative works have been published [8, 10] to determine the most accurate approach. Both conclude that the HOG-based approach outperforms existing methods.

More recently, a benchmark between sixteen state-of-the-art pedestrian detectors was presented [9]. The authors conclude that part-based HOG detectors still achieve the highest accuracy, while the FPDW is one order of magnitude faster with only small loss in accuracy. Based on these conclusions, we performed extensive benchmark experiments with both pedestrian detectors to determine the most optimal one for our framework. These results, and more in-depth information on how both pedestrian detectors work are given in Sect. 3.1.

Concerning speed, several GPU optimisations were proposed. Prisacariu and Reid [22] proposed a fast GPU implementation of the standard HOG model. In [21], Pedersoli et al. presented a pedestrian detection system using a GPU implementation of the part-based HOG model. Benenson et al. [3] proposed work in which they perform model rescaling instead of image rescaling, and combined with their stixel world approximation [2] they achieve fast pedestrian detection. Recently the authors proposed their *Roerei* detector [1]. Based on a single rigid model they achieve excellent accuracy results. However, in real-life applications using embedded systems such high-end GPU computing devices are often not available. Therefore our algorithm focuses on real-time performance, while maintaining high accuracy, on standard hardware.

Speed optimisation is also achieved using pedestrian tracking algorithms, of which several are proposed in the literature. They often rely on a fixed camera, and use a form of background modelling to achieve tracking [23, 26]. Since in our application we have to work with moving camera images, this cannot be used. Pedestrian tracking algorithms based on moving cameras mostly use a forward-looking view [11] or employ disparity information [15]. Cho et al. [4] proposed a pedestrian tracking framework related to our work, exploiting scene constraints to achieve real-time detection. However, they use a basic ground-plane assumption whereas our approach is much more flexible and generic. Moreover, our specific datasets are much more challenging due to the severe distortion.

We significantly differ from all of the previously mentioned approaches. We aim to develop a monocular multi-pedestrian tracking framework with a challenging backwards/sideways looking view, targeting high accuracy at real-time performance. Furthermore, most of these classic sliding window approaches assume only object scale variation. Other geometrical variations (e.g. rotation [16] and aspect ratio [19]) are usually covered by an exhaustive search approach. Our proposed warping approach offers a solution that can even cope with perspective distortion. In fact, without our warping window paradigm it would be unfeasible in practice to perform such an exhaustive search in a perspective distortion space.

3   Algorithm Overview

As mentioned above, existing pedestrian detectors employ a sliding window approach. Across all positions and scales in the image the features are calculated and evaluated, making it almost impossible to meet the stringent real-time demands needed in most safety applications. To achieve real-time detection speeds with high accuracy we propose our novel *perspective warping window* approach.

Our idea is mainly based on the following observation. Looking at an example frame from our dataset (see Fig. 1) one clearly notices that the pedestrians appear rotated, scaled and perspectively transformed. This is due to the specific position and the wide-angle lens of our blind spot camera. The crux of the matter is that this transformation only depends on the position in the image. Thus each pixel coordinate $\mathbf{x} = [x, y]$ uniquely defines the transformation at that specific position. If at each pixel position this transformation is known, we can dramatically speed-up pedestrian detection. Based on this transformation we can locally warp each region of interest to upright pedestrians at a fixed height, and run a single-scale pedestrian detector on each warped ROI image patch. This approach effectively eliminates the need to construct a scale-rotation-transformation-space pyramid, and thus is very fast. Moreover, this approach is easily generalisable to other applications where such distortion occurs due to non-standard camera viewpoints and/or wide-angle lens distortions (e.g. surveillance cameras). To determine this transformation at each pixel coordinate a one-time calibration step is needed. To further increase both accuracy and speed, we integrate this warping window approach into an efficient tracking-by-detection framework. We use temporal information to predict future positions of pedestrians, thus further reducing the search space. Below we describe these steps in more detail. In Subsect. 3.1 we describe how our new perspective warping approach models the transformation, and motivate important algorithmic design choices such as the pedestrian detector, and the optimal scale parameter. In Subsect. 3.2 we then show how we integrate each of these steps into our total framework, thus describing our complete algorithm.

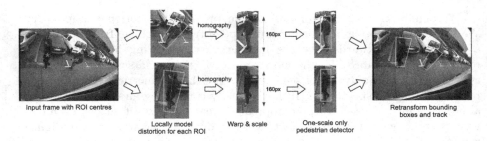

Fig. 3. Illustration of our novel perspective warping window approach. At each position in the image we locally model the distortion, warp the ROIs to a standard scale and use a one-scale only pedestrian detector.

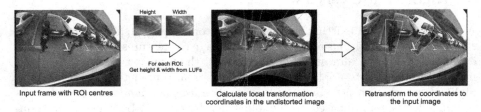

Fig. 4. The transformation is modeled as a perspective transformation, calculated in the undistorted image.

3.1   Warp Approach

Figure 3 illustrates our perspective warping window approach. Starting from input images as given in Fig. 1, pedestrians appear rotated, scaled and perspectively distorted. If we assume a flat groundplane, these transformation parameters only depend on the specific position in the image. If we know the transformation we can model the perspective distortion for that ROI, extract and warp the ROI image patch to a fixed-scale (160 pixels - motivated further in this work) and perform pedestrian detection on a single scale only. We thus eliminate the need to construct a scale-space pyramid. Note that, although we perform detection on a single scale only, the pedestrian model still provides some invariance with respect to the pedestrian height. However, if large deviations from the standard height (e.g. children) need to be detected, an extra scale needs to be evaluated. After detection, the coordinates of the detected bounding boxes are retransformed and fed into our tracking framework. Next we describe further details of our algorithm: how this position-specific transformation is mathematically modeled and how the cali-bration is performed. We further motivate the choice of our baseline pedestrian detector and determine the optimal fixed-height parameter.

Transformation Modelling. Figure 4 illustrates how the transformation is locally modeled. We use a perspective distortion model in the lens-distortion-corrected image. At each position, the height and width (at the ground) are

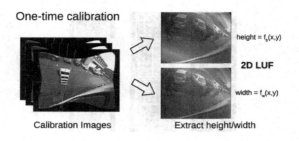

One-time calibration

Calibration Images

Extract height/width

height = $f_h(x,y)$

2D LUF

width = $f_w(x,y)$

Fig. 5. A one-time calibration is needed to determine the local perspective distortion.

known after a one-time calibration step (see further). These are visualised as two heat maps (the so-called look-up-functions or LUFs) in Fig. 4. The transformation coordinates are determined as follows. Each ROI centre coordinate (indicated with the red asterisk in the leftmost image) is first transformed into the undistorted image. This lens undistortion is simply based on the traditionally used radial lens distortion model:

$$\mathbf{x'} = \mathbf{x}(1 + k_1 r^2 + k_2 r^4) \tag{1}$$

$$r^2 = x^2 + y^2 \tag{2}$$

Here, $\mathbf{x'}$ denotes the corrected pixel coordinate, $\mathbf{x}$ the input coordinate and k_1 and k_2 indicate the radial distortion coefficients.

Next we calculate the vantage line through this ROI centre in the undistorted image, and determine the height and width (at the bottom) from the two LUFs. Based on these data we construct the perspective model in the undistorted image. The rotation of the image patch is determined from the angle of the vantage line, and the length ratio between the top and bottom is calculated based on the distance to the vantage point (visualised in the middle of Fig. 4). We thus locally model the pedestrians as if they are planar objects standing upright, faced towards the camera (that is, perpendicular to the optical axis of our blind spot camera). Our experiments show that this is a valid approximation for pedestrians. These coordinates are then retransformed to the distorted input image. Note that evidently only the coordinates are transformed, the middle image displayed here is only used for visualisation purposes. Based on the coordinates in the distorted image, and the known calibration data we apply a homography on the ROI image patch, thereby effectively undoing the local perspective distortion (visualised in Fig. 3).

Calibration. To obtain these two LUFs, a one-time calibration step is needed. To achieve this, we manually annotated about 200 calibration images. We utilised a planar calibration board of 0.5×1.80 m, and captured calibration positions homogeneously spread over the entire image (Fig. 5). The labeling was performed in the undistorted image. These images yield the vantage point, and the height and width of a pedestrian (at the ground) at each position for that image.

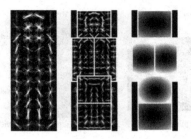

Fig. 6. The part-based pedestrian model [14]. (L) root filter (M) part filters (R) score distribution over parts.

Next we interpolated these datapoints using two-dimensional second order polynomial functions for both the height and the width: $f_h(x, y)$ and $f_w(x, y)$ with:

$$f_i(x, y) = p_0 + p_1 x + p_2 y + p_3 x^2 + p_4 xy + p_5 y^2 \tag{3}$$

Both functions are displayed as heat maps in Fig. 5: for each pixel coordinate they effectively give the height and width of the calibration pattern at that location. If for some reason the position of the camera w.r.t. the ground place changes, a recalibration needs to be performed. This is highly unlikely though, due to the robust camera mounting on the truck. Thus to summarise, detection is composed of four different steps: calculate the local perspective distortion model at each ROI centre, perform a homography and transform the pedestrians to an undistorted, upright position at a fixed height of 160 pixels, run a pedestrian detector at one scale, and finally retransform the coordinates of the detected bounding boxes to the original input image.

Pedestrian Detector. Based on the comparative works given in Sect. 2 we conclude that, since we aim for high accuracy, two approaches towards pedestrian detection are most suited for our application: the deformable part-based HOG models (DPM), and the rigid model-based FPWD. The FPDW has only slightly lower accuracy on established datasets [9] and is much faster. However, since we need to evaluate only one scale, no feature pyramid is constructed, thus this speed advantage is here not relevant. Selecting the most appropriate pedestrian detector for integration in our framework thus boils down to the selection of the most accurate detector on our dataset. For this we performed accuracy measurements for both detectors. To perform a fair comparison, let us briefly discuss how both pedestrian detectors work.

The DPM detector uses a pretrained model, consisting of HOG features (see Fig. 6). It consists of a root filter and a number of part filters representing the head and limbs of the pedestrian. To use this model, first a scale-space pyramid is constructed, using repeated smoothing and subsampling. For each pyramid layer the HOG features are computed. Then, for a specific scale the response of the root filter and the feature map is combined with the response of the part filters to calculate a final detection score. On our 640 × 480 resolution images this detector off-the-shelf needs an average of 2.3 s per frame, while their cascaded version needs on average 0.67 s per frame.

As opposed to the deformable part-based detector, the FPDW detector utilises a rigid model, not making use of deformable parts. Again, a scale-space pyramid is constructed. Next, features are calculated on transformed versions of the original input image, called *channels*. Examples are color channels and gradient channels. The most basic features are a sum over a rectangle region in one of the channels. These features are combined into higher-order features, and classification is performed using a depth two decision tree boosted classifier (AdaBoost). Fast rejection of candidate windows is possible through the use of a *soft-cascade* approach. Essentially, this detector uses ICF as a baseline detector, and achieves a speed-up through a more efficient construction of this feature pyramid; intermediate feature scales are approximated using scales nearby, avoiding the need to compute all feature scales. Out-of-the-box calculation time for ICF on average equals 451 ms per frame (on 640 × 480 images), while using the FPDW approach the calculation time drops to 147 ms per frame.

We altered both detectors into single-scale detectors to utilise in our framework. We need to determine the most optimal scale for each detector; the rescale height at which the maximal accuracy is reached.

Determining the Optimal Scale Factor. As mentioned, we rescale the pedestrians to a fixed height in order to reduce the calculation time. For this, an optimal value needs to be determined. To achieve this, we labeled and extracted 6000 pedestrian images at different locations in the images from our dataset, and performed the warp operation as given above. These pedestrians were warped to fixed heights, and we then performed accuracy measurements with both single-scale pedestrian detectors to determine the optimal height for each of them. Besides our *perspective transformation model* presented in this paper, we also warped the pedestrians using the *similarity transformation model* as explained in [25], simply consisting of a rotation and scaling operation (see Fig. 2 for a qualitative comparison). This was done to analyse the benefit of our more complex

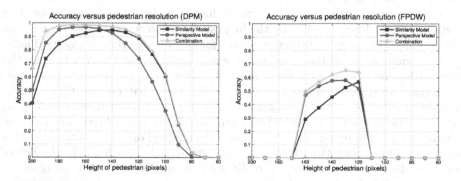

Fig. 7. Determining the optimal scale parameter. **Left:** results for the deformable-part detector. **Right:** results for the FPDW detector.

perspective model. Figure 7 displays our results for both pedestrian detectors (left: DPM, right: FPDW). Besides the individual transformations, we also give the combined accuracy. Evidently, optimal accuracy is reached when pedestrians in the rescaled image patches approximate the height of the detection models. Note that the one-scale DPM detector achieves much better accuracy results for all transformation models as compared to the FPDW detector. The reason for this significant difference is found in the design methodology of both detectors. Due to the part-based approach, the DPM detection model is much more flexible, making this detector invariant to slight deviations in height between the pedestrians that need to be detected, and the actual pedestrian model. The FPDW is much more sensitive for this due to the rigidness of their detection model. Since in our image patches slight differences between the actual and estimated pedestrian height exist (due to small calibration errors and the inherent height differences between pedestrians), more search scales would be needed to obtain higher accuracy with the FPDW approach. Further note that for both detectors the optimal resolution of the perspective and similarity transformation model differs. Concerning DPM, for the first transformation model the optimal height lies at 160 pixels, whereas the latter reaches its optimum at 140 pixels. As can be seen, the perspective model has a clear accuracy advantage over the similarity model. If both models were combined, an even higher accuracy is achieved. This, however, would double the calculation time. Although at lower accuracy, a similar trend is noticeable for the FPDW detector.

These insights favor the use of the deformable part-based model over the rigid FPDW approach. However, these experiments exclude the influence of the position in the image w.r.t. the detection accuracy. Specifically for our images, the position inherently defines the scale and the amount of transformation. For example, specific positions in the image require significant upscaling to transform these patches to the fixed height. Furthermore, research indicates that rigid models perform better on low-resolution pedestrian patches, whereas part-based models achieve higher accuracy on large pedestrian patches [20]. Thus, a combined approach—where the *best* detector is selected using the spatial location in the image—may increase accuracy.

We performed such experiments to evaluate this hypothesis as follows. The *best* detector is defined as having the highest detection score on the same image patch. Since the range of detection scores for both pedestrian detectors differs, first a normalisation of the detection scores is applied. For this, we warped the 6000 pedestrian image patches mentioned above using the perspective transformation to the optimal height of 160 pixels. These patches are then equally divided in a *training* and a *test set*. Next, both pedestrian detectors were evaluated on the test set, and their detection scores were normalised (subtract the average and divide by standard deviation - determined on the trainingset). This allows for a fair comparison between both detectors. For each patch, the detector with the highest detection score (if both found a detection) is assigned as optimal detector for this patch. These results are visualised in Fig. 8 in function of the position in the image. The colored dots indicate where which pedestrian

Fig. 8. Performance of both pedestrian detectors in function of the position in the image. Colored dots indicate which pedestrian detector performed best. Yellow: DPM. Magenta: FPDW (Color figure online).

Fig. 9. Performance of the two transformation models in function of the position. Colored dots indicate which model performed the detection. Red: perspective model. Blue: similarity model. Green: both models. Yellow indicates missed detections (Color figure online).

detector performed best (yellow: DPM, magenta: FPDW). As visualised, no specific image location is favored by any detector. Therefore, currently we do not perform such a combination, and utilise a single pedestrian detector.

Based on all experimental results mentioned above, we thus opted for the cascaded deformable part-based models as baseline detector in our framework: it achieves excellent accuracy results, lends itself perfect to perform *true* single scale detections and, due to this single scale approach, achieves excellent speed results (as shown further). Figure 9 shows (only for the deformable part-based model) where each transformation model performs best in function of the position in the image. Red dots indicate where the perspective model worked, blue where the similarity model worked and green were both models found the detection. Yellow indicates a missed detection. The perspective model obviously performs much better than the similarity model. The similarity model performs slightly better only at the image border, due to the small calibration error there.

The perspective model performs better close to the truck because of the large amount of viewpoint distortion there. Note that if we analyse positions where both models found the detection, the perspective model achieves the best detection score in 69 % of these cases, further indicating its clear advantage over the similarity transformation model.

3.2 Tracking Framework

To further improve the accuracy and detection speed we integrated our warping window approach in a tracking-by-detection framework. This is implemented as follows. Instead of a full frame search, we use initialisation coordinates (which define transformation ROIs) at the border of the image, and initially only perform detection there. See Fig. 10 for an example. If a pedestrian is detected, a linear Kalman filter is instantiated for this detection. As a motion model we use a constant velocity assumption. Our experiments indicate that this assumption holds for our application. The state vector x_k consists of the centre of mass of each detection and the velocity: $x_k = \begin{bmatrix} x\ y\ v_x\ v_y \end{bmatrix}^T$. Based on the update equation $\hat{x}_k^- = A\hat{x}_{k-1}$ we estimate the next position of the pedestrian. Here, $\hat{x}_k^-$ indicates the *a priori* state estimate and $\hat{x}_k$ indicates the *a posterior* state estimate, at timestep k.

The process matrix A thus equals:

$$A = \begin{bmatrix} 1\ 0\ 1\ 0 \\ 0\ 1\ 0\ 1 \\ 0\ 0\ 1\ 0 \\ 0\ 0\ 0\ 1 \end{bmatrix} \tag{4}$$

Based on this motion model we predict the position (that is, the centre of mass) of the pedestrian in the next frame. Each estimated new pixel coordinate is then used as input for our warping window approach: we calculate the transformation model, warp the ROI and perform pedestrian detection on this ROI. For each

Fig. 10. Example of five initialisation coordinates together with their corresponding transformation ROIs.

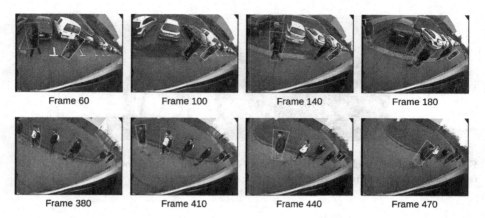

Frame 60   Frame 100   Frame 140   Frame 180

Frame 380   Frame 410   Frame 440   Frame 470

Fig. 11. Qualitative tracking sequences over two of our datasets (top and bottom row) - see http://youtu.be/gbnysSoSR1Q for a video.

pedestrian that is being tracked, our algorithm verifies if a new detection is found. This is evaluated by constructing a circular region around the estimated coordinate based on the scale of that tracked instance. If a new detection is found in this region, the Kalman filter is updated and the new position is predicted. If multiple detections are found, we associate the closest based on the Euclidean distance. The bounding box coordinates of tracked instances are averaged to assure smooth transitions between frames. If for tracked pedestrians no new detection is found, the Kalman filter is updated based on the estimated position. In this case we apply a dynamic score strategy, and lower the detection threshold for that instance (within certain boundaries). This ensures that pedestrians which are difficult to detect (e.g. partially occluded or a temporarily low HOG response) can still be tracked. If no detection is found for multiple frames in a row, the tracker is discarded. Evidently, if a new detection is found for which no previous tracker exists, tracking starts from there on. Figure 11 qualitatively illustrates tracking sequences on two of our datasets.

4   Experiments and Results

We performed extensive experiments concerning both speed and accuracy. Our datasets consists of simulated dangerous blind spot scenarios, recorded with a real truck. We used a commercial blind spot camera (Orlaco CCC115°) with a resolution of 640×480 at 15 frames per second. This camera has a 115 degree wide-angle lens. See Fig. 12 for the exact position of the camera. Five different scenarios were recorded, each in which the truck driver makes a right turn and the pedestrians react differently (e.g. the truck driver lets the pedestrians pass, or the truck driver keeps on driving, simulating a near-accident). This resulted in a total of about 11000 frames. For our accuracy and speed experiments we labelled around 3200 pedestrians. Our implementation is CPU-based only, and

Fig. 12. Our test truck with the mounted commercial blind spot camera (circled in red) (Color figure online).

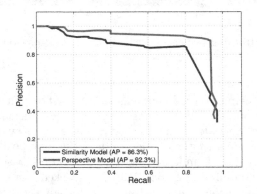

Fig. 13. Precision-recall graph over our dataset.

the hardware consists of an Intel Xeon E5 CPU which runs at 3.1 GHz. Note that all speed experiments are performed on a single core. The algorithm is mainly implemented in Matlab, with time-consuming parts (such as the homography) in OpenCV, using *mexopencv*.

4.1   Accuracy Results

Figure 13 displays the precision-recall graph of our algorithm as calculated over our datasets. The red PR curve indicates our novel perspective transformation approach, while the blue PR curve represents our previous similarity transformation approach. They are calculated as follows. For each detected pedestrian in our algorithm, we look for a labeled instance in a circular region (based on the scale) around the centre of our detection. If such an instance is found, this is counted as being a *true positive*. If this is not the case, this detection is counted as being a *false positive*. Each labeled pedestrian which is not detected accounts for a *false negative*. The PR-graph is then determined as: $precision = \frac{TP}{TP+FP}$ and $recall = \frac{TP}{TP+FN}$. We notice that, although both achieve very good accuracy results, our novel perspective warping window approach has a clear accuracy advantage over our similarity warping window approach. Indeed, the average precision (AP) for the similarity model equals 86.3 %, whereas for the perspective model $AP = 92.3\%$. With the perspective model, at a recall rate of 94 %, we still achieve a precision of 90 %. Such high accuracy results are due to our warping window approach. Since we know the scale at each position, the number of false positives is minimized. Furthermore this allows us to use a sensitive pedestrian detection threshold.

4.2 Speed Results

As mentioned in Sect. 3.1, if used out-of-the-box the baseline pedestrian detector takes 670 ms (i.e. 1.5 fps). Since in our framework we only need to perform detection at a single scale and ROI, the calculation time drastically decreases. For each default search region and tracked pedestrian in the image we need to perform a warp operation and detection. Thus, the total calculation time evidently depends on the number of tracked pedestrians per image. Figure 14 displays the average calculation time per ROI. Note that if a detection is found, the average calculation time equals 18.3 ms, while if no detection is found the average calculation time drops to 10.8 ms. This calculation time per region is independent of the position in the image. The average detection time per ROI is subdivided into five steps: the calculation of the warp coordinates, the time needed to perform the warp operation, calculation of the HOG features, evaluation of the pedestrian model, and finally the retransformation of the detected coordinates to the input image. The total warp time (*calc. warp coord.* and *perform warping*) only equals about 3 ms. Most time is spent on the actual pedestrian detection. The time needed to perform the retransformation of the coordinates is negligible. Figure 15 displays the frames per second as a function of the number of tracked pedestrians we reached on our datasets. If no pedestrians are tracked we achieve 28.2 fps. On average we achieve 13.0 fps (with an average of 3.4 pedestrians), while our worst-case framerate equals 7.0 fps.

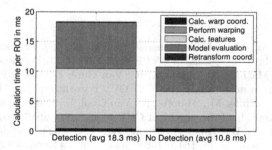

Fig. 14. Calculation time per ROI.

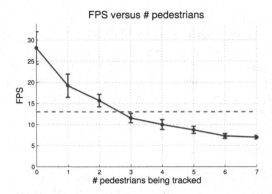

Fig. 15. Speed performance versus the number of tracked pedestrians (dotted red line indicates the average fps) (Color figure online).

5   Conclusions and Future Work

In this work we proposed a multi-pedestrian tracking framework achieving excellent accuracy and speed results on a single-core CPU implementation. The algorithm is based on our novel perspective warping window approach. We proposed this approach to allow for efficient pedestrian detection on the challenging, highly distorted camera images from a blind-spot camera, with minimal CPU resources. However, this approach is easily generalisable to other applications with non-standard camera-viewpoints.

In the future we plan to further extend our framework to multi-class detection: we aim to develop a complete vulnerable road users detection system, starting with bicyclists. Additionally, an efficient combination of multiple pedestrian detectors to further increase the accuracy could be evaluated. Furthermore we aim to investigate if the inclusion of other features (e.g. motion information) could further increase the robustness of our framework.

References

1. Benenson, R., Markus, M., Tuytelaars, T., Van Gool, L.: Seeking the strongest rigid detector. In: Proceedings of CVPR, pp. 3666–3673, Portland, Oregon (2013)
2. Benenson, R., Mathias, M., Timofte, R., Van Gool, L.: Fast stixels computation for fast pedestrian detection, In: ECCV, CVVT workshop, pp. 11–20 (2012)
3. Benenson, R., Mathias, M., Timofte, R., Van Gool, L.: Pedestrian detection at 100 frames per second. In: Proceedings of CVPR, pp. 2903–2910 (2012)
4. Cho, H., Rybski, P., Bar-Hillel, A., Zhang, W.: Real-time pedestrian detection with deformable part models. In: IEEE Intelligent Vehicles Symposium, pp. 1035–1042 (2012)
5. Dalal, N., Triggs, B.: Histograms of oriented gradients for human detection. Proc. CVPR **2**, 886–893 (2005)
6. Dollár, P., Belongie, S., Perona, P.: The fastest pedestrian detector in the west. In: Proceedings of BMVC, pp. 68.1-68.11 (2010)

7. Dollár, P., Tu, Z., Perona, P., Belongie, S.: Integral channel features. In: Proceedings of BMVC, pp. 91.1-91.11 (2009)
8. Dollár, P., Wojek, C., Schiele, B., Perona, P.: Pedestrian detection: a benchmark. In: Proceedings of CVPR, pp. 304–311 (2009)
9. Dollár, P., Wojek, C., Schiele, B., Perona, P.: Pedestrian detection: an evaluation of the state of the art. IEEE PAMI **34**, 743–761 (2012)
10. Enzweiler, M., Gavrila, D.M.: Monocular pedestrian detection: survey and experiments. IEEE PAMI **31**, 2179–2195 (2009)
11. Ess, A., Leibe, B., Schindler, K., Van Gool, L.: A mobile vision system for robust multi-person tracking. In: Proceedings of CVPR, pp. 1–8 (2008)
12. EU (22 february 2006). Commision of the european communities, european road safety action programme: mid-term review
13. Felzenszwalb, P., Girschick, R., McAllester, D.: Cascade object detection with deformable part models. In: Proceedings of CVPR, pp. 2241–2248 (2010)
14. Felzenszwalb, P., McAllester, D., Ramanan, D.: A discriminatively trained, multi-scale, deformable part model. In: Proceedings of CVPR (2008)
15. Gavrila, D., Munder, S.: Multi-cue pedestrian detection and tracking from a moving vehicle. IJCV **73**, 41–59 (2007)
16. Huang, C., Ai, H., Li, Y., Lao, S.: Vector boosting for rotation invariant multi-view face detection. In: ICCV, pp. 446–453 (2005)
17. Lampert, C., Blaschko, M., Hoffmann, T.: Efficient subwindow search: a branch and bound framework for object localization. IEEE PAMI **31**, 2129–2142 (2009)
18. Martensen, H.: Themarapport vrachtwagenongevallen 2000–2007 (BIVV) (2009)
19. Mathias, M., Timofte, R., Benenson, R., Van Gool, L.: Traffic sign recognition - how far are we from the solution? In: ICJNN (2013)
20. Park, D., Ramanan, D., Fowlkes, C.: Multiresolution models for object detection. In: Daniilidis, K., Maragos, P., Paragios, N. (eds.) ECCV 2010, Part IV. LNCS, vol. 6314, pp. 241–254. Springer, Heidelberg (2010)
21. Pedersoli, M., Gonzalez, J., Hu, X., Roca, X.: Toward real-time pedestrian detection based on a deformable template model. In: IEEE ITS (2013)
22. Prisacariu, V., Reid, I.: fastHOG - a real-time gpu implementation of HOG. Technical report, Department of Engineering Science, Oxford University (2009)
23. Seitner, F., Hanbury, A.: Fast pedestrian tracking based on spatial features and colour. In: Proceedings of CVWW, pp. 105–110 (2006)
24. Van Beeck, K., Goedemé, T., Tuytelaars, T.: Towards an automatic blind spot camera: robust real-time pedestrian tracking from a moving camera. In: Proceedings of MVA, Nara, Japan (2011)
25. Van Beeck, K., Tuytelaars, T., Goedemé, T.: A warping window approach to real-time vision-based pedestrian detection in a truck's blind spot zone. In: Proceedings of ICINCO (2012)
26. Viola, P., Jones, M., Snow, D.: Detecting pedestrians using patterns of motion and appearance. IJCV **63**, 153–161 (2005)

Comparing Different Labeling Strategies in Anomalous Power Consumptions Detection

Fernanda Rodríguez$^{(\boxtimes)}$, Federico Lecumberry, and Alicia Fernández

Facultad de Ingeniería, Instituto de Ingeniería Eléctrica, Universidad de la República,
J. Herrera y Reissig 565, 11300 Montevideo, Uruguay
{fernandar,fefo,alicia}@fing.edu.uy

Abstract. Detecting anomalous events is a complex task, specially when it should be performed manually and for several hours. In the case of electrical power consumptions, the detection of non-technical losses also has a high economic impact. The diversity and big number of consumption records, makes it very important to find an efficient automatic method for detecting the largest number of frauds. This work analyses the performance of a strategy based on learning from expert labeling: suspect/no-suspect, with one using inspection labels: fraud/no-fraud. Results show that the proposed framework, suitable for imbalance problems, improves performance in terms of the $F_{measure}$ with inspection labels, avoiding hours of experts labeling.

Keywords: Electricity fraud · Support vector machine · Optimum Path Forest · Unbalance class problem · Combining classifier · UTE

1 Introduction

Non-technical loss detection is a huge challenge for electric power utility. In Uruguay the national electric company (henceforth UTE) faces the problem by manually monitoring a group of customers. A group of experts inspect at the monthly consumption curve of each customer and indicates those with some kind of suspicious behavior. This set of customers, initially classified as suspects are then analyzed taking into account other factors (such as fraud history, electrical energy meter type, etc.). Finally a subset of customers is selected to be inspected by an UTE's employee, who confirms (or not) the irregularity. The procedure is illustrated in Fig. 1. The procedure described before, has major drawbacks, mainly, the number of customers that can be manually controlled is small compared with the total amount of customer (around 500.000 only in Montevideo).

Different machine learning aproaches have addressed the detection of non-technical losses, both supervised or unsupervised. Leon et al. review the main research works found in the area between 1990 and 2008 [1]. Here we present a brief review that builds on this work and wide it with new contributions published between 2008 and 2013. Several of these approaches consider unsupervised classification using different techniques such as fuzzy clustering [2], neural

© Springer International Publishing Switzerland 2015
A. Fred et al. (Eds.): ICPRAM 2014, LNCS 9443, pp. 196–205, 2015.
DOI: 10.1007/978-3-319-25530-9_13

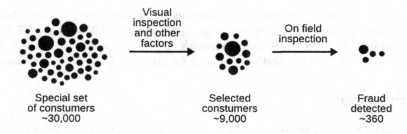

Fig. 1. Manual fraud detection scheme.

networks [3,4], among others. Monedero et al. use regression based on the correlation between time and monthly consumption, looking for significant drops in consumption [5]. Then they make a second stage where suspicious customers are eliminated if the consumption of these depend on the economy of the moment or the year's season. Only major customers were inspected and 38 % were detected as fraudulent. Similar results (40 %) were obtained in [6] using a tree classifier and customers who had been inspected in the past year. In [7,8] SVM is used. In the latter, Modified Genetic Algorithm is employed to find the best parameters of SVM. In [9], is compared the methods Back-Propagation Neural Network (BPNN), Online-sequential Extreme Learning Machine (OS-ELM) and SVM. Biscarri et al. [10] seek for outliers, Leon et al. [1] use Generalized Rule Induction and Di Martino et al. [11] combine CS-SVM classifiers, One class SVM, and C4.5 OPF using various features derived from the consumption. Different kinds of features are used among this works, for examples, consumption [8,10], contracted power and consumed ratio [12], Wavelet transformation of the monthly consumption [13], amount of inspections made to each client in one period and average power of the area where the customer resides [2], among others.

On the other hand, Romero proposes [14] a method to estimate and reduce non-technical losses, such as advanced metering infrastructure, fraud deterrence prepayment systems, system remote connection and disconnection, etc. Lo et al. based on real-time measurements, design [15] an algorithm for distributed state estimation in order to detect irregularities in consumption.

To improve the efficiency of fraud detection and resource utilization, in [16] was implemented a tool that automatically detects suspicious behavior analyzing customers historical consumption curve. This approach has the drawback of requiring a base previously tagged by the experts, in order to use it in the training stage.

In this work we set out to analyze the behavior of the proposed framework to fraud classification and compare it by using labels based on the inspection results instead of labels defined by experts. This new approach does not require that the company personnel conduct a manual study of the customers' consumption curve, since it use labels resulting from inspections in the past. We investigate performance improvement originated by training with individual algorithms and their combinations with labels of fraud and no fraud (based on inspections) and the importance of choosing the appropriate performance measure to solve the problem.

This paper is an extension of our previous work presented in the International Conference on Pattern Recognition Application and Methods (ICPRAM 2014) [17]; in Sect. 2 describes the framework and the strategies to be compare. Section 3 presents the obtained results and, finally Sect. 4 concludes the work.

2   Framework

The system presented consists basically on three modules: Pre-Processing and Normalization, Feature Extraction and Selection, and Classification. Figure 2 shows the system configuration. The system input corresponds to the last three years of the monthly consumption curve of each costumer.

The first module, Pre-Processing and Normalization, modifies the input data so that they all have normalized mean and implements some filters to avoid peaks from billing errors. A feature set was proposed taking into account UTE's technician expertize in fraud detection by manual inspection and recent papers on non technical loss detection [18–20]. Di Martino et al. use a list of the features extracted from the monthly consumption records [16]. In this work, we use the framework illustrated in Fig. 2 and a subset of the same set of features used in [16] but doing a selection of them taking into account the label type (based on inspection or expertise's criterion).

It is well known that finding a small set of relevant features can improve the final classification performance; this is why we implemented a feature selection stage. We used two types of evaluation methods: filter and wrapper. Filters methods looks for subsets of features with low correlation between them and high

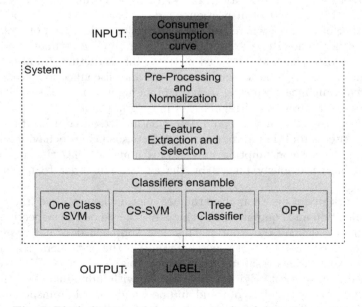

Fig. 2. Block diagram of the automatic fraud detection system.

correlation with the labels, while wrapper methods evaluate the performance of a given classifier for the given subset of features. In the wrapper methods, we used as performance measure the $F_{measure}$, also, the evaluations were performed using 10 fold cross validation over the training set.

As searching method, we used *Bestfirt*, for which we found in this application a good balance between performance and computational costs.

2.1 Classifiers

In this section we describe the classifiers used in this work. The authors of [21] proposed a new classifier, Optimum Path Forest (OPF), to apply to the problem of fraud detection in electricity consumption, showing good results. It consist of creating a graph with the training dataset, associating a cost to each path between two elements, based on the similarity between the elements of the path. This method assumes that the cost between elements of the same class is lower than those belonging to different classes. Next, a representative is chosen for each class, called prototypes. A new element is classified as the class that has lower cost with the corresponding prototype. Since OPF is very sensitive to class imbalance, we change class distribution of the training dataset by undersampling the majority class.

The decision tree proposed by Ross Quinlan: C4.5 is used as another classifier. It is widely utilized because it is a very simply method that obtain good results. However, it is very unstable and highly dependent on the training set. Thus, a later stage of AdaBoost was implemented, accomplishing a more robust results. Just as with the previews classifier, it was needed an resamplig stage to manage the dependency of the C4.5 with the class distribution.

The other two classifiers consider the widely used method, SVM, cost-sensitive learning (CS-SVM) and one-class classifier (O-SVM). In the former different cost were assigned to the misclassification of the elements of each class, in order to tackle the unbalanced problem. The second one considers the minority class as the outliers.

Finally, we also consider another method, that performs an optimal combination of the before mentioned classifiers. Taking the labels of that each classifier find, the following functions are define:

$$g_p(x) = w_1^p d_{OPF}^p + w_2^p d_{Tree}^p + w_3^p d_{CS-SVM}^p + w_4^p d_{O_SVM}^p$$

$$g_n(x) = w_1^n d_{OPF}^n + w_2^n d_{Tree}^n + w_3^n d_{CS-SVM}^n + w_4^n d_{O_SVM}^n$$

where $d_i^j(x) = 1$ if the classifier j labels the sample as i and 0 otherwise. Then if $g_p(x) > g_n(x)$ the sample is assigned to the positive class, if $g_n(x) > g_p(x)$ the sample is assigned to the negative class. The choice of combination's weights (w_i^j) is done exhaustively in order to maximize the $F_{measure}$, over a predefined grid and was evaluated with a 10-fold cross validation.

2.2 The Class Imbalance Problem and the Choice of Performance Measure

When working on fraud detection problems, we can not assume that the number of people who commit fraud are near the same than those who do not, usually they are a minority class. This situation is known as class imbalance problem, and it is particularly important in real world applications where it is costly to misclassify examples from the minority class. In this cases, standard classifiers tend to be overwhelmed by the majority class and ignore the minority class, hence obtaining suboptimal classification performance. In order to confront this type of problem, different strategies can be used on different levels: (i) changing class distribution by resampling; (ii) manipulating classifiers; (iii) and on the ensemble of them, as proposed in [16].

Another problem which arises when working with imbalanced classes is that the most widely used metrics for measuring the performance of learning systems, such as *Accuracy* and *ErrorRate*, are not appropriate because they do not take into account misclassification costs, since they are strongly biased to favor the majority class [22]. Then others measures have to be considered:

- *Recall* is the percentage of correctly classified positive instances, in this case, the fraud samples.

$$Recall = \frac{TP}{TP + FN}$$

- *Precision* is defined as the proportion of labeled as positive instances that are actually positive.

$$Precision = \frac{TP}{TP + FP}$$

Where TP, FN and FP are defined in Table 1.
- The combination of this two measurements, the $F_{measure}$, represents the geometric mean between them, weighted by the parameter β,

$$F_{measure} = \frac{(1 + \beta^2) Recall \times Precision}{\beta^2 Recall + Precision} \tag{1}$$

Depending on the value of β we can prioritize *Recall* or *Precision*. For example, if we have few resources to perform inspections, it can be useful to prioritize *Precision*, so the set of samples labeled as positive has high density of true positive.

Table 1. Confusion matrix.

	Labeled as	
	Positive	Negative
Positive	TP (True Positive)	FN (False Negative)
Negative	FP (False Positive)	TN (True Negative)

When working with inspection labels the imbalance problem is worst, in terms of unbalance, than dealing with experts labels. In the experts labels method, the ratio of suspect to no suspect is near 10 %, while in the one based on inspection labels, the ratio is near 0.4 %.

3   Experiments and Results

In this work we used a data set of 456 industrial profiles obtained from the UTE's database. Each profile is represented by the customers monthly consumption in the last 36 months and has two labels, one dictated manually by technicians previous the inspection and another based on the inspection results. Training was done considering both labels separately and performance evaluation was done given the inspection labels, using a 10-fold cross validation scheme.

3.1   Features Selection

Different feature subsets were selected from the original set of 28 features proposed in [16], for the different classifiers and for both approaches. For example, for the experts' labels approach and the classifier CS-SVM, the features are:

- Consumption ratio for the last 3, 6 and 12 months and the average consumption (feature 1, 2 and 3).
- Difference between fourth Wavelet coefficient from the last and previous years (feature 11).
- Euclidean distance of each customer to the *mean customer*, where the *mean customer* is calculated by taking the mean for each month between all the customers (feature 20).
- Rate between the mean variance and the variance in the last year of the consumption curve (feature 21).
- Module of the first two Fourier coefficients (feature 23 and 24).
- Slope of the straight line that fits the consumption curve (feature 28).

While for inspection label approach and CS-SVM, the features are:

- Consumption ratio for the last 3 months and the average consumption (feature 1).
- Norm of the difference between the expected consumption and the actual consumption (feature 4).
- Difference between the third, fourth and fifth Wavelet coefficient from the last and previous years (feature 11, 12 and 13).
- Euclidean distance of each customer to the *mean customer* (feature 20).
- Ratio between the mean variance of each costumer and the mean variance of all the costumers, of the consumption curve (feature 22).
- Slope of the straight line that fits the consumption curve (feature 28).

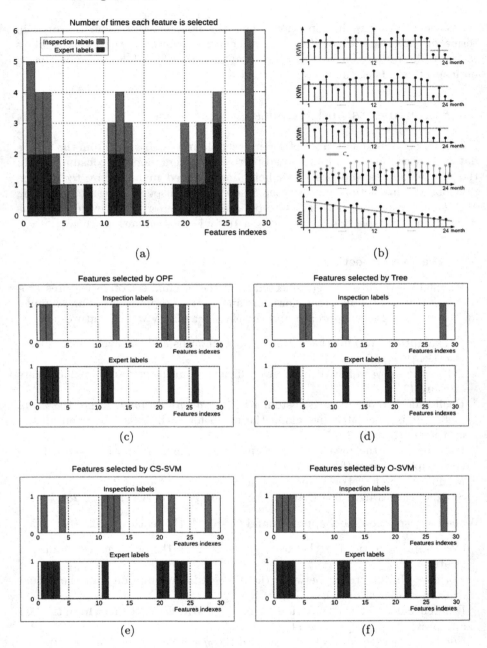

Fig. 3. Represents the features selection. (a) the abscissa indicates the 28 features and the ordinate is the number of times each feature is selected by the 4 classifiers and both methods (inspection label in red and expert label in blue). (b) outlines the features 1, 2, 3, and 28. (c), (d), (e) and (f) shows the features selected by each classifier and both methods separately (Color figure online).

Table 2. Fraud detection with experts label training.

Description	Recall (%)	Precision (%)	$F_{measure}$ (%) [$\beta = 1$]
OPF	39	27	32
Tree (C4.5)	38	23	29
O-SVM	51	22	30
CS-SVM	35	20	26
Iterative combination	77	22	35

Table 3. Fraud detection with inspection label training.

Description	Recall (%)	Precision (%)	$F_{measure}$ (%) [$\beta = 1$]
OPF	36	34	35
Tree (C4.5)	33	37	35
O-SVM	71	31	44
CS-SVM	74	33	46
Iterative combination	77	33	46

We observed that some features were selected for both approaches. Figure 3(d) represents the list indicated above, and Fig. 3(c), (d) and (f) indicates the features selected by OPF, Tree C.4.5 and O-SVM respectively, using the two different labels. Figure 3(a) shows the features selected by the four classifiers and both approaches all together. We can see that some features were never selected, other were selected only by one method and other by both method. Notice that each feature can be selected at the most eight times. The feature 28, represented by the fifth image in Fig. 3(b), is the most selected one (six times). Figure 3(b) also represents the feature 1, 2, 3 and 20. We conclude that, although the best subset of features depends on the classifiers and the type of label, some features are more representative than others.

3.2 Performance Analysis

Tables 2 and 3 shows the results obtained when experts and inspection labels are used to train the different classifiers respectively. The results for the method performed manually by experts, i.e. validating the expert labels with inspection labels, are $Recall = 38\%$, $Precision = 51\%$ and $F_{measure} = 44\%$.

In Table 2, we observe that the Iterative Combination technique with expert label training obtains the best result for fraud detection clearly overpassing the other methods, however the number false positive (FP) is relatively high, since

$$\frac{FP}{TP} = \frac{1}{Precision} - 1 \approx 4.$$

On the other hand, as Table 3 illustrates, if we use the inspection labels, the Iterative Combination also obtains the best results for fraud detection, but reducing in a half the number of FP ($\frac{FP}{TP} \approx 2$).

If we compare both approaches, we see that learning from the inspection labels could get better results (in the $F_{measure}$ sense) than learning from the labels set by experts. The former has the additional advantage of not requiring that the experts made the manual labeled of the training base.

Comparing the F_{masure} obtained manually by the experts (44%) and automatically by the Iterative Combination (46%) both are similar. However, the former consider other features as the history's fraud detection, contracted power, number of estimated readings, etc. and not only the monthly consumption, as the automatic one.

4   Conclusions and Future Work

In this work we compare the performance of a strategy based on learning from expert labeling: suspect/no-suspect, with one using inspection labels: fraud/no-fraud. In the $F_{measure}$ sense with all the tested classifiers the classification with inspection label obtains better results than using experts labels. Among them the Iterative Combination obtains the best result and also better than the manual method.

In future work we propose to include new categorical attributes as the history's fraud detection, contracted power, number of estimated readings, etc. We also want to explore a semi-supervised approach that allows to learn from data with and without previous inspection labels.

Acknowledgements. This work was supported by the program Sector Productivo CSIC UTE. Authors would like to thank UTE, especially Juan Pablo Kosut and Fernando Santomauro, for providing datasets and share fraud detection expertise.

References

1. Leon, C., Biscarri, F.X.E.L., Monedero, I.X.F.I., Guerrero, J.I., Biscarri, J.X.F.S., Millan, R. X.E.O.: Variability and trend-based generalized rule induction model to NTL detection in power companies (2011)
2. dos Angelos, E., Saavedra, O., Corts, O., De Souza, A.: Detection and identification of abnormalities in customer consumptions in power distribution systems (2011)
3. Markoc, Z., Hlupic, N., Basch, D.: Detection of suspicious patterns of energy consumption using neural network trained by generated samples (2011)
4. Sforna, M.: Data mining in power company customer database. Electr. Power Syst. Res. **55**(3), 201–209 (2000)
5. Monedero, I., Biscarri, F., León, C., Guerrero, J.I., Biscarri, J., Millán, R.: Using regression analysis to identify patterns of non-technical losses on power utilities. In: Setchi, R., Jordanov, I., Howlett, R.J., Jain, L.C. (eds.) KES 2010, Part I. LNCS, vol. 6276, pp. 410–419. Springer, Heidelberg (2010)
6. Filho, J.R., Gontijo, E.M., Delaiba, A.C., Mazina, E., Cabral, J.E., Pinto, J.O.P.: Fraud identification in electricity company customers using decision tree (2004)
7. Depuru, S.S.S.R., Wang, L., Devabhaktuni, V.: Support vector machine based data classification for detection of electricity theft (2011)
8. Yap, K.S., Hussien, Z., Mohamad, A.: Abnormalities and fraud electric meter detection using hybrid support vector machine and genetic algorithm (2007)
9. Yap, K.S., Tiong, S.K., Nagi, J., Koh, J.S.P., Nagi, F.: Comparison of supervised learning techniques for non-technical loss detection in power utility (2012)
10. Biscarri, F., Monedero, I., Leon, C., Guerrero, J.I., Biscarri, J., Millan, R.: A data mining method based on the variability of the customer consumption - a special application on electric utility companies. In: Volume AIDSS, pp. 370–374. Inst. for Syst. and Technol. of Inf. Control and Commun. (2008)
11. Di Martino, J., Decia, F., Molinelli, J., Fernández, A.: Improving electric fraud detection using class imbalance strategies. In: 1st International Conference in Pattern Recognition Aplications and Methods, vol. 2, pp. 135–141 (2012)

12. Galvn, J., Elices, E., Noz, A.M., Czernichow, T., Sanz-Bobi, M.: System for detection of abnormalities and fraud in customer consumption (1998)
13. Jiang, R., Tagaris, H., Laschusz, A.: Wavelets based feature extraction and multiple classifiers for electricity fraud detection (2002)
14. Romero, J.: Improving the efficiency of power distribution system through technical and non-technical losses reduction (2012)
15. Lo, Y.L., Huang, S.C., Lu, C.N.: Non-technical loss detection using smart distribution network measurement data. In: 2012 IEEE Innovative Smart Grid Technologies - Asia (ISGT Asia), pp. 1–5 (2012)
16. Di Martino, M., Decia, F., Molinelli, J., Fernández, A.: A novel framework for nontechnical losses detection in electricity companies. In: Latorre Carmona, P., Sánchez, J.S., Fred, A.L.N. (eds.) Pattern Recognition - Applications and Methods. AISC, vol. 204, pp. 109–120. Springer, Heidelberg (2013)
17. Rodriguez, F., Lecumberry, F., Fernndez, A.: Non technical loses detection: experts labels vs. inspection labels in the learning stage (2014)
18. Alcetegaray, D., Kosut, J.: One class SVM para la detección de fraudes en el uso deenergía eléctrica. Trabajo Final Curso de Reconocimiento de Patrones, Dictado por el IIE- Facultad de Ingeniería- UdelaR (2008)
19. Muniz, C., Vellasco, M., Tanscheit, R., Figueiredo, K.: Ifsa-eusflat 2009 a neurofuzzy system for fraud detection in electricity distribution (2009)
20. Nagi, J., Mohamad, M.: Nontechnical loss detection for metered customers in power utility using support vector machines. IEEE Trans. Power Deliv. 25(2), 1162–1171 (2010)
21. Ramos, C., de Sousa, A.N., Papa, J., Falcao, A.: A new approach for nontechnical losses detection based on optimum-path forest. IEEE Trans. Power Syst. (2010)
22. Garcia, V., Sanchez, J., Mollineda, R.: On the suitability if numerical performance evaluation measures for class imbalance problems. In: 1st International Conference in Pattern Recognition Aplications and Methods, vol. 2, pp. 310–313 (2012)

An Efficient Shape Feature Extraction, Description and Matching Method Using GPU

Leonardo Chang[1,2]($\boxtimes$), Miguel Arias-Estrada[2], José Hernández-Palancar[1], and L. Enrique Sucar[2]

[1] Advanced Technologies Application Center (CENATAV),
7A # 21406, Siboney, Playa, CP 12200 Havana, Cuba
{lchang,jpalancar}@cenatav.co.cu
[2] Instituto Nacional de Astrofísica, Óptica y Electrónica (INAOE),
Luis Enrique Erro # 1, Tonantzintla,
CP 72840 Heróica Puebla de Zaragoza, Puebla, Mexico
{ariasmo,esucar,lchang}@ccc.inaoep.mx

Abstract. Shape information is an important cue for many computer vision applications. In this work we propose an invariant shape feature extraction, description and matching method for binary images, named LISF. The proposed method extracts local features from the contour to describe shape and these features are later matched globally. Combining local features with global matching allows us to a obtaining a trade-off between discriminative power and robustness to noise and occlusion in the contour. The proposed extraction, description and matching methods are invariant to rotation, translation, and scale and present certain robustness to partial occlusion. The conducted experiments in the Shapes99, Shapes216, and MPEG-7 datasets support the mentioned contributions, where different artifacts were artificially added to obtain partial occlusion as high as 60 %. For the highest occlusion levels LISF outperformed other popular shape description methods, with about 20 % higher bull's eye score and 25 % higher accuracy in classification. Also, in this paper, we present a massively parallel implementation in CUDA of the two most time-consuming stages of LISF, i.e., the feature extraction and feature matching steps; which achieves speed-ups of up to 32x and 34x, respectively.

Keywords: Shape matching · Invariant shape features · Shape occlusion · Efficient feature extraction · Efficient feature matching · GPU

1 Introduction

Shape descriptors have proven to be useful in many image processing and computer vision applications (e.g., object detection [1,2], image retrieval [3,4], object categorization [5,6], etc.). However, shape representation and description remains as one of the most challenging topics in computer vision. The shape representation problem has proven to be hard because shapes are usually more

© Springer International Publishing Switzerland 2015
A. Fred et al. (Eds.): ICPRAM 2014, LNCS 9443, pp. 206–221, 2015.
DOI: 10.1007/978-3-319-25530-9_14

complex than appearance. Shape representation inherits some of the most important considerations in computer vision such as the robustness with respect to the image scale, rotation, translation, occlusion, noise and viewpoint. A good shape description and matching method should be able to tolerate geometric intra-class variations, but at the same time should be able to discriminate from objects of different classes.

In this work, we describe object shape locally, but global information is used in the matching step to obtain a trade-off between discriminative power and robustness. The proposed approach has been named Invariant Local Shape Features (LISF), as it extracts, describes, and matches local shape features that are invariant to rotation, translation and scale. LISF, besides closed contours, extracts and matches features from open contours making it appropriate for matching occluded or incomplete shape contours. Conducted experiments showed that while increasing the occlusion level in shape contour, the difference in terms of bull's eye score, and accuracy of the classification gets larger in favor of LISF compared to other state of the art methods.

Another important requirement for a promising shape descriptor is computational efficiency. Several applications demand real time processing or handling large image datasets. General-Purpose Computing on Graphics Processing Units (GPGPU) is the utilization of GPUs to perform computation in applications traditionally handled by a CPU, having obtained considerable speed-ups in many computing tasks. In this work, we also propose a massively parallel implementation in GPUs of the two most time consuming stages of LISF, namely, the feature extraction and feature matching stages. Our proposed GPU implementation achieves a speed-up of up to 32x and 34x for the feature extraction and matching steps, respectively.

The rest of the paper is organized as follows. Section 2 discusses some shape description and matching approaches. Section 3.1 presents the local shape feature extraction method. The feature descriptor is presented in Sect. 3.2. Its robustness and invariability to translation, rotation, scale, and its locality property are discussed in Sect. 3.3. Section 4 describes the proposed feature matching schema. The performed experiments and discussion are presented in Sect. 6. Finally, Sect. 7 concludes the paper with a summary of our proposed methods, main contributions, and future work.

2   Related Work

Some recent works where shape descriptors are extracted using all the pixel information within a shape region include Zernike moments [7], Legendre moments [8], and generic Fourier descriptor [9]. The main limitation of region-based approaches resides in that only global shape characteristics are captured, without taking into account important shape details. Hence, the discriminative power of these approaches is limited in applications with large intra-class variations or with databases of considerable size.

Curvature scale space (CSS) [10], multi-scale convexity concavity (MCC) [11] and multi-scale Fourier-based descriptor [12] are shape descriptors defined in a multi-scale space. In CSS and MCC, by changing the sizes of Gaussian kernels in contour convolution, several shape approximations of the shape contour at different scales are obtained. CSS uses the number of zero-crossing points at these different scale levels. In MCC, a curvature measure based on the relative displacement of a contour point between every two consecutive scale levels is proposed. The multi-scale Fourier-based descriptor uses a low-pass Gaussian filter and a high-pass Gaussian filter, separately, at different scales.

The main drawback of multi-scale space approaches is that determining the optimal parameter of each scale is a very difficult and application dependent task.

Geometric relationships between sampled contour points have been exploited effectively for shape description. Shape context (SC) [13] finds the vectors of every sample point to all the other boundary points. The length and orientation of the vectors are quantized to create a histogram map which is used to represent each point. To make the histogram more sensitive to nearby points than to points farther away, these vectors are put into log-polar space. The triangle-area representation (TAR) [14] signature is computed from the area of the triangles formed by the points on the shape boundary. TAR measures the convexity or concavity of each sample contour point using the signed areas of triangles formed by contour points at different scales. In these approaches, the contour of each object is represented by a fixed number of sample points and when comparing two shapes, both contours must be represented by the same fixed number of points. Hence, how these approaches work under occluded or uncompleted contours is not well-defined. Also, most of these kind of approaches can only deal with closed contours and/or assume a one-to-one correspondence in the matching step.

In addition to shape representations, in order to improve the performance of shape matching, researchers have also proposed alternative matching methods designed to get the most out of their shape representations. In [15], the authors proposed a hierarchical segment-based matching method that proceeds in a global to local direction. The locally constrained diffusion process proposed in [16] uses a diffusion process to propagate the beneficial influence that offer other shapes in the similarity measure of each pair of shapes. Authors in [17] replace the original distances between two shapes with distances induced by geodesic paths in the shape manifold.

Shape descriptors which only use global or local information will probably fail in presence of transformations and perturbations of shape contour. Local descriptors are accurate to represent local shape features, however, are very sensitive to noise. On the other hand, global descriptors are robust to local deformations, but can not capture the local details of the shape contour. In order to balance discriminative power and robustness, in this work we use local features (contour fragments) for shape representation; later, in the matching step, in a global manner, the structure and spatial relationships between the extracted local features are taken into account to compute shapes similarity. To improve matching performance, specific characteristics such as scale and orientation of the extracted features are used. The extraction, description and matching processes are invariant to rotation, translation and scale changes. In addition, there is not restriction about only dealing with closed contours or silhouettes, i.e. the method also extract features from open contours.

The shape representation method used to described our extracted contour fragments is similar to that of shape context [13]. Besides locality, the main difference between these descriptors is that in [13] the authors obtain a histogram for each point in the contour, while we only use one histogram for each contour fragment, i.e. our representation is more compact. Unlike our proposed method, shape context assumes a one-to-one correspondence between points in the matching step, which makes it more sensitive to occlusion.

The main contribution of this paper is a local shape feature extraction, description and matching schema that (i) is invariant to rotation, translation and scaling, (ii) provides a balance between distinctiveness and robustness thanks to the local character of the extracted features, which are later matched using global information, (iii) deals with either closed or open contours, and (iv) is simple and easy to compute. An additional contribution is a massively parallel implementation in GPUs of the proposed method.

3 Proposed Local Shape Feature Descriptor

Psychological studies [18, 19] show that humans are able to recognize objects from fragments of contours and edges. Hence, if the appropriate contour fragments of an object are selected, they are representative of it.

Straight lines are not very discriminative since they are only defined by their length (which is useless when looking for scale invariance). However, curves provide a richer description of the object as these are defined, in addition to its length, by its curvature (a line can be seen as a specific case of a curve, i.e., a curve with null curvature). Furthermore, in the presence of variations such as changes in scale, rotation, translation, affine transformations, illumination and texture, the curves tend to remain present. In this paper we use contour fragments as repetitive and discriminant local features.

3.1 Feature Extraction

The detection of high curvature contour fragments is based on the method proposed by Chetverikov [20]. Chetverikov's method inscribes triangles in a segment of contour points and evaluates the angle of the median vertex which must be smaller than α_{max} and bigger than α_{min}. The sides of the triangle that lie on the median vertex are required to be larger than d_{min} and smaller than d_{max}:

$$d_{min} \leq ||p - p^+|| \leq d_{max}, \tag{1}$$

$$d_{min} \leq ||p - p^-|| \leq d_{max}, \tag{2}$$

$$\alpha_{min} \leq \alpha \leq \alpha_{max}, \tag{3}$$

d_{min} and d_{max} define the scale limits, and are set empirically in order to avoid detecting contour fragments that are known to be too small or too large. α_{min} and α_{max} are the angle limits that determine the minimum and maximum sharpness accepted as high curvature. In our experiments we set $d_{min} = 10$ pixels, $d_{max} = 300$ pixels, $\alpha_{min} = 5°$, and $\alpha_{max} = 150°$.

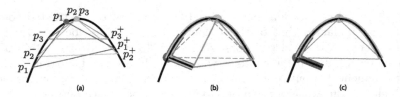

Fig. 1. Detection of contour fragments. (a) Are candidates contour fragments those contour fragments where it is possible to inscribe a triangle with aperture between α_{min} and α_{max}, and adjacent sides with lengths between d_{min} and d_{max}. If several triangles are found on the same point or near points, the sharpest triangle in a neighborhood is selected. (b) Noise can introduce false contour fragments (the contour fragment in yellow). (c) To counteract the false contour phenomenon we add another restriction, candidate triangles will grow until another corner is reached (Color figure online).

Several triangles can be found over the same point or over adjacent points at the same curve, hence it is selected the point with the highest curvature. Each selected contour fragment i is defined by a triangle (p_i^-, p_i, p_i^+), where p_i is the median vertex and the points p_i^- and p_i^+ define the endpoints of the contour fragment. See Fig. 1(a).

The Chetverikov's corners detector has the disadvantage of not being very stable to noisy contours or highly branched contours, which may cause that false corners are selected. For example, see Fig. 1(b). In order to deal with this problem, another restriction is added to the Chetverikov's method. Each candidate triangle (p_k^-, p_k, p_k^+) will grow while the points p_k^- and p_k^+ do not match any p_j point of another corner. Figure 1(c) shows how this restriction overcome the false detection in the example in Fig. 1(b).

Then, each feature ς_i extracted from the contour is defined by $\langle P_i, T_i \rangle$, where $T_i = (p_i^-, p_i, p_i^+)$ is the triangle inscribed in the contour fragment and $P_i = \{p_1, ..., p_n\}, p_j \in \mathbb{R}^2$ is the set of n points which form the contour fragment ς_i, ordered so that the point p_j is adjacent to the point p_{j-1} and p_{j+1}. Points $p_1, p_n \in P_i$ match with points $p_i^-, p_i^+ \in T_i$, respectively.

3.2   Feature Description

The definition of contour fragment given by the extraction process (specifically the triangle (p_i^-, p_i, p_i^+)) provides a compact description of the contour fragment as it gives evidence of amplitude, orientation and length; however, it has low distinctiveness due to the fact that different curves can share the same triangle.

In order to give more distinctiveness to the extracted features, we represent each contour fragment in a polar space of origin p_i, where the length r and the orientation θ of each point are discretized to form a two-dimensional histogram of $n_r \times n_\theta$ bins:

$$H_i(b) = |\{w \in P_i : (w - p_i) \in \text{bin}(b)\}|. \tag{4}$$

Note that for a sufficiently large number of n_r and n_θ this is an exact representation of the contour fragment.

3.3   Robustness and Invariability Considerations

In order to have a robust and invariant description method, several properties must be met:

Locality: The locality property is met directly from the definitions of interest contour fragment and its descriptor given in Sects. 3.1 and 3.2. A contour fragment and its descriptor only depend on a point and a set of points in a neighborhood much smaller than the image area, therefore, in both the extraction and description processes, a change or variation in a portion of the contour (produced, for example, by noise, partial occlusion or other deformation of the object), only affects the features extracted in that portion.

Translation Invariance: By construction, both the feature extraction and description processes are inherently invariant to translation since they are based on relative coordinates of the points of interest.

Rotation Invariance: The contour fragment extraction process is invariant to rotation by construction. An interest contour fragment is defined by a triangle inscribed in a contour segment, which only depends on the shape of the contour segment rather than its orientation. In the description process, it is possible to achieve rotation invariance by rotating each feature coordinate systems until alignment with the bisectrix of the vertex p_i.

Scale Invariance: This could be achieved in the extraction process by extracting contour fragments at different values of d_{min} and d_{max}. In the description process it is achieved by sampling contour fragments (i.e., P_i) to a fixed number M of points or by normalizing the histograms.

4   Feature Matching

In this section we describe the method for finding correspondences between LISF features extracted from two images. Let's consider the situation of finding correspondences between N_Q features $\{a_i\}$, with descriptors $\{H_i^a\}$, extracted from the query image and N_C features $\{b_i\}$, with descriptors $\{H_i^b\}$, extracted from the database image.

The simplest criterion to establish a match between two features is to establish a global threshold over the distance between the descriptors, i.e., each feature a_i will match with those features $\{b_j\}$ which are at distance $D(a_i, b_j)$ below a given threshold. Usually, matches are restricted to nearest neighbors in order to limit multiple false positives. Some intrinsic disadvantages of this approach limit its use; such as determining the number of nearest neighbors depends on the specific application and type of features and objects. The mentioned approach obviates the spatial relations between the parts (local features) of objects, which is a determining factor. Also, it fails in the case of objects with multiple occurrences of the structure of interest or objects with repetitive parts (e.g. buildings of several equal windows). In addition, the large variability of distances between the descriptors of different features makes the task of finding an appropriate threshold a very difficult task.

To overcome the previous limitations, we propose an alternative for feature matching that takes into account the structure and spatial organization of the features. The matches between the query features and database features are validated by rejecting casual or wrong matches.

4.1   Finding Candidate Matches

Let's first define the scale and orientation of a contour fragment.

Let the feature ς_i be defined by $\langle P_i, T_i \rangle$, its scale s_{ς_i} is defined as the magnitude of the vector $\mathbf{p_i^+} + \mathbf{p_i^-}$, where $\mathbf{p_i^+}$ and $\mathbf{p_i^-}$ are the vectors with initial point in p_i and terminal points in p_i^+ and p_i^-, respectively, i.e.,

$$s_{\varsigma_i} = |\mathbf{p_i^+} + \mathbf{p_i^-}|. \tag{5}$$

The orientation ϕ_{ς_i} of the feature ς_i is given by the direction of vector $\mathbf{p_i}$, which we will call orientation vector of feature ς_i, and is defined as the vector that is just in the middle of vector $\mathbf{p_i^+}$ and vector $\mathbf{p_i^-}$, i.e.,

$$\mathbf{p_i} = \hat{\mathbf{p}}_i^+ + \hat{\mathbf{p}}_i^-, \tag{6}$$

where $\hat{\mathbf{p}}_i^+$ and $\hat{\mathbf{p}}_i^-$ are the unit vectors with same direction and origin that $\mathbf{p_i^+}$ and $\mathbf{p_i^-}$, respectively.

We already defined the terms scale and orientation of a feature ς_i. In the process of finding candidate matches, for each feature a_i, its K nearest neighbors $\{b_j^K\}$ in the candidate image are found by comparing their descriptors (in this work we use χ^2 distance to compare histograms). Our method tries to find among the K nearest neighbors the best match (if any), so K can be seen as an accuracy parameter. To provide the method with rotation invariance the feature descriptors are normalized in terms of orientation. This normalization is performed by rotating the polar coordinate system of each feature by a value equal to $-\phi_{\varsigma_i}$ (i.e., all features are set to orientation zero) and calculated their descriptors. The scale and translation invariance in the descriptors is accomplished by construction (for details see Sect. 3.2).

4.2   Rejecting Casual Matches

For each pair $\langle a_i, b_j^k \rangle$, the query image features $\{a_i\}$ are aligned according to the correspondence $\langle a_i, b_j^k \rangle$:

$$a_i' = (a_i \cdot s + \mathbf{t}) \cdot R(\theta(a_i, b_j^k)),$$

where $s = s_{a_i}/s_{b_j^K}$ is the scale ratio between the features a_i and b_j^k, $\mathbf{t} = p_{a_i} - p_{b_j^k}$ is the translation vector from point p_{a_i} to point $p_k^{b_j}$, $R(\theta(a_i, b_j^k))$ is the rotation matrix for a rotation, around point p_{a_i}, equal to the direction of the orientation vector of feature a_i with respect to the orientation of b_j^k, (i.e., $\phi_{a_i} - \phi_{b_j^k}$).

Once aligned both images (same scale, rotation and translation) according to correspondence $\langle a_i, b_j^k \rangle$, for each feature a_i' its nearest neighbor b_v in $\{b_j^k\}$ is found. Then, vector $\mathbf{m}$ defined by (l, φ) is calculated, where l is the distance from point p_{b_v} of feature b_v to a reference point $p_\bullet$ in the candidate object (e.g., the object centroid, the point p of some feature or any other point, but always the same point for every candidate image) and φ is the orientation of feature b_v with respect to the reference point $p_\bullet$, i.e., the angle between the orientation vector $\mathbf{p_{b_v}}$ of feature b_v and the vector $\mathbf{p_\bullet}$, the latter defined from point p_{b_v} to point $p_\bullet$,

$$l = ||p_{b_v} - p_\bullet||, \tag{7}$$

$$\varphi = \arccos\left(\frac{\mathbf{p_{b_v}} \cdot \mathbf{p_\bullet}}{||\mathbf{p_{b_v}}|| \; ||\mathbf{p_\bullet}||}\right). \tag{8}$$

Having obtained $\mathbf{m}$, the point p_o, given by the point at a distance l from point $p_{a_i'}$ of feature a_i' and orientation φ respect to its orientation vector $\mathbf{p_{a_i'}}$, is found,

$$p_o^x = p_{a_i'}^x + l \cdot \cos(\phi_{a_i'} + \varphi), \tag{9}$$

$$p_o^y = p_{a_i'}^y + l \cdot \sin(\phi_{a_i'} + \varphi). \tag{10}$$

Intuitively, if $\langle a_i, b_j^k \rangle$ is a correct match, most of the points p_o should be concentrated around the point $p_\bullet$. This idea is what allows us to accept or reject a candidate match $\langle a_i, b_j^k \rangle$. With this aim, we defined a matching measure Ω between features a_i and b_j^k as a measure of dispersion of points p_o around point $p_\bullet$,

$$\Omega = \sqrt{\frac{\sum_{i=1}^{N_Q} ||p_o^i - p_\bullet||^2}{N_Q}}. \tag{11}$$

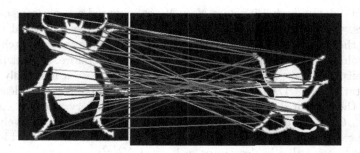

Fig. 2. Matches between local shape descriptors in two images. It can be seen how these matches were found even in presence of rotation, scale and translation changes.

Using this measure, Ω, we can determine the best match for each feature a_i of the query image in the candidate image, or reject any weak match having Ω above a given threshold λ_Ω. A higher threshold means supporting larger deformations of the shape, but also more false matches. In Fig. 2, the matches between features extracted from silhouettes of two different instances of the same object class are shown, the robustness to changes in scale, rotation and translation can be appreciated.

5   Efficient LISF Feature Extraction and Matching

In this section, we present a massively parallel implementation in GPUs of the two most time-consuming stages of LISF, i.e., the feature extraction and the feature matching steps.

5.1   Implementation of Feature Extraction Using CUDA

As mentioned in Sect. 3.1, in the feature extraction step, for each point p_i in the contour, up to P triangles are evaluated, where P is the contour size. Each one of these evaluations are independent from each other, so there is a great potential for parallelism. We present a massively parallel implementation in CUDA of this stage by obtaining in parallel the candidate triangle of each point p_i in the contour.

All the triangles of a point p_i are evaluated in a block. The constraints of each triangle (Eqs. 1–3) are evaluated in a thread; triangles that fulfill these constrains, i.e., candidate triangles, are tiled into the shared memory in order to increase data reutilization and decrease global memory accesses. Later, in each block the highest curvature candidate triangle of corresponding point p_i is selected. The final step, i.e., the selection of the shaper triangle in the neighborhood, is performed in the host. As there are only a few candidate triangles in a neighborhood, this is a task which is more favored to be performed in the CPU.

5.2   Implementation of Feature Matching Using CUDA

Finding candidate matches involves $N_Q \times N_C$ chi-squared comparisons of feature descriptors, where N_Q and N_C are the number of features extracted from the query and the database images, respectively. Also, rejecting casual matches needs $N_Q \times N_C$ chi-squared comparisons after alignment. Therefore, a great potential for parallelism is also present in these stages. We propose a massively parallel implementation in CUDA for the chi-squared comparison of $N_Q \times N_C$ descriptors.

Given the sets of descriptors extracted from the query and the candidate image, i.e., $Q = \{q_1, q_2, ..., q_{N_Q}\}$ and $C = \{c_1, c_2, ..., c_{N_C}\}$, respectively, where the size of each descriptor is given by $n_r \times n_\theta$. To perform $N_Q \times N_C$ chi-squared comparisons each value in descriptor q_i is used N_C times. In order to increase data reutilization and decrease global memory accesses, Q and C are tiled into the shared memory. In each device block the chi-squared distances between every

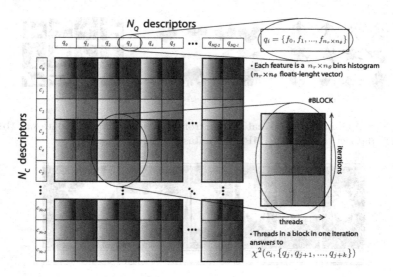

Fig. 3. Overview of the proposed feature comparison method in GPU.

pair of descriptors in $a \subset Q$ and $b \subset C$ are computed, where $|a| \ll N_Q$ and $|b| \ll N_C$. Then, all the comparison are obtained in $|b|$ iterations, where in the jth iteration the threads in the block compute the chi-squared distance of the jth descriptor in b against every descriptor in a. Figure 3 shows a graphical representation.

For values of N_Q and N_C such that the features and comparison results do not fit in the device global memory, the data could be partitioned and the kernel lunched several times.

6   Experimental Results

Performance of the proposed LISF method has been evaluated on three different well-known datasets. The first dataset is the Kimia Shapes99 dataset [21], which includes nine categories and eleven shapes in each category with variations in form, occlusion, articulation and missing parts. The second dataset is the Kimia Shapes216 dataset [21]. This database consists of 18 categories with 12 shapes in each category. The third dataset is the MPEG-7 CE-Shape-1 dataset [22], which consists of 1400 images (70 object categories with 20 instances per category). In the three datasets, in each image there is only one object, defined by its silhouette, and at different scales and rotations. Example shapes are shown in Fig. 4.

6.1   Shape Retrieval and Classification Experiments

In order to show the robustness of the LISF method to partial occlusion in the shape, we generated another 15 datasets by artificially introducing occlusion of different magnitudes (10 %, 20 %, 30 %, 45 % and 60 %) to the Shapes99,

a) Shapes99   b) Shapes216   c) MPEG-7

Fig. 4. Example images and categories from (a) the Shapes99 dataset, (b) the Shapes216 dataset, and (c) the MPEG-7 dataset.

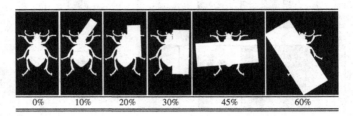

Fig. 5. Example image from the MPEG-7 dataset with different levels of occlusion (0 %, 10 %, 20 %, 30 %, 45 % and 60 %) used in the experiments.

Shapes216 and MPEG-7 datasets. Occlusion was added by randomly choosing rectangles that occlude the desired portion of the shape contour. A sample image from the MPEG-7 dataset at different occlusion levels is shown in Fig. 5.

As a measure to evaluate and compare the performance of the proposed shape matching schema in a shape retrieval scenario we use the so-called bull's eye score. Each shape in the database is compared with every other shape model, and the number of shapes of the same class that are among the $2N_c$ most similar is reported, where N_c is the number of instances per class. The bull's eye score is the ratio between the total number of shapes of the same class and the largest possible value.

The results obtained by LISF ($n_r = 5$, $n_\theta = 10$, $\lambda_\Omega = 0.9$) were compared with those of the popular shape context descriptor (100 points, $n_r = 5$, $n_\theta = 12$) [13], the Zernike moments (using 47 features) [23] and the Legendre moments (using 66 features) [8]. Rotation invariance can be achieve by shape context, but it has several drawbacks, as mentioned in [13]. In order to perform a fair comparison between LISF (which is rotation invariant) and shape context, in our experiments the non-rotation invariant implementation of shape context is used, and images used by shape context were rotated so that the objects had the same rotation.

Motivated by efficiency issues, for the MPEG-7 CE-Shape-1 dataset we used only 10 of the 70 categories (selected randomly) with its 20 samples each. The bull's eye score implies all-against-all comparisons and experiments had to be done across the 18 datasets for the LISF, shape context, Zernike moments and Legendre moments methods. There is no loss of generality in using a subset of the MPEG-7 dataset since the aim of the experiment is to compare the behavior of the LISF method against other methods, across increasing levels of occlusion.

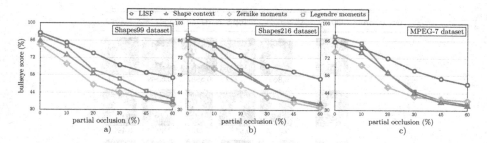

Fig. 6. Bull's eye score comparison between LISF, shape context, Zernike moments and Legendre moments in the (a) Shapes99, (b) Shapes216 and (c) MPEG-7 datasets with different partial occlusions (0 %, 10 %, 20 %, 30 %, 45 % and 60 %) (Color figure online).

As a similarity measure of image a with image b, with local features $\{a_i\}$ and $\{b_j\}$ respectively, we use the ratio between the number of features in $\{a_i\}$ that found matches in $\{b_j\}$ and the total number of features extracted from a.

Figure 6 shows the behavior of the bull's eye score of each method while increasing partial occlusion in the Shapes99, Shapes216 and MPEG-7 datasets. Bull's eye score is computed for each of the 18 datasets independently.

As expected, the LISF method outperforms the shape context, Zernike moments and Legendre moments methods. Moreover, while increasing the occlusion level, the difference in terms of bull's eye score gets bigger, with about 15–20 % higher bull's eye score across highly occluded images; which shows the advantages of the proposed method over the other three.

Figure 7 shows the top 5 retrieved images and its retrieval score for the *beetle-5* image with different occlusions. Top 5 retrieved images are shown for each database at different occlusion levels, respectively (MPEG-7 with 0 % to 60 % partial occlusion). The robustness to partial occlusion of the LISF method can be appreciated. Retrieval score of images that do not belong to the same class as the query image are depicted in bold italic.

In a second set of experiments, the proposed method is tested and compared to shape context, Zernike moments and Legendre moments in a classification task also under varying occlusion conditions. A 1-NN classifier was used, i.e., we assigned to each instance the class of its nearest neighbor. The same data as in the first set of experiments is used. In order to measure the classification performance, the accuracy measure was used. Accuracy measures the percentage of data that are correctly classified. Figure 8 shows the results of classification under different occlusion magnitudes (0 %, 10 %, 20 %, 30 %, 45 % and 60 % occlusion).

In this set of experiments, a better performance of the LISF method compared to previous work can also be appreciated. As in the shape retrieval experiment, while increasing the occlusion level in the test images, the better is the performance of the proposed method with respect to shape context, Zernike moments and Legendre moments, with more than 25 % higher results in accuracy.

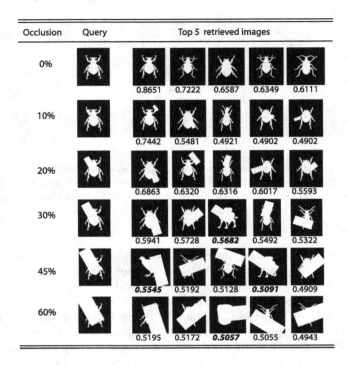

Fig. 7. Top 5 retrieved images and similarity score. In each row retrieval results for the *beetle-5* image in the six MPEG-7 based databases. Red retrieval scores represent images that do not belong to the same class of the query image (Color figure online).

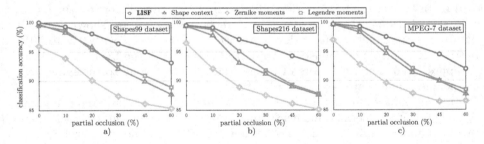

Fig. 8. Classification accuracy comparison between LISF, shape context, Zernike moments and Legendre moments in the (a) Shapes99, (b) Shape 216, and (c) MPEG-7 dataset, with different partial occlusions (0 %, 10 %, 20 %, 30 %, 45 % and 60 %) (Color figure online).

6.2   Efficiency Evaluation

The computation time of LISF has been evaluated and compared to other methods. Table 1 shows the comparison of LISF computation time against shape context, Legendre moments, and Zernike moments. The reported times correspond to the average time needed to describe and match two shapes of the

Table 1. Average feature extraction and matching time for two images of the MPEG7 database, in seconds.

Method	Computation time (s)
Shape context	2.66
Legendre moments	7.48
Zernike moments	26.47
LISF_CPU	**0.47**
LISF_GPU	**0.16**

MPEG-7 database over 500 runs. The LISF_CPU, shape context, Legendre and Zernike moments results were obtained on a single thread of a 2.2 GHz processor and 8 GB RAM PC, and the LISF_GPU results were obtained on a NVIDIA GeForce GT 610 GPU. As can be seen in Table 1, both implementations of LISF are the least time-consuming compared with shape context, Legendre moments, and Zernike moments.

In order to show the scalability of our proposed massively parallel implementation in CUDA, we reported the time and achieved speed-up while increasing the contour size and the number of features to match for the feature extrac-

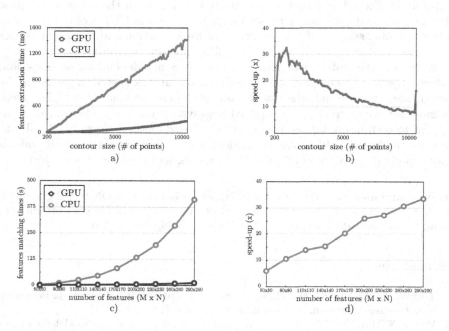

Fig. 9. Computation time and achieve speed-up by the proposed massively parallel implementation in CUDA wrt. the CPU implementation for the (a,b) feature extraction and (c,d) feature matching stages of LISF (Color figure online).

tion and feature matching stages, respectively. These results were obtained on a NVIDIA GeForce GTX 480 GPU and compared with those obtained in a single threaded Intel CPU Processor at 3.4 GHz.

As it can be seen in Figs. 9(a) and 9(b), tested on contours of sizes ranging from 200 to 10 000 points, the proposed feature extraction implementation on GPU achieves up to a 32x speed-up and a 16x average speed-up. For the feature matching step (see Figs. 9(c) and 9(d)), the proposed GPU implementation were tested for comparing from 50 vs. 50 to 290 vs. 290 features. The GPU implementation showed linear scaling against exponential scaling of the CPU implementation and obtained a 34x speed-up when comparing 290 vs. 290 LISF features.

7   Conclusions and Future Work

As a result of this work, a method for shape feature extraction, description and matching, invariant to rotation, translation and scale, have been developed. The proposed method allows us to overcome the intrinsic disadvantages of only using local or global features by capturing both local and global information. The conducted experiments supported the mentioned contributions, showing larger robustness to partial occlusion than other methods in the state of the art. It is also more efficient in terms of computational time than the other techniques. Also, we proposed a massively parallel implementation in CUDA of the two most time-consuming stages of LISF, i.e., the feature extraction and feature matching steps, which achieves speed-ups of up to 32x and 34x, respectively.

Moreover, the feature extraction process does not depend on accurate and perfect object segmentation since the features are extracted from both the contour and the internal edges of the object. Therefore, the method has great potential for use in "real" images (RGB or grayscale images) and also, as a complement to certain limitations of appearance based methods (e.g., SIFT, SURF, etc.); particularly in object categorization, where shape features usually offer a more generic description of objects. Future work will focus on this subject.

Acknowledgements. This project was supported in part by CONACYT grant Ref. CB-2008/103878 and by Instituto Nacional de Astrofísica, Óptica y Electrónica. L. Chang was supported in part by CONACYT scholarship No. 240251.

References

1. Toshev, A., Taskar, B., Daniilidis, K.: Shape-based object detection via boundary structure segmentation. Int. J. Comput. Vis. **99**, 123–146 (2011)
2. Wang, X., Bai, X., Ma, T., Liu, W., Latecki, L. J.: Fan shape model for object detection. In: CVPR, pp. 151–158. IEEE (2012)
3. Shu, X., Wu, X.J.: A novel contour descriptor for 2D shape matching and its application to image retrieval. Image and Vision Computing **29**, 286–294 (2011)

4. Yang, X., Bai, X., Köknar-Tezel, S., Latecki, L.: Densifying distance spaces for shape and image retrieval. J. Math. Imaging Vis. **46**, 12–28 (2013)
5. Trinh, N.H., Kimia, B.B.: Skeleton search: category-specific object recognition and segmentation using a skeletal shape model. Int. J. Comput. Vis. **94**, 215–240 (2011)
6. Gonzalez-Aguirre, D.I., Hoch, J., Röhl, S., Asfour, T., Bayro-Corrochano, E., Dillmann, R.: Towards shape-based visual object categorization for humanoid robots. In: ICRA, pp. 5226–5232. IEEE (2011)
7. Kim, W.Y., Kim, Y.S.: A region-based shape descriptor using zernike moments. Signal Process. Image Commun. **16**, 95–102 (2000)
8. Chong, C.W., Raveendran, P., Mukundan, R.: Translation and scale invariants of legendre moments. Pattern Recogn. **37**, 119–129 (2004)
9. Zhang, D., Lu, G.: Shape based image retrieval using generic fourier descriptors. Signal Process. Image Commun. **17**, 825–848 (2002)
10. Mokhtarian, F., Bober, M.: Curvature Scale Space Representation: Theory, Applications, and MPEG-7 Standardization. Kluwer, Dordrecht (2003)
11. Adamek, T., O'Connor, N.E.: A multiscale representation method for nonrigid shapes with a single closed contour. IEEE Trans. Circ. Syst. Video Technol. **14**, 742–753 (2004)
12. Direkoglu, C., Nixon, M.: Shape classification via image-based multiscale description. Pattern Recogn. **44**, 2134–2146 (2011)
13. Belongie, S., Malik, J., Puzicha, J.: Shape matching and object recognition using shape contexts. IEEE Trans. Pattern Anal. Mach. Intell. **24**, 509–522 (2002)
14. Alajlan, N., Rube, I.E., Kamel, M.S., Freeman, G.: Shape retrieval using triangle-area representation and dynamic space warping. Pattern Recogn. **40**, 1911–1920 (2007)
15. McNeill, G., Vijayakumar, S.: Hierarchical procrustes matching for shape retrieval. In: CVPR (1), pp. 885–894. IEEE Computer Society (2006)
16. Yang, X., Köknar-Tezel, S., Latecki, L.J.: Locally constrained diffusion process on locally densified distance spaces with applications to shape retrieval. In: Proceedings of the IEEE Conference on Computer Vision and Pattern Recognition (CVPR) (2009)
17. Bai, X., Yang, X., Latecki, L.J., Liu, W., Tu, Z.: Learning context-sensitive shape similarity by graph transduction. IEEE Trans. Pattern Anal. Mach. Intell. **32**, 861–874 (2010)
18. Biederman, I., Ju, G.: Surface versus edge-based determinants of visual recognition. Cogn. Psychol. **20**, 38–64 (1988)
19. De Winter, J., Wagemans, J.: Contour-based object identification and segmentation: stimuli, norms and data, and software tools. Behav. Res. Methods Instrum. Comput. J. Psychon. Soc. Inc. **36**, 604–624 (2004)
20. Chetverikov, D.: A simple and efficient algorithm for detection of high curvature points in planar curves. In: Petkov, N., Westenberg, M.A. (eds.) CAIP 2003. LNCS, vol. 2756, pp. 746–753. Springer, Heidelberg (2003)
21. Sebastian, T.B., Klein, P.N., Kimia, B.B.: Recognition of shapes by editing their shock graphs. IEEE Trans. Pattern Anal. Mach. Intell. **26**, 550–571 (2004)
22. Latecki, L.J., Lakämper, R., Eckhardt, U.: Shape descriptors for non-rigid shapes with a single closed contour. In: CVPR, pp. 1424–1429. IEEE Computer Society (2000)
23. Khotanzad, A., Hong, Y.H.: Rotation invariant pattern recognition using zernike moments. In: 9th International Conference on Pattern Recognition, vol. 1, pp. 326–328 (1988)

Utilization of Multiple Sequence Analyzers for Bibliographic Information Extraction

Atsuhiro Takasu[1][✉] and Manabu Ohta[2]

[1] National Institute of Informatics, Tokyo, Japan
takasu@nii.ac.jp
[2] Okayama University, Okayama, Japan
ohta@de.cs.okayama-u.ac.jp

Abstract. This paper discusses the problems of analyzing title page layouts and extracting bibliographic information from academic papers. Information extraction is an important function for digital libraries to offer, providing versatile and effective access paths to library content. Sequence analyzers, such as those based on a conditional random field, are often used to extract information from object pages. Recently, digital libraries have grown and can now handle a large number and wide variety of papers. Because of the variety of page layouts, it is necessary to prepare multiple analyzers, one for each type of layout, to achieve high extraction accuracy. This makes rule management important. For example, at what stage should we invest in a new analyzer, and how can we acquire it efficiently, when receiving papers with a new layout? This paper focuses on the detection of layout changes and how we learn to use a new sequence analyzer efficiently. We evaluate the confidence metrics for sequence analyzers to judge whether they would be suited to title page analysis by testing three academic journals. The results show that they are effective for measuring suitability. We also examine the sampling of training data when learning how to use a new analyzer.

Keywords: Page layout analysis · Information extraction · Digital libraries · Conditional random field

1 Introduction

The digitization of documents has infiltrated our society and one piece of evidence is the rapid spread of electronic book reading devices such as iPad and Kindle. What we really need in such circumstances is not just the digitization of books, but digitization of all the printed or written documents in our society, which would create an information archive accessible from all over the world. Needless to say, digital libraries (DLs) would be one such type of information archive. Recently, some universities and research institutions have set up web-accessible archives as their own institutional repositories. Although metadata such as bibliographic information about documents are indispensable for the efficient access to and utilization of digital documents, techniques for creating

© Springer International Publishing Switzerland 2015
A. Fred et al. (Eds.): ICPRAM 2014, LNCS 9443, pp. 222–236, 2015.
DOI: 10.1007/978-3-319-25530-9_15

digital documents with appropriate metadata are not yet mature enough to be used in real applications. Extracting information that includes bibliographic data from documents is a key technology for realizing such information archives as intellectual legacies because it will enable the extraction of various kinds of metadata and will provide the users of such archives with full access to rich information sources.

For documents such as the academic papers studied here, the important bibliographic information will include the title, author information, and journal name. Extracting such bibliographic information from an academic paper is useful in creating or reconstructing metadata. For example, it could be used to link identical records stored in different DLs and for faceted retrieval. Although many researchers have studied bibliographic information extraction from papers and documents [2,12,15], it remains an active research area, with several competitions having been held[1].

For accurate information extraction, researchers have developed various rule-based methods that can exploit both logical structure and page layout. However, document archives such as DLs usually handle several different types of document. Because formulating and managing them requires effective and efficient methods, the rules used should be tailored to suit each type of document. Rule management becomes harder as the system grows and contains more papers with more varieties of types of layout. For example, when receiving a fresh set of documents, we must determine whether we should generate a new set of rules or use the existing rules. In addition, the rules should be properly updated because the layout of a particular type of document may sometimes change over time. To maintain such document archives, we require a rule management facility that can measure the suitability of rules and recompile sets of rules when required.

As reported previously, we have been developing a DL system for academic papers [9,10,15]. We are especially interested in extracting bibliographic information such as authors and titles. In previous studies, we applied a conditional random field (CRF) [5] to extract bibliographic information from the title pages of academic papers. In these studies, we observed that rule-based methods, whereby a CRF exploits several rules as a form of feature vector, can extract metadata with high accuracy. However, we had to use multiple CRFs, choosing the one to use according to the page layout of the target journal. In other words, we had to access sufficient homogeneously laid-out pages to be able to identify a CRF that could analyze the pages with high accuracy for this metadata extraction task.

The use of multiple CRFs for metadata extraction from documents requires rule management functions such as choosing the appropriate CRF for a particular document or deciding when to make a new CRF to handle a new or changed page layout. This led us to study management rules for page layout analysis and bibliographic information extraction from title pages [16].

In this paper, we first examine the effectiveness of multiple CRFs for information extraction from pages with various page layouts. We compare the labeling

---

[1] http://www.icdar2013.org/program/competitions.

performance of a single CRF and multiple CRFs in active learning and show that multiple CRFs perform better when analyzing the title pages of multiple journals. We then propose a method that uses confidence metrics calculated for the CRFs to measure the suitability of each CRF to a particular page layout among the layouts used for training. Our experimental results show that the metric's value decreases significantly when the CRF is applied to the title page of a journal that is different from the one used for learning the CRF.

This result indicates that the confidence metrics are effective in detecting page layout changes. We also examine the effectiveness of the metrics in selecting training samples during active sampling when learning a CRF for a new page layout.

The remainder of the paper is organized as follows. Section 2 defines the problem addressed by this study. Section 3 proposes a rule management method for title page analysis. Our experimental results are given in Sect. 4.

2 Problem Definition

Information extraction from academic papers has been studied by the document image analysis community [6,14]. In early work, researchers aimed to extract information from scanned document images. To realize highly accurate extraction, they have recently developed extraction methods specific to document components such as mathematical expressions [18], figures [1], and tables [17]. This paper deals with the extraction of bibliographic information such as titles, authors, and abstracts [12], which is one of the fundamental extraction tasks. Although bibliographic information may appear in various parts of the document, including title pages and reference sections, this paper focuses on title page analysis.

To extract bibliographic components such as the title and authors from title pages, we first extract tokens and apply a sequence analyzer to label each token with its type of bibliographic component. A character, word, or line can be regarded as a token. We choose lines as tokens because they achieved higher extraction accuracy in our preliminary experiments. Figure 1 shows an example

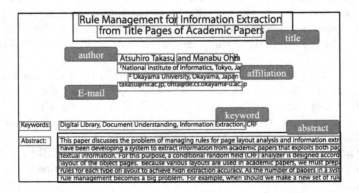

Fig. 1. Example of a page layout.

of a title page. The red rectangles are tokens extracted from the portable-document-format (PDF) file of the paper. In the figure, note that some bibliographic components are separated into tokens, such as the abstract. In addition, lines themselves can be split into multiple tokens, such as the title of the paper. The sequence analyzer merges the tokens comprising a single bibliographic component by labeling them as the same component. As a result, we obtain a set of bibliographic components, each of which comprises one or more tokens, as shown in Fig. 1.

Because bibliographic components are located in a two-dimensional (2D) space, some researchers have proposed rules that can analyze components of a page based on a page grammar [3] or a 2D CRF [7,19]. Others have proposed applying sequential analysis after serializing components of the page in a preceding step. For example, Peng et al. [12] proposed a CRF-based method for extracting bibliographic components from the title pages and reference sections of academic papers in PDF format. Councill et al. [2] developed a CRF-based toolkit for page analysis and information extraction. We adopt their approach and use a linear-chain CRF [5] as a sequence analyzer.

If the layout of the title page is different from that used when learning the CRF, the accuracy of extraction of the bibliographic components can be degraded. For large DLs, which will usually contain a variety of journals, the system will need to prepare multiple CRFs and choose one suitable for the target title page. The DL system will sometimes receive title pages for a new journal or for one with a redesigned layout. For these changes of layout, the system may not be able to analyze the title pages accurately. To address these difficulties, this paper considers the following problems:

- measuring the suitability of a CRF for a type of title page,
- detecting title pages that cannot be analyzed accurately with existing CRFs, and
- learning a new CRF so as to analyze pages efficiently for a new layout.

3   Layout Change Detection and CRF Learning

3.1   System Overview

We are developing a DL system to handle the variety of journals published in Japan. Because their bibliographic information is stored in multiple databases, the system creates linkages by locating identical papers in the multiple databases. It also provides a testbed for scholarly information studies such as citation analysis and paper recommendation.

The system aims to handle both newly published papers and papers published previously but not yet included in the system. As stated in the previous section, we use multiple CRFs in extracting information from the variety of journals. The system chooses a CRF according to the journal title and then applies it to that journal's papers to extract bibliographic information.

Whenever the layout of a paper changes or a new journal is incorporated, we must judge whether we can use a CRF already in the system or should build a new CRF. The system supports rule maintenance by:

- checking the suitability of a CRF for a given set of papers and alerting the user if the CRF does not analyze them with high confidence, and
- supporting the labeling of training data when a new CRF is generated.

3.2 The CRF

As described above, we have adopted a linear-chain CRF for the extraction of bibliographic information from the title pages of academic papers. Let L denote a set of labels. For a token sequence $\boldsymbol{x} := x_1 x_2 \cdots x_n$, a linear-chain CRF derives a sequence $\boldsymbol{y} := y_1 y_2 \cdots y_n$ of labels, i.e., $\boldsymbol{y} \in L^n$. A CRF M defines a conditional probability by:

$$P(\boldsymbol{y} \mid \boldsymbol{x}, M) = \frac{1}{Z(\boldsymbol{x})} \exp \left\{ \sum_{i=1}^{n} \sum_{k=1}^{K} \lambda_k f_k(y_{i-1}, y_i, \boldsymbol{x}) \right\}, \tag{1}$$

where $Z(\boldsymbol{x})$ is the partition constant. The feature function $f_k(y_{i-1}, y_i, \boldsymbol{x})$ is defined over consecutive labels y_{i-1} and y_i and the input sequence $\boldsymbol{x}$. Each feature function is associated with a parameter λ_k that gives the weight of the feature.

In the learning phase, the parameter λ_k is estimated from labeled token sequences. In the prediction phase, the CRF assigns the label sequence $\boldsymbol{y}^*$ to the given token sequence $\boldsymbol{x}$ that maximizes the conditional probability in Eq. (1).

3.3 Change Detection

To detect a layout change in a token sequence, we use metrics that give the likelihood that the token sequence was generated from the model. This problem is similar to the sampling problem in active sampling [13].

In Eq. (1), the CRF calculates the likelihood based on the transition weight from y_{i-1} to y_i and the correlation between a feature vector x_i and a hidden label y_i. A change of page layout may affect the transition weight between hidden labels in addition to the layout features in x_i. This will lead to a decrease in the likelihood $P(y^* \mid x, M)$ of the optimal label sequence y^* given by Eq. (1). A natural way to measure the model suitability is to use this likelihood. The CRF calculates the hidden label sequence $\boldsymbol{y}^*$ that maximizes the conditional probability given by Eq. (1). A higher $P(\boldsymbol{y}^* \mid \boldsymbol{x}, M)$ means a more confident assignment of labels, whereas a lower $P(\boldsymbol{y}^* \mid \boldsymbol{x}, M)$ means that the token sequence will make it hard for the current CRF model to assign labels.

The conditional probability is affected by the length of the token sequence $\boldsymbol{x}$. We therefore use the normalized conditional probability for the model suitability measure, as follows:

$$C_l(\boldsymbol{x}) := \frac{\log \left(P(\boldsymbol{y}^* \mid \boldsymbol{x}) \right)}{|\boldsymbol{x}|}. \tag{2}$$

Here, $|\boldsymbol{x}|$ denotes the length of the token sequence $\boldsymbol{x}$. We refer to the metric given by Eq. (2) as the *normalized likelihood*. The normalized likelihood is a type of confidence measure for when the model assigns labels to all tokens in the sequence $\boldsymbol{x}$.

A second measure is based on the confidence in assigning labels to a single token in the sequence. For a sequence $\boldsymbol{x}$, let Y_i denote a random variable for assigning a label to the ith token in $\boldsymbol{x}$. For label l in a set L of labels, $P(Y_i = l)$ denotes the marginal probability that label l is assigned to the ith token. If the token has feature values clearly supporting a specific label $l \in L$, $P(Y_i = l)$ must be significantly high and $P(Y_i = l')$ $(l' \neq l)$ must be low. The following entropy value can therefore quantify a token-level confidence:

$$e(\boldsymbol{x}, i) := \sum_{l \in L} -P(Y_i = l) \log(P(Y_i = l)) .\tag{3}$$

A low entropy value signifies that the label of token x_i is likely to be l. For the sequence analysis, an analyzer is regarded as succeeding in its analysis only if it assigns the correct label to every token. In other words, its confidence should be measured by the most difficult token to label. According to this perspective, we can use the *maximum entropy* of a token sequence $\boldsymbol{x}$ as another model of confidence, as follows:

$$C_e(\boldsymbol{x}) := - \max_{1 \leq i \leq |\boldsymbol{x}|} e(\boldsymbol{x}, i).\tag{4}$$

As opposed to the normalized likelihood, the maximum entropy can be regarded as a worst-case token-level metric.

The third metric is similar to the maximum entropy, but it measures the token-level confidence in terms of the maximum probability of label assignment to the token. It measures the confidence in the CRF for a token sequence $\boldsymbol{x}$ as the minimum of the token-level maximum probabilities over the sequence. It is defined formally as follows:

$$C_m(\boldsymbol{x}) := \min_{1 \leq i \leq |\boldsymbol{x}|} \max_{l \in L} P(Y_i = l) .\tag{5}$$

It is referred to as the *min-max label probability*.

Suppose that CRF M is used to label a token sequence obtained from a title page. There is more than one way to define the change detection problem, but the most basic definition is as follows. Given a new token sequence $\boldsymbol{x}$, determine whether the sequence is from the same information source as that from which the current CRF M was learned.

A token sequence $\boldsymbol{x}$ is judged to be a token sequence from the same information source if $C(\boldsymbol{x}) > \sigma$ holds for a predefined threshold σ, where C is C_l C_e, or C_m. Otherwise, the layout is regarded as having changed.

Because an issue of a journal will usually contain multiple papers, the change detection problem can be solved by detecting a change of title page layout adopted in journals when given a set of token sequences.

3.4 Learning a CRF for a New Layout

If we detect papers with a page layout that is different from those already known, we must derive a new CRF for these papers. We apply an active sampling technique [13], as follows:

1. Gather a significant number of papers T without labeling.
2. Choose an initial small number of papers T_0 from T, label them, and learn an *initial CRF* M_0 using the labeled papers.
3. At the tth iteration:
   (a) Let $\bar{T}$ be $T - \cup_{i=0}^{t-1} T_i$.
   (b) Calculate a metric described in Sect. 3.3 for each page in $\bar{T}$ using the CRF M_{t-1} obtained in the previous iteration.
   (c) Choose the bottom-k papers T_t from $\bar{T}$ according to the metric.
   (d) Label the papers T_t manually.
   (e) Learn CRF M_t using the labeled papers $\cup_{i=0}^{t} T_i$.

The aim of active sampling is to reduce the cost of labeling required to learn the CRF. Note that we need to delay learning a new CRF until we have gathered enough papers in the new layout (Step 1).

In active sampling, the sampling strategy for the initial CRF (Step 2) and for updating the CRF (Step 3(c)) is important. For the initial CRF, we choose k papers in T with the lowest values for the metric C introduced in Sect. 3.3, where C is calculated using the CRFs that we have at that time. This strategy means that we choose training papers T_0 that differ most in layout from those that we have so far.

In the tth update phase, we choose k training papers from $T - \cup_{i=0}^{t-1} T_i$ with the lowest values for the metric C, where C is calculated using the CRF M_{t-1} that we obtained in the previous step. This strategy means that we choose training papers with a different layout from those in $\cup_{i=0}^{t-1} T_i$.

4 Experimental Results

This section examines empirically the metrics for model fitness described in Sect. 3 by evaluating their effectiveness in detecting layout changes and in selecting training samples incrementally in active sampling.

4.1 Dataset

For this experiment, we used the same three journals as in our previous study [8], as follows:

- Journal of Information Processing by the Information Processing Society of Japan (IPSJ): We used papers published in 2003 in this experiment. This dataset contains 479 papers, most of them written in Japanese.

- English IEICE Transactions by the Institute of Electronics, Information and Communication Engineers in Japan (IEICE-E): We used papers published in 2003. This dataset contains 473 papers, all written in English.
- Japanese IEICE Transactions by the Institute of Electronics, Information and Communication Engineers in Japan (IEICE-J): We used papers published between 2000 and 2005. This dataset contains 964 papers, most of them written in Japanese.

As in [8], we used the following labels for the bibliographic components:

- Title: We used separate labels for Japanese and English titles because Japanese papers contained titles in both languages.
- Authors: We used separate labels for author names in Japanese and English as in the title.
- Abstract: As with the title and authors, we used separate labels for Japanese and English abstracts.
- Keywords: Only Japanese keywords are marked up in the IEICE-J.
- Other: Title pages usually contain paragraphs such as introductory paragraphs that are not classified into any of the above bibliographic components. We assigned the label "other" to the tokens in these paragraphs.

Note that different journals have different bibliographic components in their title pages.

Because we used the chain-model CRF, the tokens must be serialized. We therefore used lines extracted via OCR as tokens and serialized them according to the order generated by the OCR system. We labeled each token for training and evaluation manually.

4.2 Features of the CRF

As in [8], 15 feature templates were adopted. Of these, 14 were unigram features, i.e., the feature function $f_k(y_{i-1}, y_i, x)$ in Eq. (1) is calculated independently of the label y_{i-1}. There was one remaining bigram feature, i.e., the feature function $f_k(y_{i-1}, y_i, x)$ is calculated independently of the token sequence x. The unigram feature templates were further categorized into two kinds of features. Some involved layout features such as location, size, and gaps between lines. Others involved linguistic features such as the proportions of several kinds of characters in the tokens and the appearance of characteristic keywords that often appear in a particular bibliographic component such as "Institute" in affiliations. Table 1 summarizes the set of feature templates. Their values were calculated automatically from the token and label sequences.

An example of the bigram feature template $< y(-1), y(0) >$ is:

$$f_k(y_{i-1}, y_i, x) = \begin{cases} 1 \text{ if } y_{i-1} = \text{title}, y_i = \text{author} \\ 0 \text{ otherwise} \end{cases}. \tag{6}$$

Table 1. Feature templates of CRF for bibliographic component labeling [8].

Type	Feature	Description
Unigram	$< i(0) >$	Current line ID
	$< x(0) >$	Current line abscissa
	$< y(0) >$	Current line ordinate
	$< w(0) >$	Current line width
	$< h(0) >$	Current line height
	$< g(0) >$	Gap between current and preceding lines
	$< cw(0) >$	Median of character widths in the current line
	$< ch(0) >$	Median of character heights in the current line
	$< \#c(0) >$	Number of characters in the current line
	$< ec(0) >$	Proportion of alphanumerics in the current line
	$< kc(0) >$	Proportion of kanji in the current line
	$< jc(0) >$	Proportion of hiragana and katakana in the current line
	$< s(0) >$	Proportion of symbols in the current line
	$< kw(0) >$	Presence of predefined keywords in the current line
Bigram	$< y(-1), y(0) >$	Previous and current labels

This bigram feature indicates whether an author name follows a title in a label sequence, with the corresponding parameter λ_k showing how likely it is that an author name follows a title. CRF++ 0.58[2] [4] was used to learn and to label the token sequence for the title pages of each journal.

4.3   Bibliographic Component Labeling Accuracy

We first examined the bibliographic component labeling accuracy of CRFs. The purpose of this experiment was to evaluate the effectiveness of:

– the metrics described in Sect. 3.3 for active sampling, and
– multiple CRFs, each of which was learned for a particular layout of title pages.

We applied fivefold cross-validation. For each of the IPSJ, IEICE-E, and IEICE-J journals, we randomly split manually labeled title pages into five equal-sized groups of pages. In each round of the cross-validation, we examined the active sampling described in Sect. 3.4, choosing 10 papers randomly as the initial training data and the 10 least-confidently labeled papers according to the normalized likelihood at each iteration of the active sampling process.

We measured the accuracy of a learned CRF using the test token sequences for the same journal as the training data in each round of the cross-validation.

---

[2] https://code.google.com/p/crfpp/.

The accuracy was measured by:

$$\frac{\#\text{successfully labeled sequences}}{\#\text{test sequences}}. \tag{7}$$

Note that a CRF was only regarded as having succeeded in labeling if it assigned correct labels to all tokens in the token sequence. In other words, if a CRF assigned an incorrect label to one token despite correctly labeling all other tokens in a sequence x, it was regarded as having failed.

Figures 2 (a), (b), and (c) show the accuracy of CRFs for the IPSJ, IEICE-E, and IEICE-J journals, respectively. Each graph in the figure plots the accuracy of the CRF with respect to the size of training samples obtained at each iteration of the active sampling process. We first plotted the results of *normalized likelihood*, shown as a green curve in Fig. 2 and labeled as *nlh*. As shown in the graph, the accuracy increases as the active sampling proceeds. It converges when the training data size reaches about 50 for IPSJ and IEICE-E, whereas IEICE-J required about 250 training pages.

To evaluate the effectiveness of the metric for active sampling, we measured the accuracy of CRFs obtained for various quantities of randomly chosen training data. The purple curves labeled as *random* show the average accuracy of the random sampling with respect to the size of the set of training data. As shown in Fig. 2, we need much more training data to achieve an accuracy competitive with active sampling. In fact, we needed about 250 training pages for IPSJ and IEICE-E and more than 500 training pages for IEICE-J. In summary, using active sampling with the normalized likelihood described in Sect. 3.3 significantly reduces the training data required for learning CRFs to be used in bibliographic component labeling.

To evaluate the effectiveness of learning a separate CRF for each journal, we measured the accuracy of CRFs that were learned from merged training data for the three journals. More precisely, we merged the training data of the three journals to form the T used for the active sampling described in Sect. 3.4. We call the resultant CRF a *general-purpose CRF*. As in the evaluation of active

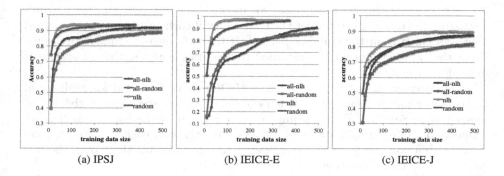

| (a) IPSJ | (b) IEICE-E | (c) IEICE-J |

Fig. 2. Bibliographic component labeling accuracy.

sampling, we chose 10 initial training pages and the 10 least-confidently labeled pages from the merged training data and measured the accuracy of the learned CRFs with respect to the test data for each journal separately. This experiment corresponds to the case of a single CRF being used to analyze the title pages of three journals.

The blue curves labeled as *all-nlh* in Fig. 2 show the average accuracy of the general-purpose CRFs when they were tested with the papers of the journals IPSJ in Fig. 2 (a), IEICE-E in Fig. 2 (b), and IEICE-J in Fig. 2 (c), respectively. Note that the convergence of the general-purpose CRFs differs according to the test journal used. For example, the accuracy converges at about 400 training papers for IPSJ papers, whereas more than 450 training papers were required for IEICE-J and IEICE-E papers. By comparing *nlh* and *all-nlh*, we can see that the separate CRFs converge for a smaller training dataset than that for a general-purpose CRF. In a fairer comparison, the separate CRFs require 50, 50, and 250 training papers for IPSJ, IEICE-E, and IEICE-J journals, respectively. This means that we would require 350 training papers in all to learn the separate CRFs, whereas we required more than 500 training papers for the general-purpose CRF.

The separate CRFs are more accurate than the general-purpose CRF. At convergence, the increased accuracy for the separate CRFs, as compared to the general-purpose CRF, are 2.2 % for IPSJ, 8.9 % for IEICE-E, and 2.6 % for IEICE-J. We also learned CRFs using randomly chosen papers from the merged training data. The red curves labeled as *all-random* show the average accuracy with respect to the training dataset size. We observed that using active sampling is effective, as with the separate CRFs, except for the IEICE-E case. Even in this exceptional case, active sampling achieved better accuracy for larger training datasets. In general, the metric described in Sect. 3.3 is effective in enabling active sampling to learn CRFs for bibliographic component labeling.

4.4 Change Detection Performance

This section evaluates the sensitivity of the confidence metrics with respect to layout change. For this purpose, we first learned a CRF by using training data for each journal. In the test phase, we merged the test sets for two journals including the training set and let the CRF judge if each title page in the merged test sets came from the training journal. If the title page was judged to come from the training journal, we regarded the page as *positive*. Otherwise it was regarded as *negative*.

The receiver operating characteristic (ROC) curve was used for evaluation. That is, the merged sets of test title pages were ranked according to the metric described in Sect. 3.3. By regarding the top k pages in the list as *positive*, we calculated the true-positive and false-positive fractions for each k. For each false-positive fraction, we plotted the averaged true-positive fractions over the five trials in the cross-validation as the ROC curve.

Figure 3 shows the ROC curves for each pair of journals. Each panel contains the ROC curves for the normalized likelihood labeled as *nlh*, for the maximum

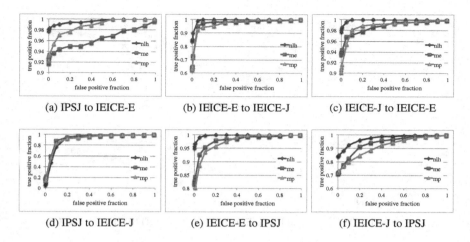

(a) IPSJ to IEICE-E   (b) IEICE-E to IEICE-J   (c) IEICE-J to IEICE-E

(d) IPSJ to IEICE-J   (e) IEICE-E to IPSJ   (f) IEICE-J to IPSJ

Fig. 3. Change detection performance.

entropy labeled as *me*, and for the min-max label probability labeled as *mp*. For example, the ROC curves in Fig. 3 (a) are the results of detecting title pages of IEICE-E from those of IPSJ using the CRF learned by labeled IPSJ training sequences. Similarly, the ROC curves in panel (b) are the results of detecting title pages of IEICE-J from those of IEICE-E using the CRF learned from labeled IEICE-E training data.

Two conclusions can be drawn from these results. First, the ROC curves show that the three metrics are very effective for detecting a test page different from the journal used for learning. Among the three metrics, the normalized likelihood is most effective for this detection. Note that both the maximum entropy and the min-max label probability can estimate the worst-case token-level confidence. This result indicates that focusing on the least-confident token in the sequence is not a good strategy for layout-change detection. We did not observe a significant difference between the maximum entropy and the min-max label probability. The latter was effective for detecting IEICE-E from IPSJ, as shown in Fig. 3 (a), whereas the former was effective for detecting IPSJ from IEICE-J, as shown in Fig. 3 (f).

Second, the journal used for learning affects the ability to detect changes. For example, compare panels (d) and (f) in Fig. 3. In both panels, we could discriminate IPSJ and IEICE-J. However, IPSJ was used for training in panel (d), whereas IEICE-J was used for training in panel (f). The panels show that the CRF learned by IEICE-J is better than the CRF learned by IPSJ.

4.5 CRF Learning Using Additional Training Data

In Sect. 4.3, we evaluated the effectiveness of the active sampling when applied to each journal independently. When learning a new CRF for pages whose layout is different from those in the system, we can utilize the labeled training data for

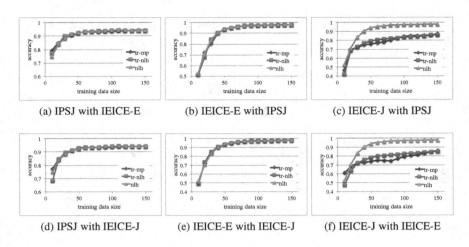

(a) IPSJ with IEICE-E   (b) IEICE-E with IPSJ   (c) IEICE-J with IPSJ

(d) IPSJ with IEICE-J   (e) IEICE-E with IEICE-J   (f) IEICE-J with IEICE-E

Fig. 4. Learning a new CRF.

journals already stored in the system. It may further reduce the cost of preparing the training data, as in the case of transfer learning [11]. To examine the effect of labeled data associated with other journals, we modified the active sampling procedure described in Sect. 4.3. That is, we added 100 randomly chosen pages from the labeled training title pages of another journal to the 10 initial training title pages. The accuracy of the CRF was measured using Eq. (7).

Figure 4 shows the accuracy of the CRFs for journals IPSJ, IEICE-E, and IEICE-J. Each graph in the figure plots the accuracy of the CRF with respect to the size of the training sample dataset chosen according to the normalized likelihood and the min-max label probability. For example, panel (a) shows the accuracy of the CRF learned from title pages of IPSJ with 100 additional title pages from IEICE-E.

The red curve labeled as *tr-nlh* depicts the accuracy for the case of training samples being chosen according to the normalized likelihood. The blue curve labeled as *tr-mp* depicts the accuracy for the case of training samples being chosen according to the min-max label probability. For comparison, the green curve labeled as *nlh* depicts the accuracy of the CRF that was learned without additional training data and sampled according to the normalized likelihood. Note that the green curves are the same as *nlh* in Fig. 2.

We observe that the normalized likelihood and the min-max label probability have similar performance with respect to the training dataset size, for all the cases in the experiment. One remarkable point is that the min-max label probability performed better than the normalized likelihood at the initial step, where 10 training datasets and 100 additional training datasets were used for training.

By comparing the curves *nlh* and *tr-mp*, we observe that *tr-mp* tends to perform better than *nlh* for a training dataset size of 10. For larger training datasets, *tr-mp* performs as well as *nlh* for IPSJ and IEICE-E, whereas it performs less well than *nlh* for IEICE-J. This result indicates that using additional training datasets from different journals is effective for the initial CRF learning, but it may become less effective as more training datasets for the target journal is used. This could be improved by introducing a weighting to the pages of the target and other journals according to the training dataset size.

5   Conclusions

We examined three confidence measures derived from a linear-chain CRF for detecting layout changes in the title pages of academic papers. We applied the measures to the active sampling process used in learning CRFs. Our experiments revealed that the confidence measures are very effective in detecting layout changes and that the measures can be used for active sampling, which will reduce the labeling cost for the training data.

We plan to extend this study in several directions. First, we will study methods that might make the best use of the data accumulated in the system so far. In our experiments, we observed that additional training data is effective when obtaining an initial CRF in the active sampling process. We will look for effective ways to utilize additional training data, as occurs in transfer learning. Second, we will study methods for making clusters of title pages for learning separate CRFs. In this paper, we split the data according to the journal type, when learning the separate CRFs. In future work, we will seek optimal ways of splitting the data for learning separate CRFs.

References

1. Choudhury, S.R., Mitra, P., Kirk, A., Szep, S., Pellegrino, D., Jones, S., Giles, C.L.: Figure metadata extraction from digital documents. In: International Conference on Document Analysis and Recognition (ICDAR 2013), pp, 135–139 (2013)
2. Councill, I.G., Giles, C.L., Kan, M.-Y.: Parscit: An open-source CRF reference string parsing package. In: Language Resources and Evaluation Conference (LREC 2008), p. 8 (2008)
3. Krishnamoorthy, M., Nagy, G., Seth, S.: Syntactic segmentation and labeling of digitized pages from technical journals. IEEE Comput. **25**(7), 10–22 (1992)
4. Kudo, T., Yamamoto, K., Matsumoto, Y.: Applying conditional random fields to Japanese morphological analysis. In: Empirical Methods in Natural Language Processing (EMNLP 2004) (2004)
5. Lafferty, J., McCallum, A., Pereira, F.: Conditional random fields: Probabilistic models for segmenting and labeling sequence data. In: Proceedings of 18th International Conference on Machine Learning, pp. 282–289 (2001)
6. Nagy, G., Seth, S., Viswanathan, M.: A prototype document image analysis system for technical journals. IEEE Comput. **25**(7), 10–22 (1992)
7. Nicolas, S., Dardenne, J., Paquet, T., Heutte, L.: Document image segmentation using a 2D conditional random field model. In: International Conference on Document Analysis and Recognition (ICDAR 2007), pp. 407–411 (2007)
8. Ohta, M., Inoue, R., Takasu, A.: Empirical evaluation of active sampling for CRF-based analysis of pages. In: IEEE International Conference on Information Reuse and Integration (IRI 2010), pp. 13–18 (2010)

9. Ohta, M., Takasu, A.: CRF-based authors' name tagging for scanned documents. In: Joint Conference on Digital Libraries (JCDL 2008), pp. 272–275 (2008)

10. Ohta, M., Takasu, A., Adachi, J.: Empirical evaluation of CRF-based bibliography extraction from reference strings. In: IAPR International Workshop on Document Analysis Systems (DAS 2014), pp. 287–292 (2014)

11. Pan, S.J., Yang, Q.: A survey on transfer learning. IEEE Trans. Knowl. Data Eng. **20**(10), 1345–1359 (2010)

12. Peng, F., McCallum, A.: Accurate information extraction from research papers using conditional random fields. In: Human Language Technologies; Annual Conference on the North American Chapter of the Association for Computational Liguistics (NAACL HLT), pp. 329–336 (2004)

13. Saar-Tsechansky, M., Provost, F.: Active sampling for class probability estimation and ranking. Mach. Learn. **54**(2), 153–178 (2004)

14. Story, G.A., O'Gorman, L., Fox, D., Schaper, L.L., Jagadish, H.V.: The rightpages image-based electronic library for alerting and browsing. IEEE Comput. **25**(9), 17–26 (1992)

15. Takasu, A.: Bibliographic attribute extraction from erroneous references based on a statistical model. In: Joint Conference on Digital Libraries (JCDL 2003), pp. 49–60 (2003)

16. Takasu, A., Ohta, M.: Rule management for information extraction from title pages of academic papers. In: 3rd International Conference on Pattern Recognition Applications and Methods (ICPRAM 2014), pp. 438–444 (2014)

17. Wang, Y., Phillips, I.T., Robert, R.M., Haralick, M.: Table structure understanding and its performance evaluation. Pattern Recogn. **37**(7), 1479–1497 (2004)

18. Zanibbi, R., Blostein, D.: Recognition and retrieval of mathematical expressions. Int. J. Doc. Anal. Recogn. **15**(4), 331–357 (2012)

19. Zhu, J., Nie, Z., Wen, J.-R., Zhang, B., Ma, W.-Y.: 2D conditional random fields for web information extraction. In: International Conference on Machine Learning (ICML 2005) (2005)

Statistically Representative Cloud of Particles for Crowd Flow Tracking

Patrick Jamet[1,3]([✉]), Stephen Chai Kheh Chew[1], Antoine Fagette[1,3], Jean-Yves Dufour[2], and Daniel Racoceanu[3]

[1] Thales Solutions Asia Pte Ltd, 28 Changi North Rise, Singapore 498755, Singapore
patrick.c.jamet@gmail.com, tephen.chai@asia.thalesgroup.com,
antoine.fagette@thalesgroup.com
[2] ThereSIS - Vision Lab, Thales Services - Campus Polytechnique, Palaiseau, France
[3] CNRS IPAL UMI 2955 Joint Lab, Singapore, Singapore

Abstract. This paper deal with the flow tracking topic applied to dense crowds of pedestrians. Using the estimated density, a cloud of particles is spread on the image and propagated according to the optical flow. Each particles embedding physical properties similar to those of a pedestrian, this cloud of particles is considered as statistically representative of the crowd. Therefore, the behavior of the particles can be validated with respect to the behavior expected from pedestrians and potentially optimized if needed. Three applications are derived by analysis of the cloud behavior: the detection of the entry and exit areas of the crowd in the image, the detection of dynamic occlusions and the possibility to link entry areas with exit ones according to the flow of the pedestrians. The validation is performed on synthetic data and shows promising results.

Keywords: Particle video · Crowd · Flow tracking · Entry-exit areas detection · Occlusions · Entry-exit areas linkage

1 Introduction

At the 2010 Love Parade in Duisburg, a mismanagement of the flows of pedestrians led to the death of 21 participants. To put it in a nutshell, on a closed area, the exit routes had been closed while the entry ones remained open. With people still coming in and none coming out, the place ended up overcrowded with a density of population unbearable for human beings who suffocated. This tragedy is one among others where the crowd itself is its own direct cause of jeopardy. Setting up video-surveillance systems for crowd monitoring, capable of automatically raising alerts in order to prevent disasters is therefore one of the new main topics of research in computer vision.

Tracking the flow of pedestrians appears here as an interesting feature for a system monitoring areas welcoming large streaming crowds. With the capability of understanding where the different flows enter the scene, where they exit it and at what rate, comes the capability of predicting how the density of population is

© Springer International Publishing Switzerland 2015
A. Fred et al. (Eds.): ICPRAM 2014, LNCS 9443, pp. 237–251, 2015.
DOI: 10.1007/978-3-319-25530-9_16

going to evolve within the next minutes. In terms of environment management, it means being able to close or open the right doors at the right moment in order the avoid a situation similar to the one of Duisburg.

This paper is presenting an original method to detect the entry and exit areas in a video stream taken by a video-surveillance camera. By linking these areas one with another according to the flows of pedestrians, it is also able to give an estimation of the trajectory followed by these pedestrians and therefore indicate the most used paths. It is based on the use of particles initialized according to an instantaneous measure of the density and driven by the optical flow. These particles are embedding physical properties similar to those of a regular pedestrian in order to perform optimization computations regarding the trajectory and the detection of incoherent behaviors.

This article first presents in Sect. 2 the existing state of the art in terms of crowd tracking. We are then presenting an overview of the method in Sect. 3 as well as our results in Sect. 4. Finally, in Sect. 5, conclusions are given followed by a discussion on the possibilities for future developments.

2   State of the Art

Tracking in crowded scenes is an important problem in crowd analysis. The goal can be to track one specific individual in the crowd in order to know his whereabouts. But, as noted previously, the objective can also be to track the different flows in order to monitor the global behavior of the crowd. There are therefore two approaches to tackle this topic. By segmenting each pedestrian of the crowd on the video and by tracking them individually, the system can keep track of some particular designated pedestrians. However, this method becomes costly as soon as the number of pedestrians rises and it may be impossible to use it when the density is too high or the image resolution too low due to the problems of occlusions. On the other hand, by considering the crowd as a whole and by applying holistic methods such as those used in climatology and fluid mechanics, it is possible to keep track of the different flows. The system builds a model that is statistically representative of the crowd. The drawback of such methods is that it does not allow the system to keep track of specific individuals designated by the operator, it can only give a probability of presence within the crowd.

Methods belonging to the first paradigm, the object tracking approach, are numerous and a classification is proposed in [1]. Yilmaz *et al.* point out that the taxonomy of tracking methods is organized in three branches: point tracking (such as the use of SIFT as in [2]), appearance tracking (such as the Viola-Jones algorithm used in [3]) and silhouette tracking (such as the CONDENSATION method developed in [4]). Based on this survey and following this classification, Chau *et al.* present a most recent and very complete overview of the existing tracking algorithms in [5]. From this overview, one can notice that these algorithms are all requiring the detection in the image of points or regions of interest (PoI/RoI) in order to perform the tracking. They characterize these PoI/RoI using features such as the HOG, Haar or LBP ones or detect them using algorithms such as FAST or GCBAC. The tracking part is then often optimized using tools such as Kalman or particle filters. As the number of pedestrians in the crowd grows, the quality of the extracted features is downgraded due mainly to the occlusions.

Regarding the holistic methods applied to the crowds it appears in the literature that the computer vision field is mostly inspired by fluid mechanics approaches used in climatology for example as in [6] or [7]. In particular, the use of optical flow algorithm applied to crowd analyses has been explored in [8] to detect abnormal events. However, such a method provides an instantaneous detection of events but does not allow long-term tracking of the flow and predicting models that would detect emerging hazardous situations. Sand and Teller in [9] introduce the Particle Video algorithm. The interest of such an algorithm is that, instead of detecting at each frame the points of interest to be tracked, the algorithm sets its own points of interest to follow: the particles driven by the optical flow. Ali and Shah in [10] are using particles to track the flows of pedestrians in dense crowds. Using the Lagrangian Coherent Structure revealed by the Finite Time Lyapunov Exponent field computed using the Flow Map of the particles, the authors can detect the instabilities and therefore the problems occurring in the crowd. Mehran *et al.* in [11] are also setting a grid on the image. The nodes of the grid are the sources for particle emission. At each frame, each node emits a particle and the authors use their trajectories to detect streaklines in the flows of pedestrians. The consistency between streaklines and pathlines is a good indicator of flow stability. Another method of crowd analysis implying particles is using the Social Force Model built by Helbing and Molnár in [12]. Mehran *et al.* are using particles, initialized on a grid as well, to compute the social forces applied on a crowd in a video footage [13]. As a result, they can link the output of their algorithm with the model's and detect abnormal behaviors quite efficiently.

At the cross-roads between the discrete and the holistic approaches, Rodriguez *et al.* in [14] are giving a review of the algorithms for crowd tracking and analysis. Subsequently, they propose to combine results from holistic methods with outputs from discrete ones. For example, their density-aware person detection and tracking algorithm is combining a crowd density estimator (holistic) with a head detector (discrete) to find and track pedestrians in the middle of a crowd.

The method presented in this article belongs to the holistic approach and is inspired by the work of Rodriguez *et al.* in the use of density. It is using the Particle Video algorithm in a new way that is described in Sect. 3.

3 Overview of the Method

The method described hereafter is following the same pattern proposed by Sand and Teller in [9]: the particles are set on the image, they are propagated following the optical flow, their positions are optimized according to our own criteria and they are removed when they are no longer relevant. We call this process the BAK (Birth Advection Kill) Process. As opposed to Sand and Teller, the particles, as much as the pedestrians in the crowd, are considered as independent from each other and are therefore not linked. This assumption allows an efficient parallel architecture implementation such as a GPU-implementation for example in our case. That way, we can deal with the great number of particles involved while achieving real time computation.

Unlike most of the methods using the Particle Video algorithm, we do not initialize our particles on a grid, nor do we add any via sources distributed on a grid. As we want the cloud of particles to be statistically representative of the crowd, the first initialization and the subsequent additions of particles are done according to the density of the crowd.

Moreover, in order to comply with the idea that our particles are a representation of the pedestrians, we match the behavior of each of them with the one expected from a regular human being wandering in the crowd.

3.1   Pre-processing

At each iteration, the algorithm updates the positions of the particles using the result of an optical flow algorithm. For this study, we are using Farnebäck's algorithm [15] as it is implemented in OpenCV.

Plus, as explained previously, the initialization and additions of particles is performed with respect to the estimated density of pedestrians. Therefore, we need to feed our algorithm with such an estimation. For this study, as we are using synthetic datasets, we rely on the ground truth to provide the estimation of density.

Finally, it is to be noted that several steps of the algorithm require to compute image coordinates into 3D coordinates and vice versa. Therefore, the camera parameters are mandatory as an input for these orthorectification processes.

3.2   The BAK Process

The Birth-Advection-Kill (BAK) Process that is described in this subsection is the process handling the whole life cycle of a particle. As explained previously, all the particles are independent from each others. This assertion means that once they are born, all the particles can be handled in parallel, hence the GPU-implementation.

The Birth part takes care of the addition of new particles. It assesses whether there is a need for new particles and the number of these to add.

The Advection part propagates the particles with respect to the optical flow provided to the algorithm. This part also evaluates the validity of the displacement ordered by the optical flow with respect to the displacement expected from a human being. If needed, the proper corrections and optimization are made also in this part of the algorithm.

The Kill part removes the particles that are no longer relevant: those that are out of the scene or those that have had an inconsistent behavior for too long (*i.e.* beyong possible corrections).

This process is summarized on Fig. 1. The Advection part, for which the particles are handled truly independently one from each other, is the part bearing

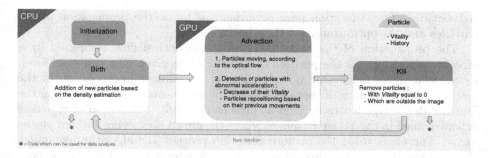

Fig. 1. The BAK process.

most of the processing time. It is the one implemented in GPU. The Birth part is implemented in CPU because it is not done independently from the state of the other particles. As for the Kill part, the killing decision is taken on the GPU-side of the implementation however, it is performed in CPU for technical reasons.

Birth. The Birth part depends exclusively on the density of pedestrians present in the scene and the density of particles set on the image. The goal is to have the density of particles meet the density of pedestrians up to a scale factor λ. Therefore, where the density of particles is too low, the algorithm adds new particles.

The scale factor λ represents the number of particles per pedestrians. This value is to be set by the user, but concretely it will mostly depend on the position of the camera with respect to the crowd. If the pedestrians appear too small, setting λ too high will only induce the creation of many particles at the same location and therefore lots of particles will be redundant. On the other hand, if the pedestrians appear quite large and λ is set too low, the particles may be fixed on parts of the body whose motion is not representative of the global motion of a pedestrian (arm, leg, etc.).

A cloud of particles generated on a crowd with a scale factor λ can be interpreted as λ representative observations of that monitored crowd.

The operation to add particles is performed using the density map provided by the ground truth in our case or by any crowd density estimation algorithm. This map is giving for each pixel a value of the density. It is subsequently divided into areas of m-by-n pixels. For each area, its actual size is computed by orthorectification. By multiplying by the average density of the $m \times n$ densities given in this area by the density map, one can find the estimated number of pedestrians. Multiplied by the scale factor λ, this gives the number of particles that are required. The particles are then added randomly in the m-by-n area in order to meet the required number. Should the required number be lower than the actual number of particles, nothing is done.

Advection. The Advection part performs two tasks: the propagation of the particles and the optimization of their positions.

The propagation of the particles is using the optical flow computed by a separate algorithm. As the position of a particle p is given at a sub-pixel level, its associated motion vector $\mathbf{u_p}$ is computed by bilinear interpolation of the motion vectors given by the optical flow on the four nearest pixels and using the fourth-order Runge-Kutta method as presented in [16]. The notion of closeness is here defined in the 3D environment where the crowd is evolving and not on the image. Therefore, using the orthorectification process, the algorithm computes the positions in the 3D environment of the particle to propagate as well as those of the four nearest pixels in the image.

The new computed position may not be valid with the expected behavior of a regular pedestrian (speed or acceleration beyond human limits). These abnormal behaviors for the particles may happen mostly for two reasons: the noise in the optical flow and occasional occlusions of the entity the particle is attached to.

Therefore, from $\mathbf{u_p}$ and the previous positions, the validity of the position can be assessed and, if needed, optimized. The optimization is performed only when the displacement $\mathbf{u_p}$ generates a speed or an acceleration that is not expected for a pedestrian. When these kinds of event occur, the particle is tagged as abnormal and the algorithm tries to find its most probable position according to its history. Each particle therefore holds a history of its N previous positions, N a number that can be set in the algorithm. These N positions are subsequently used to extrapolate the position that the particle is most likely to occupy.

As the position of a particle has to be optimized, the reliability of the computed trajectory decreases. Indeed, the particle is no longer driven by the optical flow and its computed moves are only an estimation with an associated probability. The more a trajectory requires optimized positions, the less reliable this trajectory is. This leads to another parameter attached to each particle: its vitality. This vitality, which represents the reliability of the trajectory, decreases each time the position of the particle needs to be optimized. However, as soon as the particle manages to follow the optical flow without triggering the optimization process, its vitality is reset to its maximum. A particle whose vitality decreases down to zero or below is considered as dead and will be removed in the Kill part.

Kill. Once the particles have been advected, the Kill part removes all the particles that are not bringing relevant information. These particles are the ones outside of the image, the ones with a vitality equal to or below 0 or the ones that are in an area where the density of particles is too high.

Indeed some particles can move outside of the image because the optical flow drives them out of the field of view of the camera. These particles become useless and are removed.

Some other particles have their vitality dropping down to 0. As this means that they kept having an incoherent behavior for too long, it is reasonable to think that these particles lost track of the object they were attached to. They are therefore removed. The Birth part will solve the potential imbalance between the density of particles and the density of pedestrians induced by these removals. Finally, particles can accumulate at some location in the image. This happens when the optical flow points at such a location with a norm decreasing down to zero. We identified two causes for these kinds of situation: static occlusions and crowd stopping.

Static occlusions are elements in the image that are not part of the crowd but belong to the environment and can hide the crowd (pillars, walls, trash bins, etc.). A crowd moving behind a static occlusion generates an optical flow dropping to zero. Due to the image resolution and the precision of the optical flow, the change is usually not sudden. Therefore, the velocity of the particles decreases gradually and no abnormal accelerations (dropping the speed from $1m \cdot s^1$ to $0m \cdot s^1$ in $1/25$ s would generate an acceleration of $25m \cdot s^{-2}$) are detected. Therefore they accumulate in these areas where there is no pedestrians.

Nevertheless, this can also be due to the crowd coming to a halt. In this case there would be no density issue so no particles would be removed. However if there is a density mismatch with too much particles then the algorithm will remove as much particles as needed to match the correlated amount of pedestrian. This latter case can happen for instance when pedestrians gathered in a large crowd are queuing before stepping on an escalator. Depending on the angle of the camera, the disappearance of the pedestrians might not be visible. However, the density would be quite stable, hence the number of required particles to remain roughly the same through time and therefore older particles being removed.

3.3   Entry-Exit Areas Detection

Over the course of the video, pedestrians keep entering and exiting the camera field of view. They either come in and out of the boundaries of the image or pop up from and get hidden by static occlusions. These limits where the crowd appears and disappears in the image are called respectively entry and exit areas. The Birth and Kill parts described in Subsect. 3.2 handle the appearances and disappearances of the particles on the image according to the optical flow and the crowd density estimation. Due to the noise in the optical flow and to the precision of the crowd density estimation, it is expected to have particles appearing and disappearing even in the middle of the crowd, where it should not happen. However it is a reasonable assumption to expect a higher number of particles added and removed from respectively the entry and exit areas. The subpixelic position of each particle added and removed throughout the image is known. Therefore, areas in which the number of birth (respectively kill) is a local maximum is an entry (respectively exit) area.

The accumulation of data through the iterations of the algorithm enables us to define more precisely the local maxima and therefore the entry and exit areas. Nevertheless in order to be able to detect new entry and exit areas that may appear, this accumulation of data is performed only on a gliding temporal window of Δ_t frames.

To detect these maxima the image is first divided in boxes of $a \times b$ pixels. Each box is assigned δ_+ and δ_- which are respectively the number of particles that have been added and removed through all the previous iterations in the gliding temporal window. The box map is then divided in blocks of $c \times d$ boxes.

To find the boxes of a block that may form an exit area, the algorithm computes for each block μ and σ which are respectively the mean value and the standard deviation of δ_- in the block. Then, for each block, ω and Ω are defined such as:

$$\omega = \mu + \sigma \tag{1}$$
$$\Omega = \mu_\omega + \sigma_\omega \tag{2}$$

with μ_ω and σ_ω respectively the mean value and standard deviation of the values taken by ω during all the previous iterations. Finally, in each block, the boxes with δ_- higher than both ω and Ω are considered as potential candidates to form an exit area.

To find the boxes of a block that may form an entry area, the same process is used, replacing δ_- by δ_+. For these potential entry boxes, another parameter is also taken into account, the assumption being that entry areas are only producing new particles. Therefore, if a potential entry box is crossed by k particles older than f frames, it can no longer be considered as a potential entry box.

The selected boxes are then gathered in groups following a distance criteria: two boxes belong to the same group if and only if they are at a distance of d_{min} meters or less from each other. Groups with more than n_{min} boxes form the entry and exit areas. Each area is materialized by a convex hull.

For our study, given the camera parameters of our datasets, we choose $a = b = 4$ pixels, $c = d = 50$ boxes, $k = 1$ particle, $f = 15$ frames, $d_{min} = 2$ meters, $n_{min} = 5$ boxes and $\Delta_t = 150$ frames.

3.4 Dynamic Occlusions Detection

Dynamic occlusions are entities moving in the image and occluding the pedestrians (*e.g.* a car, a truck, etc.). Particles following a portion of the crowd that is being dynamically occluded tend to have an abnormal behavior. Indeed the optical flow of the object they track is replaced by the optical flow of the occluding object usually resulting in high accelerations which causes an abnormal behavior for the concerned particles. There is then a high probability that an area with a high number of abnormal particles is highlighting a dynamic occlusion.

The method used to detect the exit areas and described in Subsect. 3.3 can be adapted to detect these dynamic occlusions. Indeed we are looking for local maxima of the number of abnormal particles in the image. As opposed to entry and exit areas, dynamic occlusions are happening at a given time and moving rapidly on the image. Therefore, we do not wait for the data to accumulate along the gliding temporal window and use rather the instantaneous number of abnormal particles.

3.5   Linkage of Entry and Exit Areas

The purpose of linking the entry and exit areas is to be able, in a video footage, to know where the pedestrians coming from one area of the image are most likely to go. This can help designing clever pathways in an environment where multiple flows are crossing each others. The interest is also to keep track of the number of pedestrians simply transiting in the scene that is monitored and the number of those staying. With such figures, the system can anticipate any potential overcrowding phenomenon and therefore prevent them.

The information to link the entry and exit areas to each other is carried by the particles themselves. While the entry and exit areas are detected, a number is given to each of them. When a particle enters the scene at a specific entry area, it embeds this entry area number. Once exiting, the particles informs the system of its corresponding exit area number.

4   Validation

The validation datasets used for this study are all synthetic. The main reason to explain this choice is that for our algorithm to work, we need the camera parameters. The second reason is that we need the ground truth to assess the validity of our results. And the third reason is that, to our knowledge, among the huge amount of video sequences displaying crowds and available all over the Internet, none are providing neither the camera parameters nor the required ground truth.

The solution of the synthetic dataset justifies itself in that nowadays, simulators manage to produce crowds with a high level of realism in terms of behavior as well as in terms of rendering. We are using two datasets; a frame of each is displayed on Fig. 2. The first one, basic, was produced by our team. The flow of pedestrians is modeled by cylinders organized in two lanes moving in opposite directions. A static occlusions is represented by a large black rectangle and one of the lane is going behind it. Two dynamic occlusions are crossing the

(a)   (b)

Fig. 2. Example of images from our datasets: (a) the basic one, (b) from Agoraset.

image beside the second lane, just like vehicles on the road next to the side-walk. Although this first dataset is not photo-realistic, it is to be noted that our algorithm does not need photo-realism but rather behavior-realism. The second dataset is taken from the Agoraset simulator [17], available on the Internet. It is more elaborated, with a better rendering as well as a more realistic engine to rule each pedestrian's behavior. This second dataset comes with the ground truth for the pedestrian's positions as well as the camera parameters.

We provide in this Section the results of our algorithm on the first dataset for the Entry-Exit area detection, the dynamic occlusions detection and the linkage of the entry and exit areas. Results are also provided for the tests on the second dataset regarding the entry-exit area detection.

4.1   Entry-Exit Areas Detection

The results provided in Fig. 3 show the detected entry and exit areas compared to the ground-truth that we manually annotated. The main entry and exit areas are accurately detected. On our basic dataset, two of the exit areas are very thin but nevertheless present.

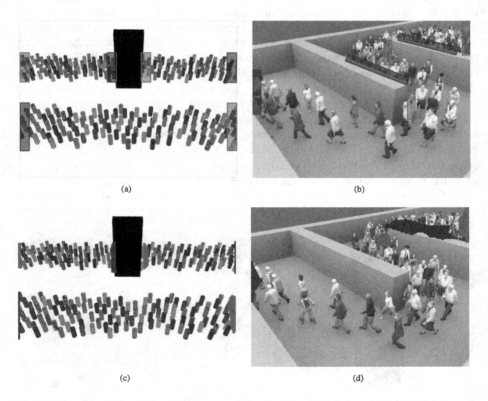

(a)   (b)

(c)   (d)

Fig. 3. Detection of entry and exit areas. The first line is the ground truth, the second line our results. The green polygons are the entry areas, the red polygons are the exit ones (Color figure online).

On Fig. 3(d), one can see that one entry area is not detected. It is an area in which some pedestrians are coming from behind a wall. The non-detection of this entry area can be explained by the fact that even though particles are being born there, and therefore boxes are labeled as potential entry boxes, these boxes cannot link to each other to form entry areas because they are crossed by older particles dragged by pedestrians who are never being hidden by the wall.

4.2   Dynamic Occlusions Detection

The Fig. 4 shows the results of our algorithm for the detection of a dynamic occlusion. One can see that it is effectively isolating the "truck" from the crowd. Due to the precision of the optical flow, the polygon embedding this dynamic occlusion is larger than the occlusion itself.

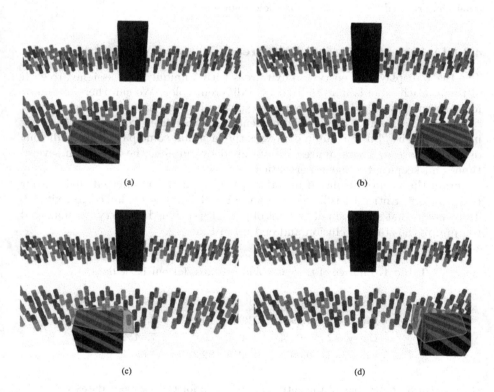

(a)   (b)

(c)   (d)

Fig. 4. Detection of a dynamic occlusion: on 4(a) and 4(b) the ground truth, at two different moment of the video, is materialized with an orange polygon. On 4(c) and 4(d) the results obtained (Color figure online).

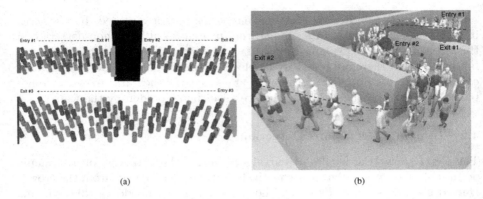

(a)   (b)

Fig. 5. Linkage of the exit areas with the entry areas on 5(a) our basic dataset and 5(b) the dataset coming from Agoraset. Each entry and exit area is assigned a unique label, materialized by a unique color (Color figure online).

4.3   Linkage of Entry and Exit Areas

The Fig. 5 displays the entry and exit areas with a unique label assigned to each of them. Each label is materialized by a different color. We can therefore know for each entry area, where the particles spawned in this area are dying.

The Tables 1 and 2 display the percentages of particles from Entry #i dying in Exit #j. The "Loss" column expresses the fact that some particles can die out of the exit areas. These figures are obtained by running the algorithm several times and keeping the mean percentages.

From the results displayed in Table 1, it can clearly be inferred that, in our basic dataset, Entry #1 is linked to Exit #1 and Entry #2 is linked to Exit #2. To be noted that a very small amount of particles spawned in Entry #1 managed to "jump" the static occlusion and end up in Exit #2.

Table 1. Linkage of the entry and exit area for our basic dataset.

Area	Exit #1	Exit #2	Exit #3	Loss
Entry #1	95.29 %	0.29 %	0.00 %	4.42 %
Entry #2	0.00 %	92.36 %	0.00 %	7.64 %
Entry #3	0.00 %	0.00 %	82.32 %	17.68 %

Table 2. Linkage of the entry and exit area for the Agoraset dataset.

Area	Exit #1	Exit #2	Loss
Entry #1	92.00 %	07.50 %	00.50 %
Entry #2	00.00 %	98.80 %	01.20 %

For the particles spawned in Entry #3 the score achieved for the rate of particles reaching Exit #3 is less than expected. This is due to the dynamic occlusion crossing the image from time to time and killing a higher number of particles than where it is not present (between Entry #1 and Exit #1 and Entry #2 and Exit #2). However, the percentage is high enough to be interpreted as a link between Entry #3 and Exit #3.

The Table 2 provides our results for the Agoraset dataset.

These results clearly indicate that, from the point of view of the camera, most pedestrians appearing in Entry #1 are getting hidden by the first wall, Exit #1. A small proportion of the pedestrians originating from Entry #1 is reaching Exit #2. They are those going along the outside of the curve imposed by the first wall. Not surprisingly, no pedestrians from Entry #2 go through Exit #1 which would be against the observed flow and most of them are reaching Exit #2. Here, the proportion of particles belonging to the "Loss" column can be explained by the fact that the exit area corresponding to the second wall (see Fig. 3(b)) is not detected. However, particles are still dying there and are therefore lost.

5   Conclusions

In this paper we have presented an adaptation of the Particle Video algorithm for crowd flow tracking. The goal was to detect where the crowd is entering the scene that is monitored and where it exits this scene. It is of a particular interest for any crowd monitoring system to be able to track the different crowd flows in order to be able to adapt the environment as efficiently as possible to the different streams of pedestrians and their strength. We showed that our algorithm can detect the different entry and exit areas of the crowd in the image and that it can also provide the route of the crowd within the image with an indication of the rate of pedestrians coming from one area and going to another. Moreover, our GPU-implementation shows that this kind of algorithm reaches real-time execution even though not fully optimized. This depends, of course, on the number of particles that are set and also on the hardware used. In our case, tests were run on a machine equipped with an Intel Core i7 @ 3.20 GHz CPU and a nVidia GeForce GTX 580 GPU. About 10^5 particles were deployed.

To conclude, we would like to point out some further directions of research that could be done in order to enhance such a system. First, regarding the algorithm itself, the conditions of abnormality that lead to a decrease of the vitality of a particle or to its killing could be improved. For the moment, they are based on physical properties linked to the pedestrians accelerations and crowd density but need to be further investigated.

Then, it is obvious that some additional applications could be added on top of those existing in order to compute more crowd features and associate them to specific behaviors. A simple example would be to perform some clustering of the activities present in the image by grouping the particles according to their velocity, direction or location in the scene. From these markers and their spatial and temporal combinations we should derive additional behavior features and be able to classify these activities. Grouping the particles according to their behavior and being able to put a label on top of these groups could help the human operator to analyze the monitored scene.

Finally, as explained in Subsect. 3.2, a cloud of particles generated on a crowd with a number of particles per pedestrian λ can be interpreted as λ representative observations of that monitored crowd. Therefore, these λ observations could be used to train crowd simulators specifically designed to reproduce the behavior of crowds at some location of interest monitored by video-surveillance. The learning of these specific behaviors would help to generate crowd models adapted to specific environments and help, once again, a human operator to design some environmental responses to events of interest.

References

1. Yilmaz, A., Javed, O., Shah, M.: Object tracking: A survey. Acm Comput. Surv. (CSUR) **38**, 1–45 (2006). Article 13
2. Zhou, H., Yuan, Y., Shi, C.: Object tracking using sift features and mean shift. Comput. Vis. Image Underst. **113**, 345–352 (2009)
3. Viola, P., Jones, M.: Rapid object detection using a boosted cascade of simple features. In: Proceedings of the 2001 IEEE Computer Society Conference on Computer Vision and Pattern Recognition, CVPR 2001, Vol. 1, p. I-511. IEEE (2001)
4. Isard, M., Blake, A.: Condensation - conditional density propagation for visual tracking. Int. J. Comput. Vision **29**, 5–28 (1998)
5. Chau, D.P., Bremond, F., Thonnat, M.: Object tracking in videos: Approaches and issues. arXiv preprint arXiv:1304.5212 (2013)
6. Corpetti, T., Heitz, D., Arroyo, G., Memin, E., Santa-Cruz, A.: Fluid experimental flow estimation based on an optical-flow scheme. Exp. Fluids **40**, 80–97 (2006)
7. Liu, T., Shen, L.: Fluid flow and optical flow. J. Fluid Mech. **614**, 1 (2008)
8. Andrade, E.L., Blunsden, S., Fisher, R.B.: Modelling crowd scenes for event detection. In: Proceedings of the 18th International Conference on Pattern Recognition ICPR 2006, vol. 1, pp. 175–178 (2006)
9. Sand, P., Teller, S.: Particle video: Long-range motion estimation using point trajectories. In: Computer Vision and Pattern Recognition, vol. 2, pp. 2195–2202 (2006)
10. Ali, S., Shah, M.: A lagrangian particle dynamics approach for crowd flow segmentation and stability analysis. In: IEEE International Conference on Computer Vision and Pattern Recognition (2007)
11. Mehran, R., Morre, B.E., Shah, M.: A streakline representation of flow in crowded scenes. In: Proceedings of the 11th European Conference on Computer Vision (2010)
12. Helbing, D., Molnár, P.: Social force model for pedestrian dynamics. Phys. Rev. E **51**, 4282 (1995)
13. Mehran, R., Omaya, A., Shah, M.: Abnormal crowd behavior detection using social force model. In: Proceedings of the IEEE International Conference on Computer Vision and Pattern Recognition (2009)
14. Rodriguez, M., Sivic, J., Laptev, I.: Analysis of crowded scenes in video. In: Intelligent Video Surveillance Systems, vol. 1, pp. 251–272 (2012)

15. Farnebäck, G.: Two-frame motion estimation based on polynomial expansion. In: Gustavsson, T., Bigun, J. (eds.) SCIA 2003. LNCS, vol. 2749, pp. 363–370. Springer, Heidelberg (2003)
16. Tan, D., Chen, Z.: On a general formula of fourth order runge-kutta method. J. Math. Sci. Math. Educ. **7**(2), 1–10 (2012)
17. Allain, P., Courty, N., Corpetti, T.: AGORASET: a dataset for crowd video analysis. In: 1st ICPR International Workshop on Pattern Recognition and Crowd Analysis, Tsukuba, Japan (2012)

Preserving Maximum Color Contrast
in Generation of Gray Images

Alex Yong-Sang Chia[1], Keita Yaegashi[2]([✉]), and Soh Masuko[2]

[1] Rakuten Institute of Technology, Singapore, Singapore
alex.a.chia@rakuten.com
[2] Rakuten Institute of Technology, Tokyo, Japan
{keita.yaegashi,so.masuko}@rakuten.com

Abstract. We propose a method to preserve maximum color contrast when converting a color image to its gray representation. Specifically, we aim to preserve color contrast in the color image as gray contrast in the gray image. Given a color image, we first extract unique colors of the image through robust clustering for its color values. We tailor a non-linear decolorization function that preserves the maximum contrast in the gray image on the basis of the color contrast between the unique colors. A key contribution of our method is the proposal of a color-gray feature that tightly couples color contrast information with gray contrast information. We compute the optimal color-gray feature, and focus the search for a decolorization function on generating a color-gray feature that is most similar to the optimal one. This decolorization function is then used to convert the color image to its gray representation. Our experiments and a user study demonstrate the superior performance of this method in comparison with current state-of-the-art techniques.

Keywords: Image processing · Image decolorization · Feature representation · Coarse-to-fine search

1 Introduction

Image decolorization refers to the process of converting a color image to its gray representation. This conversion is important in applications such as grayscale printing, single channel image/video processing, and image rendering for display on monochromatic devices such as e-book readers. Image decolorization involves dimension reduction, which inevitably results in information loss in the gray image. Therefore, the goal in image decolorization is to retain as much of the appearance of the color image as possible in the gray image. In this work, we aim to generate a gray image in such a way that the maximum of color contrast visible in the color image is retained as gray contrast in the gray image. This ensures different colored patches (both connected and non-connected) of the color image can be distinguished as different gray patches in the gray image. Given that a color image can have arbitrary colors that are randomly distributed across the

© Springer International Publishing Switzerland 2015
A. Fred et al. (Eds.): ICPRAM 2014, LNCS 9443, pp. 252–262, 2015.
DOI: 10.1007/978-3-319-25530-9_17

image, decolorization that focuses on preserving maximum contrast is a very difficult image processing problem.

A key contribution of this paper is the proposal of a novel color-gray feature, which tightly couples color contrast information in a color image with gray contrast information in its corresponding gray image. This provides several unique advantages. First, by encapsulating both color and gray contrast information in a single representation, it allows us to directly evaluate the quality of a gray image based on information available in its color image. More importantly, it provides a convenient means of defining an optimal feature that represents the maximum preservation of color contrast in the form of gray contrast in the gray image. Consequently, by searching for a color-gray feature that is most similar to the optimal feature, we can compute a gray image in which different colored regions in the color image are also distinguishable in the gray image.

We employ a non-linear decolorization function to convert a color image to a gray image. The use of a non-linear function increases the search space for the optimal gray image. To reduce computation cost, we adopt a coarse-to-fine search strategy that quickly eliminates unsuitable parameters of the decolorization function to hone in on the optimal parameter values. We show, through experimental comparisons, that the proposed decolorization method outperforms current state-of-the-art methods.

This paper is organized as follows. Immediately below, we discuss related work. We present our decolorization method in Sect. 2. Experimental evaluations against existing methods are detailed in Sect. 3. We offer our conclusions in Sect. 4.

1.1   Related Work

Traditional decolorization methods apply a weighted sum to each of the color planes to compute a gray value for each pixel. For example, MATLAB [1] eliminates the hue and saturation components from a color pixel, and retains the luminance component as the gray value for each pixel. Neumann et al. [2] conducted a large-scale user study to identify the general set of parameters that perform best on most images, and used these parameters to design their decolorization function. Their methods support very fast computation of gray values, in which the computational complexity is $O(1)$. However, given that the decolorization function is not tailored to the input image, these methods rarely maintain the maximum amount of information from the color image.

Modern approaches to convert a color image to its gray representation tailor a decolorization function to the color image. Such approaches can be classified into two main categories: local mapping and global mapping. In local mapping approaches, the decolorization function applies different color-to-gray mappings to image pixels based on their spatial positions. For example, Bala and Eschbach [3] enhanced color edges by adding high frequency components of chromaticity to the luminance component of a gray image. Smith et al. [4] used a local mapping step to map color values to gray values, and utilized a local sharpening step to further enhance the gray image. While such methods can improve the perceptual quality of the gray representation, a weakness of these methods is that they can distort the appearance of uniform color regions. This may result in haloing artifacts.

Global mapping methods use a decolorization function that applies the same color-to-gray mapping to all image pixels. Rasche et al. [5] proposed an objective function that addresses both the need to maintain image contrast and the need for consistency of the luminance channel. In their method, a constrained multi-variate optimization framework is used to find the gray image that optimizes the objective function. Gooch et al. [6] constrained their optimization to neighboring pixel pairs, seeking to preserve color contrast between pairs of image pixels. Kim et al. [7] developed a fast decolorization method that seeks to preserve image feature discriminability and reasonable color ordering. Their method is based on the observation that more saturated colors are perceived to be brighter than their luminance values would suggest. Recently, Lu et al. [8] proposed a method that first defines a bimodal objective function, then uses a discrete optimization framework to find a gray image that preserves color contrast.

One weakness of the methods proposed by Gooch et al. [6], Kim et al. [7], and Lu et al. [8] is that they optimize contrast between neighboring connected pixel pairs and do not consider color/gray differences between non-connected pixels. As a result, different colored regions that are non-connected may be mapped to similar gray values. This results in the loss of appearance information in the gray image. Our framework considers both connected and non-connected pixel pairs to find the optimal decolorization function of an image, and thus does not suffer from this shortcoming.

2   Our Approach

Figure 1 outlines our method, which consists of four main modules. We first extract unique colors from a given a color image. Thereafter, we compute the corresponding gray values of these color values using a currently considered decolorization function. Based on these color and gray values, we compute a color-gray feature to encapsulate the color and gray contrast information in a single representation. The best possible color-gray feature is computed, and we evaluate the quality of the currently considered decolorization function by comparing its color-gray feature with the optimal one. Then, a coarse-to-fine

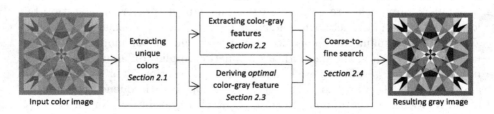

Fig. 1. Overview of our decolorization method.

search strategy is employed to search for a decolorization function that provides the color-gray feature that is the closest fit to the optimal feature. Our method explicitly drives the search toward the gray image that preserves the maximum color contrast in the form of gray contrast. The best-fit function is then used to convert the color image into its gray representation. We elaborate on these modules in the following subsections.

2.1 Extracting Unique Colors

Given an input color image, we first extract its unique color values by applying the robust mean shift clustering method proposed by Cheng [9] to its color values. Let $\{c_i\}$ denote the set of clusters formed. We do not remove weakly populated clusters, but instead consider all mean shift clusters for subsequent processing. Consequently, the color clusters do not represent only the dominant colors of the image; collectively, they represent all the unique colors of the image.

We can highlight three advantages of using these unique colors in the decolorization framework. First, use of these unique colors keeps the computational cost of our method lower than the computational cost of methods that operate on a per-pixel basis, since the number of unique colors is typically much fewer than the number of image pixels. Second, as the clusters represent all unique colors that are extracted from the image, our method does not concentrate the color-to-gray optimization on only the dominant color space of the image but instead on the entire color space represented by the image. Third, and most important, these color clusters do not include spatial information of pixels. By focusing the search for an optimal decolorization function across these color clusters, our method optimizes on a global basis rather than a local neighborhood basis. This ensures that non-connected colored regions map to distinguishable gray regions.

2.2 Extracting Color-Gray Features

Before we describe the computation for a color-gray feature, we must discuss the non-linear decolorization function that we adopt to convert a color image to its gray counterpart. Given a pixel p of a color image, we define its red, green, and blue components as $\{p_r, p_g, p_b\}$ in which the values vary in the range of 0 to 1. p_{gray} denotes the gray value of pixel p. We compute p_{gray} by the multivariate non-linear function:

$$p_{\text{gray}} = ((p_r)^x \times w_r + (p_g)^y \times w_g + (p_b)^z \times w_b)^{\frac{3}{x+y+z}}, \tag{1}$$

where $\{w_r, w_g, w_b\}$ are weight values and $\{x, y, z\}$ are power values that correspond, respectively, to the red, green, and blue components. The non-linear function increases the search space (and affords significant flexibility) to find an optimal gray representation. Our aim is to find the set of weight and power values that results in maximum retention of color contrast in the form of gray contrast in the image.

To compute a color-gray feature that tightly couples color contrast to gray contrast, our process is as follows. For each cluster c_i obtained in Sect. 2.1, we first compute mean color value from all pixels within the cluster. We compute the Euclidean distance between the mean color values of all clusters, and collect the color distances into a distance matrix Φ. Additionally, for each cluster, we compute the gray values for all pixels within the cluster based on a currently considered set of weight $\{w_r, w_g, w_b\}$ and power $\{x, y, z\}$ values using Eq. (1). We use these gray values to compute an average gray value for the cluster. Similarly, we compute the Euclidean distances between the gray values of all clusters, and collect these distances into a matrix Γ. Implicitly, each value in $\Phi(i)$ reflects the color contrast between two color clusters whose gray contrast is given in $\Phi(i)$.

Let F denote a color-gray feature, where $F(j)$ denotes the value at the jth dimension of the feature. Each dimension corresponds to a gray-distance interval $[a, b]$. We define set $\Psi_{(a,b)}$ to be the set of color-distance values whose gray distance is within the interval $[a, b]$. Mathematically, this is represented as follows:

$$\Psi_{(a,b)} = \Phi\left(\phi_{(a,b)}\right), \tag{2}$$

where set $\phi_{(a,b)}$ is the set of matrix indices whose gray distance is within the interval $[a, b]$:

$$\phi_{(a,b)} = \cup k, \quad \forall k, \ a \leq \Gamma(k) \leq b. \tag{3}$$

We compute $F(j)$ as

$$F(j) = \max\left(\Psi_{(a,b)}\right). \tag{4}$$

Figure 2 shows a toy example for computing a feature value $F(j)$. Here, three color clusters are considered, where the clusters are depicted as the color patches in the color-distance matrix Φ given in Fig. 2(a). We show the corresponding gray patches and the gray-distance matrix Γ in Fig. 2(b). Consider the computation of $F(8)$, which corresponds to gray-distance interval $[0.7, 0.8]$. We show all entries in G that belong to this interval by the blue outlines in Fig. 2(b), and the color distances that are considered by the red outlines in Fig. 2(a). Maximum color distance within this gray-distance interval is 0.47, and thus, this is the feature value of $F(8)$ (as shown in Fig. 2(c)).

We note here the following points. First, a feature dimension corresponds to a gray-distance interval, and a feature value corresponds to a color-distance value. Therefore, the proposed color-gray feature directly incorporates both color and gray information in its representation. More importantly, each feature value indicates the maximum color contrast of the color image that is represented within the currently considered gray-contrast interval, and indicates the importance of the gray-contrast interval. This information enables us to compute an optimal feature that retains maximum color contrast of the color image in the form of gray contrast, as described below.

2.3   Deriving Optimal Color-Gray Feature

We seek color-gray features in which larger feature values are present in the dimensions of the feature which are furthest to the right. Intuitively, feature

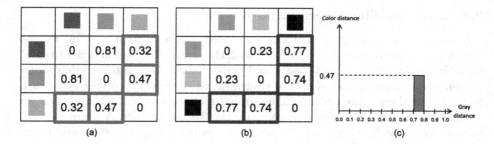

Fig. 2. Toy example for computing value $F(8)$ for three color clusters. (a) Color-distance matrix F for the three color clusters. (b) Corresponding gray-distance matrix G. (c) Feature value $F(8)$ computed at intervals $[0.7, 0.8]$. Entries in F and G used to compute $F(8)$ are represented by the red and blue outlines in (a) and (b) respectively (Color figure online).

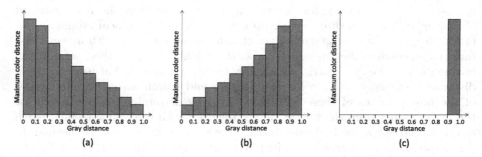

Fig. 3. Pictorial representations of (a) poor, (b) good and (c) optimal color-gray features. See text for details (Color figure online).

dimensions depict the extent clusters $\{c_i\}$ that can be distinguished in the gray space, whereas larger dimensions correspond to a higher gray contrast, and therefore more perceivable differences in the gray space. A feature value $F(j)$ indicates the importance of the feature dimension j in the color space, where larger feature values indicate that clusters exist that are readily distinguishable in the color space. Thus, a color-gray feature that has a heavy right tail corresponds to a color-to-gray mapping in which clusters whose color contrasts are readily distinguished in the color space have gray contrast that is easily perceived in the gray space. Figure 3 provides a pictorial representation of several color-gray features. The feature of Fig. 3(a) implies that clusters that are readily distinguished in the color space (i.e., have high color contrast) are weakly perceived in the gray space. Conversely, the feature in Fig. 3(b) shows that clusters that have sharp color contrast can be readily distinguished in the gray space. This indicates a better gray representation of the color clusters. By extending this reasoning, we can derive the optimal color-gray feature in Fig. 3(c), where regardless of the color differences between the clusters, these clusters have maximum contrast in the gray space.

2.4   Coarse-to-Fine Search Strategy

We compare a color-gray feature, generated by a current set of weight $\{w_r, w_g, w_b\}$ and power $\{x, y, z\}$ values, with the optimal feature in order to evaluate the quality of the set of weight and power values. Our aim is to determine the minimum computational cost of transforming the current color-gray feature to the optimal one. The feature values are normalized to sum to 1, and then we use the earth mover's distance to compare the feature vectors. A naïve method of identifying the best set of weight and power values is to perform iteration over all values, and select the values whose color-gray feature has the minimum earth mover's distance to the optimal feature. This, however, has high computational costs in the order of $O\left(n^6\right)$. Instead, we adopt a fast coarse-to-fine hierarchical search strategy to find the best set of weight and power values. Specifically, we first search across coarse ranges of the weight values, and identify a seed weight value $\{w_r^*, w_g^*, w_b^*\}$ that has the least earth mover's distance from the optimal color-gray feature. Thereafter, we search at a finer scale in the neighborhood of $\{w_r^*, w_g^*, w_b^*\}$. To ensure that we do not get trapped in a local minimum, we retain three sets of seed weight values that have the smallest earth mover's distances, and conduct the fine search across the neighborhoods of these sets. At the termination of the fine search, we identify the weight values that have the least distance from the optimal color-gray feature and search across various ranges of the power values of these weight values. The decoupling of the search for the power values from the weight values, together with the coarse-to-fine search strategy, makes our method substantially faster than the brute force method. In this paper, a coarse search is conducted in the range $\{0, 0.2, 0.4, 0.6, 0.8, 1.0\}$ and a fine search at an offset of $\{-0.10, -0.05, 0, +0.05, 0.10\}$ from three sets of seed weight values. During computation, we ignore a set of weight values if the sum of the weights exceeds 1. The search for the power values is in the range $\{0.25, 0.5, 0.75, 1.0\}$. The proposed method searches over a maximum of 435 different value settings, as opposed to 47,439 settings using the naïve search method. This yields more than $100\times$ increase in speed with our method, as compared to the naïve method.

3   Results

We compare our method with MATLAB's *rgb2gray* function, and the recent state-of-the-art methods of Lu et al. [8], Rasche et al. [5], and Smith et al. [4]. For all experiments, we evaluate the methods using the publicly available color-to-gray benchmark dataset of Čadík [10], which comprises 24 images. Image decolorization by our method on all test images is achieved with the same set of parameter settings, and takes under one minute per image on unoptimized codes. We construct color-gray features using 20 equally spaced gray intervals. Three sets of seed weight values are computed from coarse intervals of $\{0, 0.2, 0.4, 0.6, 0.8, 1.0\}$. These seed weight values are then used to initialize a fine search at an offset of $\{-0.10, -0.05, 0, +0.05, 0.10\}$. We find the best power values by searching across values $\{0.25, 0.5, 0.75, 1.0\}$.

Figure 4 show decolorization results obtained by the proposed and comparison methods across various synthetic and real images. We show the input color images in the first column. Gray images computed with MATLAB's *rgb2gray* function and the methods of Lu et al. [8], Rasche et al. [5], and Smith et al. [4] are shown in the second to fifth columns, respectively. Gray images obtained by our method are shown in the final column. As can be seen in the figure, the proposed method provides perceptually more meaningful representation of the

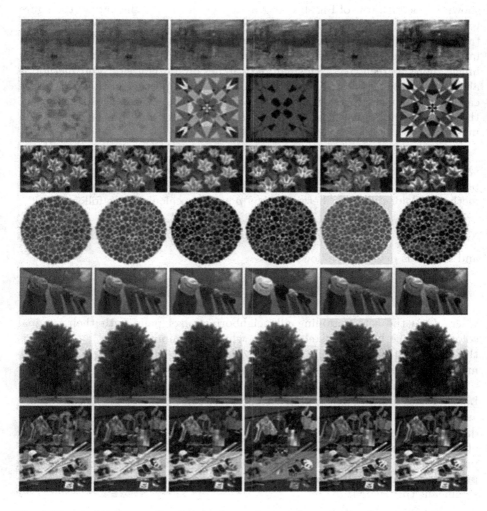

Fig. 4. Decolorization results. First column shows reference color image. Gray images obtained with MATLAB's rgb2gray function, the method of Lu et al. [8], the method of Rasche et al. [5], and the method of Smith et al. [4] are shown in the second to fifth columns, respectively. The final column shows gray images obtained with our method. This figure is best viewed with magnification (Color figure online).

color images, where color contrast present in the images is well preserved as gray contrast. For example, consider the synthetic image shown in the first row of Fig. 4. We note that our method resulted in superior representation, with the contrast between the red sun and the background being better maintained in our gray image than in the gray images of the comparison methods. Additionally, contrast in the fine scale details in the middle-right portion of the image are well preserved using our method. Decolorization results on real images also bear out the superior performance of our method. For example, consider the real image shown in the fifth row of Fig. 4. Our gray representation of the hats in the figure renders them distinguishable in the gray image, and is an improved representation as compared with representations produced by the other methods. An interesting example is shown in the sixth row of the figure, in which the color image shows a green tree with small red patches on its right side. Our method is able to generate a gray image in which red and green patches in the color image can be distinguished as different gray patches in the gray image. In contrast, these patches are indistinguishable in gray images produced by the comparison methods.

Quantitative Evaluation. We use the color contrast preserving ratio (CCPR) proposed by Lu et al. [8] to quantitatively compare our method with other methods. Color difference between pixel p and q is calculated as follows:

$$\delta_{p,q} = \sqrt{(p_r - q_r)^2 + (p_g - q_g)^2 + (p_b - q_b)^2}, \tag{5}$$

and CCPR is defined as follows:

$$\text{CCPR} = \frac{\# \{(p, q)|(p, q) \in \Omega, \ |p_{\text{gray}} - q_{\text{gray}}| \geq \tau \}}{\|\Omega\|} \tag{6}$$

where Ω is the set containing all neighboring pixel pairs with their original color difference between pixel p and q, which is computed as $\delta_{p,q} \geq \tau$. $\|\Omega\|$ is the number of pixel pairs in Ω. $\# \{(p, q)|(p, q) \in \Omega, \ |p_{\text{gray}} - q_{\text{gray}}| \geq \tau \}$ is the number of pixel pairs that are still distinctive after decolorization.

This measurement evaluates the ability of the methods to preserve contrast. Larger ratios indicate better ability to preserve color contrast in the color image as gray contrast in the gray image. We calculate the average CCPR for the whole dataset by varying τ from 1 to 15. We report the mean and standard deviation of the ratios in Table 1. The quantities indicate our method can satisfactorily preserve color distinctiveness and perform better than the other methods in preserving contrast. A t-test shows the comparison results to be statistically significant ($\rho < 10^{-7}$).

User Study. We perform a user study to qualitatively evaluate our method. We engaged 60 subjects with normal vision (whether natural or corrected) for the study. We show each subject 24 sets of images. Each set consists of a reference color image and decolorized results obtained by the proposed and comparison methods. To avoid bias, we randomly jumble the ordering of the

Table 1. Quantitative evaluation using color contrast preserving ratio (CCPR) proposed by Lu et al. [8].

	Mean	Std
MATLAB *rgb2gray*	0.7134	0.2872
Lu et al. [8]	0.8213	0.1595
Rasche et al. [5]	0.7442	0.2565
Smith et al. [4]	0.8212	0.2708
Proposed method	0.8497	0.1621

gray images before presenting them to the subjects. Subjects are instructed to identify the gray image that best represents color patches in the color image as different gray patches in the gray image. Overall, the subjects identify gray images produced by our method as the best representations 27.2 % of the time. This compares favorably to selection of the comparison methods, which are as follows: 16.1 % MATLAB, 18.5 % Lu et al. [8], 19.3 % Rasche et al. [5], and 18.9 % Smith et al. [4]. A t-test shows the comparison results to be statistically significant ($\rho < 2.0 \times 10^{-3}$).

4   Discussion

We propose an image decolorization method that retains maximum color contrast in a color image as gray contrast in the corresponding gray image. To this end, we propose a novel color-gray feature that couples color contrast and gray contrast information. This feature provides a unique ability to directly evaluate the quality of a gray image based on information available in the color image. More importantly, it provides a mechanism to drive our search to find the color-to-gray decolorization function that best preserves contrast in the gray image. A non-linear decolorization function is employed to convert a color image to its gray representation, and we reduce computational cost by using a coarse-to-fine search strategy. Experimental comparisons and a user study show the superior effectiveness of our approach. For future work, we are interested in extending our method to decolorize movie frames. We will exploit spatial and temporal cues to ensure coherence in gray representation is maintained across different movie frames.

References

1. The MathWorks, I.: MATLAB version 7.10.0 (r2010a) (2010). http://www.mathworks.com/products/matlab/
2. Neumann, L., Čadík, M., Nemcsics, A.: An efficient perception-based adaptive color to gray transformation. In: Proceedings of the Third Eurographics Conference on Computational Aesthetics in Graphics, Visualization and Imaging, Eurographics Association, pp. 73–80 (2007)

3. Bala, R., Eschbach, R.: Spatial color-to-grayscale transform preserving chrominance edge information. In: Color and Imaging Conference, vol. 2004, Society for Imaging Science and Technology, pp. 82–86 (2004)
4. Smith, K., Landes, P.E., Thollot, J., Myszkowski, K.: Apparent greyscale: a simple and fast conversion to perceptually accurate images and video. Comput. Graph. Forum **27**, 193–200 (2008)
5. Rasche, K., Geist, R., Westall, J.: Re-coloring images for gamuts of lower dimension. Comput. Graph. Forum **24**, 423–432 (2005)
6. Gooch, A.A., Olsen, S.C., Tumblin, J., Gooch, B.: Color2gray: salience-preserving color removal. ACM Trans. Graph. (TOG) **24**, 634–639 (2005)
7. Kim, Y., Jang, C., Demouth, J., Lee, S.: Robust color-to-gray via nonlinear global mapping. ACM Trans. Graph. (TOG) **28**, 161 (2009)
8. Lu, C., Xu, L., Jia, J.: Real-time contrast preserving decolorization. In: SIGGRAPH Asia 2012 Technical Briefs, p. 34. ACM (2012)
9. Cheng, Y.: Mean shift, mode seeking, and clustering. IEEE Trans. Pattern Anal. Mach. Intell. **17**, 790–799 (1995)
10. Čadík, M.: Perceptual evaluation of color-to-grayscale image conversions. Comput. Graph. Forum **27**, 1745–1754 (2008)

Real-Time Facial Analysis in Still Images and Videos for Gender and Age Estimation

Lionel Prevost[1(✉)], Philippe Phothisane[2], and Erwan Bigorgne[2]

[1] University of the French West Indies,
BP 250, 97157 Pointe à Pitre, Guadeloupe
lionel.prevost@univ-ag.fr
[2] Eikeo, 11 Rue Léon Jouhaux, 75010 Paris, France
erwan.bigorgne@eikeo.com

Abstract. Research has recently focused on human age and gender estimation because they are useful cues in many applications such as human-machine interaction, soft biometrics and demographic studies.

In this paper, we propose a real time face tracking framework that includes a sequential estimation of people's gender then age. Local binary patterns histograms extracted from facial images. A single gender estimator and several gender-specific age estimators are trained using a boosting scheme. Their decisions are combined to output a gender and an age in years.

The whole process is thoroughly tested on state-of art databases and video sets. Results on the popular FG-NET database are comparable to human perception (overall 70 % correct responses within 5 years tolerance and almost 90 % within 10 years tolerance). The age and gender estimators combined with the face tracker provide real-time estimations at 21 frames per second.

Keywords: Face analysis · LBP · Boosting · Gender estimation · Age estimation

1 Introduction

Humans can glean a wide variety of information from a face image, including identity, age, gender, and ethnicity. Despite the broad exploration of person identification from face images, there is only a limited amount of research on how to automatically and accurately estimate demographic information contained in face images such as age or gender.

Gender identification is a cognitive process learned and consolidated throughout childhood. It finally becomes mature in teenage years [1] Children can make good guesses but also use many social stereotypes such as facial features, hair type, clothes or interests. Cropped faces without these external features are generally enough for adult operators to classify properly men and women. The goal of an automatic gender estimator is to match adult human accuracy on cropped face image. As social stereotypes are too variable, they are not considered by most existing methods.

Age identification is also learned throughout life experiences. It is easier to guess someone's age when his(her) age range and ethnicity are seen frequently [2]. One's appearance age may be altered by his (her) individual growth pattern, general health,

© Springer International Publishing Switzerland 2015
A. Fred et al. (Eds.): ICPRAM 2014, LNCS 9443, pp. 263–278, 2015.
DOI: 10.1007/978-3-319-25530-9_18

ethnicity, gender, etc. All these parameters should be considered to determine someone's age with accuracy. But most of the time, facial images provide enough information about a subject to be able to estimate his(her) age range.

Automatic age and gender estimators can be used in many different applications such as human-computer interaction, security control, demographic segmentation for marketing studies, etc. Research teams report good performances on databases well spread in the FAR (Facial Analysis and Recognition) community, such as FERET or LFW [3].

In this article, we introduce a real-time algorithm for age and gender estimation. It is implemented in a real-time 3D face tracker derived from the one detailed in [4] and provide age and gender estimation (Fig. 1). The main contributions of the paper are:

- The estimation of gender using one binary boosted classifier based on fast multi-block local binary pattern (MB-LBP) histogram comparisons.
- The estimation of age using several gender specific age classifiers (using the same features) combined with a weighted sum rule to output an age.
- The comparison of human perception vs. automatic estimation of age on two benchmark databases.
- The study of our estimators in videos using 3D face tracking. Such experiments on real data appear rarely in publications focusing on age and gender estimation.

In Sect. 2, previous methods are introduced, providing state of the art performances on age and gender estimation. Section 3 describes the feature extraction process. Section 4 details boosting training and decision process for gender and age estimation. Section 5 presents various experiments on common databases and video sequences to validate our approach. Comparisons with state of the art are provided too. Finally, Sect. 6 concludes and adds some prospects.

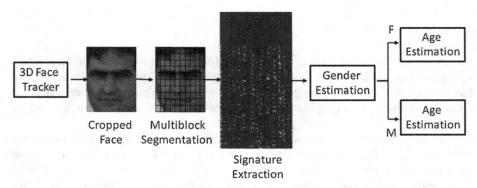

Fig. 1. Gender then age estimation process flowchart. The apparent blocks are the 64 (8 rows, 8 columns) regions of interest. The signature is the 256×59 2-uniform LBP normalized histograms matrix.

2 Previous Work

2.1 Gender Recognition

Most published methods use face cues for gender recognition. The first attempts of automatic gender estimation started in the early 90's with the SEXNET [5]. This method used a two-layer neural network trained to classify 30×30 facial images. Tests were done on 90 images (45 males, 45 females) and obtained a 8.68 % error rate. At the same time, Cottrell and Metclafe [6] used 160 images 64×64 (10 males, 10 females). Images were reduced to 40 components vectors and used to train a single layer neural network. This experiment provided a perfect recognition rate on the training database.

Recently, other methods have also used gait cues to gather more information on targeted subjects [7, 8]. As our human interaction applications are aimed at being used at close range, only faces are visible. The focus here is set on methods using only face images. Many recent papers report results on the FERET database. Moghaddam and Yang [9] achieve an overall 3.38 % error rate using a support vector machine with a RBF kernel on low resolution images. Baluja and Rowley [10] report comparable results using simple pixel comparisons on 20×20 face images. This feature extraction process is very interesting because it is not time consuming. Other studies published results on more unconstrained databases, as image sets downloaded from the web. Shan's gender estimator [11] applies SVMs to LBP histograms on 7,443 images of the LFW database, obtaining a 5.19 % error rate. Shakhnarovich et al. [12], Gao and Ai [13], Kumar et al. [14] experimented on non publicly available databases. In [12] authors use an adaBoost on Haar filters outputs applied to 30×30 images. Kumar et al. [14] obtain an 8.62 % error rate on a 1,954 images database (1,087 males, 867 females) with SVM comparable to those seen in [9].

Most studies focus on still image databases using k-fold cross-validation, and few provide cross-database results. Makinen and Raisamo [15] provide a deep comparison of some state-of-art classifiers on gender estimation (Fig. 2). These classifiers are trained on the FERET database and evaluated on a homemade "internet" database. Authors show that the mean accuracy is good (from 80 % to 90 %) and does not vary significantly from one classifier to another. Overall, only a few experiments were conducted on video sequences [16].

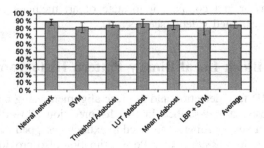

Fig. 2. Comparison of state of art methods for gender classification (reprinted from [15]).

2.2   Age Estimation

Estimating an age means to automatically assign an age to the current subject, whether in years or as an age interval. It is the reverse action of age modeling [17]. There appear to be several definitions of "age" described in [18].

- The actual age is the real age of an individual.
- The perceived age is gauged by another person.
- The appearance age, given by the person's image.
- The estimated age is given by a computer.

Age estimation can be seen as two different problems. The first is a regression problem where the estimator has to predict someone's age as closely as possible with a year precision. The second aims at classifying a face image into one of several bins. As an example, Gao and Ai [13] use a linear discriminant analysis on Gabor wavelets and classify images into 4 bins ("baby", "child", "adult", "old"). Recently, Guo et al. [19] studied both questions using bio-inspired features (BIF) and achieved a 4.77 years accuracy on the FG-NET database. Thukral et al. [20] report a mean absolute error (MAE) of 6.2 years on the whole FG-NET database using geometric features and relevance vector machines. Luu et al. report a MAE of 4.37 years on a subset of FG-NET using active appearance models and support vector machine regressors [21] and 4.12 years on the complete set with a contourlet appearance model [22]. Others report results on the whole database, using methods such as RUN (Regressor on Uncertain Nonnegative labels) [23, 24]. In [25], Guo et al. report 4.69 years accuracy on the non-publicly available Yamaha Gender and Age (YGA) database. As we can see, many studies report results on the FG-NET database which is publicly available (http://www.fgnet.rsunit.com). Most of them use a "Leave-One-Subject-Out" evaluation scheme and their MAE varies between 4 and 6 years.

2.3   Discussion

We decide to perform LBP histogram bin comparisons instead of the simplistic pixel comparison proposed in [10] because this feature extraction process is one the fastest in the literature (see Sect. 3). Then, form same reasons, we use an AdaBoost scheme [11, 29] to perform sequential gender then age comparison (see Sect. 5). We will study our method's behavior and compare it to state of art methods on standard image databases and unconstrained videos (see Sect. 5).

3   Multi-scale Block Local Binary Pattern Histograms

First, we perform 2D face detection and 3D face alignment using a real real-time face tracker and pose estimator described in [4]. Its precision allows us to track facial features accurately. Eyes coordinates are used to extract a cropped face from the source image and normalize to 128×96 pixels. Pose estimation also provides valuable information and allows to reject images of face too far from a frontal pose. This rejection is only used on our live video tests. For still image databases, no rejection is applied.

Then, we compute Multi-scale Block LBP (MBLBP) and histograms fo MBLBP. The following subsections describe this feature extraction process thoroughly.

3.1 Uniform Local Binary Patterns

LBP are commonly used local texture descriptors [26]. Their evolutions include Multi-resolution Histograms of Local Variation Patterns [27] and Multi-scale Block LBP (MB-LBP) [28] which inspired our method.

The original (scale-1) LBP operator labels the pixels of an image by thresholding the 3×3-neighborhood of each pixel with the center value p_0 and considering the result as a binary string. For each p_i, $i = \{1, \ldots, 8\}$ surrounding p_0 in a circular fashion, the boolean b_i is defined as follows (Eq. 1).

$$b_i = \begin{cases} 1 \text{ if } p_i > p_0 \\ 0 \text{ if } p_i \leq p_0 \end{cases} \tag{1}$$

Using the 8 bits, LBP_{P_0} has 256 possible values (Eq. 2). The histogram of the labels can be used as a texture descriptor.

$$LBP_{P_0} = \sum_i 2^{i-1} \times b_i \tag{2}$$

In MB-LBP, the comparison operator between single pixels in LBP is simply replaced with comparison between average gray-values of sub-regions. Each sub-region is a square block containing neighboring pixels (or just one pixel particularly). The whole filter is composed of 9 blocks. We take the size k of the filter as a parameter, and $k \times k$ denoting the scale of the MB-LBP operator. For instance, a scale-3 LBP centered on pixel p_0 uses all the pixels in the 9×9 region surrounding it. The reference region, r_0 is a 3×3 area around p_0. Each other r_i, $i = \{1, \ldots, 8\}$ is another 3×3 region encircling r_0 defined as the sum of its pixel values p. Thus, the b_i are defined according to the sum of the pixel values p inside the r_i regions (Eq. 3).

$$b_i = \begin{cases} 1 \text{ if } \sum_{p \in r} p > \sum_{p \in r_0} p \\ 0 \text{ if } \sum_{p \in r_i} p \leq \sum_{p \in r_0} p \end{cases} \tag{3}$$

In our method, scale-1, 3, 5 and 9 LBP are computed before conversion into 2-uniform LBP. The k-uniform LBP are a subset of the original LBP. The criterion used is the number of circular bit transitions: a k-uniform LBP has k or less transitions. For instance, 11001111 and 00000001 have both 2 transitions and thus are 2-uniform LBP. According to [11] and [26], 2-uniform LBP provide the majority of seen patterns. There are 58 possible values of 2-uniform LBP, the remaining values are all set as non-uniform. We use a look-up table which directly transforms LBP values into (58+1) different values. We obtain in the end four 128×96 2-uniform LBP maps for each input image (one for each scale).

3.2   Block Histograms

In order to add spatial information, in a similar fashion of [11] and [28], we divided the image into blocks of 26×20 pixels. Then, we compute histograms of 2-uniform LBP values on 4 scales (1, 3, 5 and 9). Blocks are regularly distributed on 8 rows and 8 columns. In the end, we obtain 8×8×4 59-bin histograms. Each 59-bin histogram is normalized to obtain a unit vector. These 256×59 matrix signatures are computed on each face image.

4   Boosting and Decision

The signatures are used to classify age and gender. Before training, the face databases (described in Sect. 5.1) are labeled, with the actual gender and a perceived age. Each image is mirrored to avoid asymmetrical bias during the learning process. The database thus doubles in size. The weak classifiers $f(c, j_1, j_2)$, $c=\{1,\ldots,59\}$, $j_n=\{1,\ldots,256\}$, $j_1 \neq j_2$ are simple comparisons of histogram components across blocks. For instance, the cth histogram bin value $h(c, j_1)$ block j_1 is compared to every other c^{th} histogram bin value $h(c, j_1)$ of block j_2, $j_2 \neq j_1$.

- If $h(c, j_1) > h(c, j_2)$, $f(c, j_1, j_2) = 1$
- if $h(c, j_1) \leq h(c, j_2)$, $f(c, j_1, j_2) = 0$

There are $C = 59×(256×255)/2 = 1,925,760$ weak classifiers in total. All these C weak classifiers are used to build our gender and age estimators.

4.1   Gender Estimation

Gender identification is a bi-class segmentation and age estimation is a multi-class segmentation. For the gender estimator, a single strong classifier is built up, by sequentially selecting weak classifiers using an AdaBoost training scheme [29]. The gender strong classifier S_g combines weak classifiers outputs using Eq. (4).

$$S_g = \frac{\sum_{k=1}^{K} log\left(\frac{1-e_k}{e_k}\right) \times f(c_k, j_{1k}, j_{2k})}{\sum_{k=1}^{K} log\left(\frac{1-e_k}{e_k}\right)} \tag{4}$$

We define K as the number of training iterations and e_k as the weighted error on the training database after the kth iteration. The output of the kth selected weak classifier is $f(c_k, j_{1k}, j_{2k})$. Then, the decision is taken by thresholding S_g, with δ being the decision threshold:

- if $S_g > 0.5 + \delta$, subject is a male.
- if $S_g < 0.5 - \delta$, subject is a female.

S_g values within the $[0.5 - \delta, 0.5 + \delta]$ interval are considered neutral. In our real-time implementation, we set $\delta = 0.01$ and obtain good qualitative results in live demonstrations (see Sect. 5.3).

4.2   Age Estimation

According to the results provided by [25], age estimators provide better results after being trained on specific genders. This result is intuitive as male and female facial features are not altered by age in the same way. So, males and females are segregated in the training databases in order to build two gender specific age estimators. We build a complete age estimator by training several strong classifiers S_a, $a{\in}A{=}\{10,15,20,\ldots,50,55\}$ which output a real value like the previous gender strong classifier. Each S_a is constructed using specific image selections and labels. The S_a learn to classify face images into two classes: those younger than a, and those older than a. The output of S_a is computed by using Eq. 4. Then, all the strong classifiers outputs are used to compute an over-the-ages score S_{age}. The age decision is made by finding the maximum value and associated age $k{\in}\{10,15,20,\ldots,50,55\}$ of S_{age}:

$$\mathrm{argmax}_k S_{age}(k) = \sum_{a\in A}(S_a - 0.5) \times \frac{1}{1 - e^{k-a}}$$

The sigmoid function is used to normalize the strong classifier outputs and to build a continuous over-the-ages score. Though simplistic, this decision is effective.

4.3   Real-Time Video Analysis

On still images databases used for validation, the age and gender estimations are made without any specific threshold. However, for sequential databases and the live application, the distribution of gender and age estimations associated with each target is important. Both age and gender estimators are implemented in our real-time facial analysis system, which provides accurate head pose estimation. The gender estimation is triggered when the face alignment is considered satisfactory enough, according to specific regression score thresholds on each facial landmark. According to the gender estimator's decision, the male or the female age estimator makes the age estimation.

To use the information provided by the face tracking, computed age and gender estimations are collected over the sequence. These estimations are recorded for each tracked target, building two one-dimensional votes distributions. Our observations made us choose to model these distributions with Gaussian mixture models, and to use the E-M algorithm to take decisions. After fitting the age model and the gender model, the most weighted Gaussian bell means are selected as the final decision.

For the video analysis, every frame is considered. The computation times of our C++ implementation were measured on Intel Core i7-2600 hardware: 42+/-1 ms for a MB-LBP matrix and 1.5+/-0.1 ms for 12,000 weak classifiers (age and gender). Added up, these single-threaded processes can be computed at 21 frames per second with one core while the other cores are dedicated to other tasks such as face tracking. The computational load can be lowered by reducing the MB-LBP signature generation frequency, as two consecutive frames are likely to be only slightly different.

5 Experiments

Our age and gender estimators are compared to other state of the art methods on still images databases used in the facial analysis community. The FG-NET database is used to test our method for age estimation, and LFW and FERET are used for gender recognition. Other experiments are conducted on video sequences. These databases are described in the next subsection.

5.1 Databases

Gender. Labeled Faces in the Wild (LFW) and FERET are commonly used database among the facial analysis community, particularly for face and gender recognition. We randomly select a subset of LFW to keep a balanced repartition of males and females. This final selection contains a total of 2,758 images. The FERET image selection contains the 1,696 images extracted from the *fa* and *fb* subsets. We provide results on these databases using 4-fold cross validation.

Age. The FG-NET database is an age estimation specific database. It contains 1002 images of 82 different people, with age labels going from 0 to 69 years. The "*100*" video ("*from 0 to 100 years in 150 s*") is available on youtube. During this sequence, 101 people from 0 to 100 years old tell their age while facing the camera. The first frame corresponding to each subject was extracted and labeled accordingly. This "*0 to 100*" database is used to measure age estimation errors from human operators and automatic system.

Age and Gender. We built our own age and gender face database by using images collected from the web. The objective was to gather faces with a wide variety of pose, illumination and expression for a large number of people from various origins. It contains for now 5,814 images, including 3,366 males and 2,448 females. Ten human operators labeled these images with the age they perceived. In order to measure accuracy of human age perception, all these operators participated in a dedicated experiment described in Sect. 5.2. A test-only database was also collected in order to have a constant validation set, as the size of our training dataset aims to grow in size. Even though this method is inherently subject to bias, the error margin is measured in two experiments.

The recorded video dataset uses 16 videos of 8 people, including 6 males and 2 females. The face alignment was considered good enough in 2,086 frames. Every subject was asked to look at the camera then look at specific items in order to capture a wide range of face poses. The relative low number of sequences is compensated by the quantity of images (2,086). In order to investigate the estimators' behavior towards asymmetric facial appearances, each subject was captured in two different illumination conditions: one in ambient lighting and the other with a supplemental lateral light source. The experimental results are provided in the next subsection.

5.2   Results on Still Images

For the following experiments, gender is estimated first and according to the estimator's decision, the age estimation uses either the male or the female features selection. Age estimation results are given in term of Mean Absolute Error (MAE). A preliminary study shows that AdaBoost training mainly selects weak classifiers comparing scale-3 ($\approx$42 %), then scale-5 ($\approx$25 %) and scale-7 ($\approx$23 %). On the over hand, scale-1 weak classifiers represent less than 10 % of those selected. This result shows all the interest of using multi-scale LBP for these tasks.

Gender Estimation. Experiments on gender estimation are done on the LFW and FERET databases. We proceed to four-fold cross validation tests on both databases separately, obtaining 90.7 % accuracy on our subset of LFW and 93.4 % of correct answers on the FERET database. This kind of experiments does not provide information about how the estimators behave when more than an image is provided per target, which is the case in video streams. This is why we proceeded to experiments using sequential data. They are described in Sect. 5.3 and measure the gender estimator's performance in video sequences.

Human Age Perception Experiments. Measuring human errors of age perception would help appreciate automatic age estimation results. Two distinct measurements are conducted. The first is done on a subset of the FG-NET database, and the other on the "0 to 100" dataset. Ten people participated in each experiment. The objective is to measure the accuracy of human age perception and compare it to state of art methods, including ours.

The first experiment uses an age-uniform selection of 60 clear FG-NET pictures of males and females from 0 to 69 years old. Human age perception errors are shown in Fig. 2. The MAE is 4.9 years with a standard deviation of 4.6 years. More than 65 % of the errors are below 5 years and almost 90 % of the errors are below 10 years. These values are close to the performance of the best age estimators published recently and not far from those reported by [30] on the whole FG-NET set (MAE=4.7 years).

For the second experiment, either all images (from 0 to 100 years old) or only a subset (from 0 to 60 years old) are considered. The human performance is similar to the one in the previous test using the FG-NET subset (Fig. 3). This provides information about what kind of accuracy is achievable and reveals that the performance of the best age estimators is actually close to human perception.

Age Estimation. To perform fair comparison with existing methods, a leave-one-subject-out (LOSO) scheme was used on the FG-NET database. Each strong age classifier was built with only 200 weak classifiers, as the LOSO scheme costs a lot of time. This comparison was conducted using many other age estimation methods, including [19]'s BIF or [23]'s RUN. As a large part of the database is focused on people younger than 20, we added the 5 and 10 years old strong classifiers to our initial set. The best reported result is obtained by [22] with a MAE of 4.12 years. Other report detailed results on several age subsets as shown in Table 1. We obtain results close to Guo's performance [19] with a MAE value of 4.94 years (0.04 year difference with human perception error). Our results are the best available on the [0-9] and [40+]

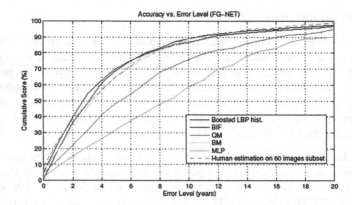

Fig. 3. Human perception and estimation errors: comparison with state-of-art works (FGNET database).

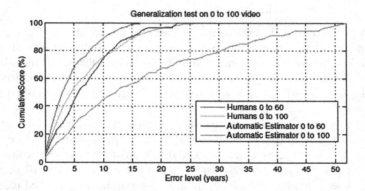

Fig. 4. Human perception errors and estimation errors (" *0 to 100*" database).

Table 1. Age estimation results on FG-NET database. Leave-One-Subject-Out training. Mean absolute error in year on every age subsets. QM and MLP methods are described in [24].

Range	#image	MBLBP	BIF	RUN	QM	MLP
0-9	371	**2.42**	2.99	2.51	6.26	11.63
10-19	339	3.92	3.39	3.76	5.85	**3.33**
20-29	144	4.95	**4.30**	6.38	7.10	8.81
30-39	70	10.56	**8.24**	12.51	11.56	18.46
40-49	46	**14.59**	14.98	20.09	14.80	27.98
50-59	15	**18.45**	20.49	28.07	24.27	49.13
60-69	8	**27.62**	31.62	42.50	37.38	49.13
0-69	1002	4.94	**4.77**	5.78	5.57	10.39

subsets and the second best on the [20-39] subset. Cumulative scores on FG-NET are available in Fig. 2. Table 1 compares mean errors for each age subset. The best age estimation methods (including ours) have results close to human perception on this database. Human perception errors and estimation errors on the "*0 to 100*" database are presented in Fig. 3. This is pure generalization as our estimators were trained on FG-NET and then tested on this new, different database. Our estimator was not trained for age boundaries above 60 years. It explains its poor performance on this generalization experiment over the full "*0 to 100*" database and its acceptable results (not far from human operators) on the "*0 to 60*" subset.

5.3   Results on Sequences

In the following experiments, the complete system is tested on our video dataset. Both gender and age estimators share the same cropped images and MB-LBP histogram signatures to predict their output. Each frame is computed to make a complete observation of the estimation outputs.

Gender Estimation. For the video set, our gender estimator outputs distributions of real values between 0 and 1. Using E-M algorithm [31], each video's most weighted distributions' mean is on the correct side of the 0.5 threshold. As the estimation score is centered on 0.5, we can study the estimator's outputs distribution considering several rejection thresholds δ. For instance, with $\delta = 0.01$, outputs within the [0.49 0.51] interval are discarded. The ROC-like graphs shown in Fig. 5 are plotted using the subsequent good and false response rates. This experiment uses two different block settings to compute the MB-LBP histograms (Fig. 5.a). Both settings use blocks of the same size, one fifth of the original 128×96 crop. The first setting uses 25 (5×5) non-overlapping 26×20 pixels blocks across the face crop and the second uses 64 (8×8) overlapping 26×20 pixels blocks. This setting (8×8 blocks) performs slightly better than the other on the lateral illumination dataset (Fig. 5.b). The use of symmetry refers to using each source frame's symmetric image, thus using two MB-LBP histogram signatures for each computed frame. As the learning database itself was mirrored, using symmetry does not dramatically improve the results. The pixel comparisons method is an implementation of Baluja's boosted gender estimator [10], using 30×30 face images. It was used as a baseline for our latest experiments. It seems more robust to illumination. Anyway, whatever the dataset, our method performs better as its classification rate is higher than 99 % even when rejecting one frame out of two (rejection rate≈50 %).

Age Estimation.   Accordingly to previous results, we choose to use the 8 × 8 blocks settings. Age estimation results over the video set are shown as cumulative scores in Fig. 6: 65.4 % of the estimations are within the 5 years threshold on the neutral illumination set. Despite having mirrored the database, we still obtain slightly better results on this dataset. These results are comparable to the human age perception experiment results on FG-NET or the "*0 to 60*" dataset seen in Fig. 4.

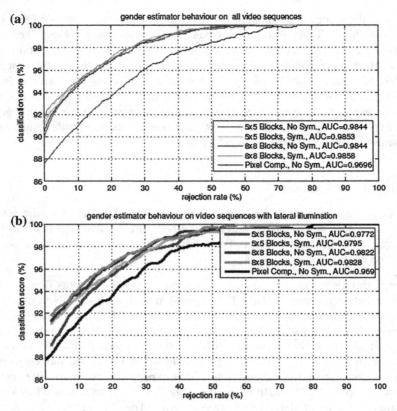

Fig. 5. Gender estimator behavior and comparison with Baluja's estimator [10] for different settings (a) and illumination (b).

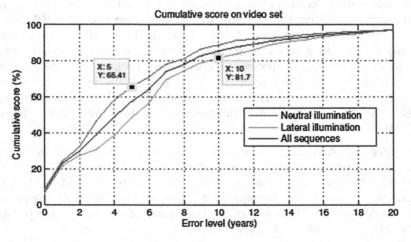

Fig. 6. Age estimation: cumulative scores on the video set with neutral or with added lateral illumination.

Fig. 7. Results in outdoor conditions: face tracking, pose estimation, gender and age estimation. test video n°1 and test video n°2.

6   Conclusions and Perspectives

An alternative method for live stream oriented age and gender estimation is provided. It uses boosted comparisons over uniform LBP histograms based facial signatures. The system provides real-time estimations and is able to track several targets simultaneously. The system's performance is compared favorably to state of the art techniques of age and gender recognition on common databases. Other experimentations are conducted on video sequences and on live streams to show the accuracy of the whole process including face tracking, pose estimation, gender and age estimation (Fig. 7).

Apart from the inevitable database collection needed to improve our training sessions, many perspectives appear. The next natural evolution would be to change the face cropping for face warping using our 3D model to be more resilient to face orientation, instead of only rejecting extreme poses. The present final age and gender decision uses simple logic over the strong classifiers. It would be interesting to build a system able to decide from all the strong classifiers outputs. Another part of this industrial project is the design of a multi-target re-identification system across multiple camera streams.

Acknowledgements.   We provide our test videos including the ground truth measures on request. Please contact us by mail to receive our data. The authors gratefully acknowledge the contribution of the Agence National de la Recherche (CIFRE N°533/2009).

References

1. Wild, H.A., Barett, S.E., Spence, M.J., O'Toole, A.J., Cheng, Y.D., Brooke, J.: Recognition and sex categorization of adults' and children's faces Examining performance in the absence of sex-stereotyped cues. J. Exp. Child Psychol. **77**(4), 269–291 (2000). Elsevier
2. Anastasi, J., Rhodes, M.: An own-age bias in face recognition for children and older adults. Psychon. Bull. Rev. **12**(6), 1043–1047 (2005). Springer
3. Huang, G.B., Ramesh, M., Berg, T., Learned-Miller, E.: Labeled faces in the wild: A database for studying face recognition in unconstrained environments. University of Massachusetts, Amherst, Technical Report 07–49 (2007)
4. Phothisane, P., Bigorgne, E., Collot, L., Prevost, L.: A robust composite metric for head pose tracking using an accurate face model. In: IEEE International Conference on Automatic Face and Gesture Recognition (FG 2011), pp. 694–699. IEEE (2011)
5. Golomb, B., Lawrence, D., Sejnowski, T.: Sexnet, a neural network identifies sex from human faces. In: NIPS-3 Proceedings of the 1990 Conference on Advances in Neural Information Processing Systems, vol. 3, pp. 572–577 (1991)
6. Cottrell, G., Metclafe, J.: Empath: face, emotion, and gender recognition using holons. In: NIPS-3 Proceedings of the 1990 Conference on Advances in Neural Information Processing Systems, vol. 3, pp. 567–571 (1990)
7. Li, X., Maybank, S., Yan, S., Tao, D., Dacheng, T.: Gait components and their application to gender recognition. IEEE Trans. on SMC-B **38**(2), 145–155 (2008). IEEE
8. Shan, C., Gong, S., McOwan, P.: Fusing gait and face cues for human gender recognition. Neuro Comput. **71**(10–12), 1931–1938 (2008). Elsevier

9. Moghaddam, B., Yang, M.: Learning gender with support faces. IEEE Trans. on PAMI **24** (5), 707–711 (2002). IEEE
10. Baluja, S., Rowley, H.: Boosting sex identification performance. Int. J. Comput. Vis. **71**, 111–119 (2007). Kluwer Academic Publishers Hingham
11. Shan, C.: Learning local binary patterns for gender classification on real-world face images. Pattern Recogn. Lett. **33**(4), 431–437 (2012). Elsevier
12. Shakhnarovich, G., Viola, P., Moghaddam, B.: A unified learning framework for real time face detection and classification. In: IEEE International Conference on Automatic Face and Gesture Recognition (FG 2002), pp. 14–21. IEEE (2002)
13. Gao, F., Ai, H.: Face age classification on consumer images with gabor feature and fuzzy LDA method. In: Tistarelli, M., Nixon, M.S. (eds.) ICB 2009. LNCS, vol. 5558, pp. 132–141. Springer, Heidelberg (2009)
14. Kumar, N., Belhumeur, P.N., Nayar, S.K.: FaceTracer: a search engine for large collections of images with faces. In: Forsyth, D., Torr, P., Zisserman, A. (eds.) ECCV 2008, Part IV. LNCS, vol. 5305, pp. 340–353. Springer, Heidelberg (2008)
15. Makinen, E., Raisamo, R.: An experimental comparison of gender classification methods. Pattern Recogn. Lett. **29**(10), 1544–1556 (2008). Elsevier
16. Hadid, A., Pietikainen, M.: Combining motion and appearance for gender classification from video sequences. In: International Conference on Pattern Recognition (ICPR 2008), pp. 1–4. IEEE (2008)
17. Ramanathan, N., Chellappa, R.: Face verification across age progression. IEEE Trans. Image Proces. **15**(11), 3349–3361 (2006). IEEE
18. Fu, Y., Guo, G., Huang, T.: Age synthesis and estimation via faces: A survey. IEEE Trans. PAMI **32**(11), 1955–1976 (2010). IEEE
19. Guo, G., Mu, G., Fu, Y., Huang, T.: Human age estimation using bio inspired features. In: IEEE Conference on Computer Vision (CVPR 2009), pp. 112–119. IEEE (2009)
20. Thukral, P., Mitra, K., Chellappa, R.: A hierarchical approach for human age estimation. In: IEEE International Conference on Acoustic, Speech and Signal Processing (ICASSP 2012), pp. 1529–1532. IEEE (2012)
21. Luu, K., Ricanek, K., Bui, T., Suen, C.: Age estimation using active appearance models and support vector machine regression. In: IEEE Conference on Biometrics: Theory, Applications, and Systems (BTAS), pp. 1–5. IEEE (2009)
22. Luu, K., Seshadri, K., Savvides, M., Bui, T.D., Suen, C.: Contourlet appearance model for facial age estimation. In: IEEE International Joint Conference on Biometrics (IJCB 2011), pp. 1–8 (2011)
23. Yan, S., Wang, H., Tang, X., Huang, T.S.: Learning auto-structured regressor from uncertain nonnegative labels. In: International Conference on Computer Vision (ICCV 2007), pp. 1–8 (2007)
24. Lanitis, A., Draganova, C., Christodoulou, C.: Comparing different classiers for automatic age estimation. IEEE Trans. on SMC-B **34**(1), 621–628 (2004)
25. Guo, G., Mu, G., Dyer, D., T.S., H.: A study on automatic age estimation using a large database. In: International Conference on Computer Vision (ICCV 2009), pp. 1986–1991 (2009)
26. Ojala, T., Pietikinen, M., Maenpaa, T.: Multi-resolution gray-scale and rotation invariant texture classification with local binary patterns. IEEE Trans. PAMI **24**(7), 971–979 (2002)
27. Zhang, W., Shan, S., Zhang, H., Gao, W., Chen, X.: Multi-resolution histograms of local variation patterns (MHLVP) for robust face recognition. In: Kanade, T., Jain, A., Ratha, N. K. (eds.) AVBPA 2005. LNCS, vol. 3546, pp. 937–944. Springer, Heidelberg (2005)

28. Liao, S., Zhu, X., Lei, Z., Zhang, L., Li, S.Z.: Learning multi-scale block local binary patterns for face recognition. In: Lee, S.-W., Li, S.Z. (eds.) ICB 2007. LNCS, vol. 4642, pp. 828–837. Springer, Heidelberg (2007)
29. Freund, Y., Schapire, H.: Experiments with a new boosting algorithm. In: International Conference on Machine Learning (ICML 1996), pp. 148–156. Maurgan Kaufmann (1996)
30. Han, H., Otto, C., Jain, A.: Age estimation from face images: Human vs. machine performance. In: International Conference on Biometrics (ICB 2013), pp. 4–7. IEEE (2013)
31. Dempster, A.P., Laird, N.M., Rubin, D., et al.: Maximum Likelihood from incomplete data via the EM algorithm. J. Royal Stat. Soc. Ser. B (Methodological) 39(1), 1–38 (1977)

Author Index

Printed in the United States
By Bookmasters